Ecology, Biodiversity and Conservation

Ecology, Biodiversity and Conservation

Edited by Jeffery Clarke

SYRAWOOD
PUBLISHING HOUSE

New York

Published by Syrawood Publishing House,
750 Third Avenue, 9th Floor,
New York, NY 10017, USA
www.syrawoodpublishinghouse.com

Ecology, Biodiversity and Conservation
Edited by Jeffery Clarke

International Standard Book Number: 978-1-68286-696-2 (Hardback)

Cataloging-in-Publication Data

Ecology, biodiversity and conservation / edited by Jeffery Clarke.
    p. cm.
Includes bibliographical references and index.
ISBN 978-1-68286-696-2
1. Ecology. 2. Biodiversity. 3. Nature conservation. 4. Biodiversity conservation.
5. Ecosystem management. 6. Conservation biology. I. Clarke, Jeffery.
QH541 .E26 2019
577--dc23

TABLE OF CONTENTS

PREFACE

Ecology, biodiversity and conservation are interrelated fields of study concerned with assessing, evaluating and maintaining the ecological balance of our planet. The focus areas of ecology and biodiversity are planning and management of natural resources, biodiversity maintenance as well as sustaining genetic diversity for wildlife species conservation. The interdisciplinary branches of evolutionary ecology, biology, genetics and ethology are studied to formulate conservation methodologies. This book provides the latest research and technological advancements in the field of ecology, biodiversity and conservation. It strives to provide significant knowledge on these frontiers and help to develop a holistic understanding of these fields. The book is appropriate for students seeking detailed information in this area as well as for experts, ecologists, environmentalists and conservationists.

Significant researches are present in this book. Intensive efforts have been employed by authors to make this book an outstanding discourse. This book contains the enlightening chapters which have been written on the basis of significant researches done by the experts.

Finally, I would also like to thank all the members involved in this book for being a team and meeting all the deadlines for the submission of their respective works. I would also like to thank my friends and family for being supportive in my efforts.

Editor

Looking Beyond the Fenceline: Assessing Protection Gaps for the World's Rivers

Robin Abell[1], Bernhard Lehner[2], Michele Thieme[3], & Simon Linke[4]

[1] Global Water Program, The Nature Conservancy, Arlington, VA 22203, USA
[2] Department of Geography, McGill University, Montreal, QC H3A 0B9, Canada
[3] Freshwater Program, World Wildlife Fund, Washington, DC 20037, USA
[4] Australian Rivers Institute, Griffith University, Brisbane, QLD 4111, Australia

Keywords
Freshwater conservation; protected areas; inland waters; rivers; global.

Correspondence
Bernhard Lehner, Department of Geography, McGill University, 805 Sherbrooke Street West, Montreal, QC H3A 0B9, Canada.
E-mail: bernhard.lehner@mcgill.ca

Editor
Richard Zabel

Abstract

Protected areas are a cornerstone strategy for terrestrial and increasingly marine biodiversity conservation, but their use for conserving inland waters has received comparatively scant attention. In 2010, the Convention on Biological Diversity (CBD) included a target of 17% protection for inland waters, yet there has been no meaningful way of measuring progress toward that target. Defining and evaluating "protection" is especially complicated for rivers because their integrity is intimately linked to impacts in their upstream catchments. A new generation of global hydrographic data now enables a high-resolution, standardized assessment of how upland activities may be propagated downstream. Here, we develop and apply, globally, a river protection metric that integrates both local and upstream catchment protection. We found that "integrated" river protection is highly variable across geographies and river size classes and in most basins falls short of the 17% CBD target. Around the world, about 70% of river reaches (by length) have no protected areas in their upstream catchments, and only 11.1% (by length) achieve full integrated protection. The average level of integrated protection is 13.5% globally, yet the majority of the world's largest basins show averages below 10%. Within basins, gaps are particularly severe for larger rivers.

Introduction

The world's inland waters—rivers, lakes, springs, ground waters, and wetlands—contain exceptional numbers of species, provide critical ecosystem services, and are among the most threatened ecosystems globally (Dudgeon *et al.* 2005; Balian *et al.* 2008; Vörösmarty *et al.* 2010). Protected area (PA) coverage, in this article defined as all nationally designated PAs listed by the International Union for Conservation of Nature (IUCN), has rapidly expanded around the world in the last half century, and evidence suggests that well-managed PAs can achieve biodiversity conservation goals (Geldmann *et al.* 2013). Yet, the extent to which PAs can and do benefit inland waters has been little examined, with global assessments focusing squarely on terrestrial and marine systems (Watson *et al.* 2014).

Measuring the protection of inland waters is of more than academic interest. In 2010, for the first time, the Convention on Biological Diversity's (CBD) PA target (Aichi Target 11) required that "at least 17 percent of terrestrial and inland water areas ... are conserved through effectively and equitably managed, ecologically representative and well-connected systems of protected areas ..." (Convention on Biological Diversity 2010). Even putting aside the important qualifiers of management, representation, and connectivity, 6 years on there remains no globally comprehensive gap analysis of inland waters to provide information on where the numeric 17% target is unmet.

The extent to which existing PAs may protect inland waters has been poorly known due in large part to a lack of accurate, comprehensive spatial datasets of freshwater systems. One study has put coverage of lakes by PAs at less than 2%, but spatial lake datasets have been unreliable for some parts of the world, and lakes comprise only one type of inland water (Chape *et al.* 2003). The Millennium Ecosystem Assessment (Finlayson & D'Cruz 2005) calculated that 12% of the world's freshwaters were included in PAs by overlaying PA polygons with inland water categories of the Digital Chart of the World. This simple overlay analysis confirmed only that freshwaters had not been intentionally excluded from PAs.

Evaluating the extent of river protection has been attempted in the past (e.g., Nel *et al.* 2007; Sowa *et al.* 2007; Stein & Nevill 2011) but has proven especially difficult as hydrologic flows, which originate upstream, are of critical importance for defining the connectivity, character, and integrity of rivers (Poff *et al.* 1997). In fact, it is a core feature of fluvial systems that they are shaped and affected not only by local circumstances but also by conditions in their oftentimes remote upland areas (Johnson & Host 2010). For large-scale assessments, a second challenge is posed by the requirement of adequate information regarding river networks and analytical tools to trace their connectivity. This challenge, however, has recently been addressed in the creation of new data and modeling frameworks. In particular, the HydroSHEDS database (Hydrological data and maps based on SHuttle Elevation Derivatives at multiple Scales) now provides maps of the world's rivers at a high spatial resolution (500 m) and with associated information such as upstream topology and streamflow quantities, which enables complex analyses along river networks (Lehner & Grill 2013).

By overlaying the river and catchment information of HydroSHEDS with the World Database of Protected Areas (IUCN & UNEP-WCMC 2014), we calculated the extent to which rivers are captured within existing PAs ("local protection"), as well as the degree of upland protection for each river reach. Using this information, we propose a novel "integrated protection" metric that combines both local and upland river protection. While this first-of-its-kind metric does not encompass management effectiveness, it does offer a step toward assessing where protection gaps exist.

Methods

Data

We used the HydroSHEDS database (Lehner *et al.* 2008; Lehner & Grill 2013) to provide a consistent global river network at 15 arc-second spatial resolution (approximately 500 m pixel resolution at the equator). HydroSHEDS includes an estimate of long-term average "naturalized" discharge, derived by downscaling coarse-resolution (0.5°) discharge estimates of the global hydro-logical WaterGAP model (Döll *et al.* 2003; model version 2.2 as of 2014). We assessed all river reaches—defined as stretches of rivers between consecutive tributaries—with a minimum average discharge of 100 l/second (0.1 m³/second). The resulting dataset encompasses 6.3 million reaches worldwide with an average length of 3.9 km each, amounting to a total of 24.3 million river kilometers. Smaller rivers have been excluded from the analysis, primarily due to increasing uncertainties in the underpinning global hydrographic and streamflow data.

We used all nationally designated PAs (DESIG_TYPE = "national"; STATUS = "designated") of all IUCN categories (IUCN_CAT = "I-VI," "not reported," or "not assigned") from the October 2014 World Database on Protected Areas (IUCN & UNEP-WCMC 2014) as our source data to describe the coverage of protected land surface areas globally (~160,000 polygons representing 19.2 million km², or 14.3% of the total global land surface area, excluding Antarctica). In cases where PA sites were only given as point data (~17,000 points representing 1.1 million km²), we approximated their spatial extent as a circle with a size representing the reported area.

River protection metrics

The network topology of rivers means that upstream and downstream reaches are inherently connected and that local conditions may be influenced by upstream and upland activities. An assessment of a river reach's protection must therefore look both at local protection (whether a reach falls within a PA) and the degree of landscape protection within a reach's upstream catchment.

We propose a four-tiered approach to define the protection status of a river by calculating: (1) "local protection," which refers to all river reaches that lie within PAs; (2) "upland protection," which measures the percentage of protection of the upstream catchment area associated with each river reach; (3) "achieved target protection," which determines the deviation from a proposed upland protection threshold that represents sufficient protection; and (4) "integrated protection," which combines the requirement of local protection and achieved target protection.

Local protection

To determine local protection, we analyzed whether a river reach falls within a PA and assigned a binary protection status (0% or 100%). Reaches that cross PA borders were first split to allow for partial accounting. It is not uncommon to find rivers flowing along PA borders as they may be used to delimit the PA boundaries. We chose to consider these rivers as being inside PAs. However, the

Target upland catchment protection and measurement of protection shortfall

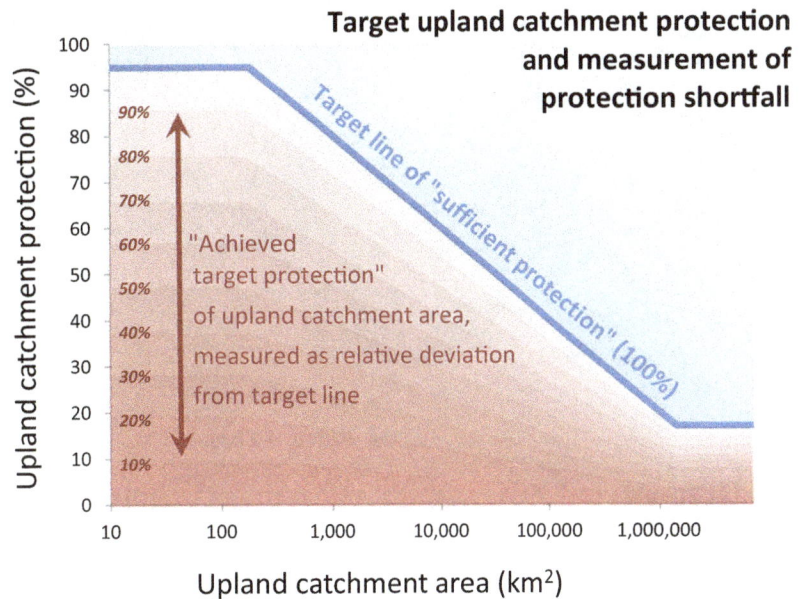

Figure 1 Calculation of "achieved target protection." Once an area-based target line has been defined (blue line), every individual river reach can be assessed as to how far it deviates from the target (based on its individual values for "upland protection" and catchment area). For example, a reach with an upland area of 10,000 km² of which 30% is protected achieves 50% of its protection target. Reaches that exceed the target line (anywhere in the blue area) are defined to achieve 100% of their protection target.

correct detection of boundary rivers is difficult as even small spatial inaccuracies inherent in the underpinning maps can lead to significant misalignments between river lines and PA polygons. To minimize errors, we added a 500 m buffer around all PAs before conducting the assessment as 500 m represents the spatial precision of the river network.

Upland protection

Using catchment delineations from the HydroSHEDS database, we quantified "upland protection" for each river reach as the percentage of PA coverage within the associated upstream catchment of the reach (0–100%). Following a particular river course from its headwaters to the ocean outlet, total upland protection can increase along the river network at confluences where tributaries with high protection ratios join, or it can decrease where tributaries with low levels of protection merge.

Achieved target protection

While "upland protection" provides a first-order proxy to characterize the degree of overall protection within a basin, the interpretation of this indicator is highly scale-dependent. For small headwater catchments, it has been shown that as little as 2% land cover change may affect the ecological status of small streams (Schueler et al. 2009; Cuffney et al. 2010). Thus, 100% upland protection is desirable and achievable for headwaters by putting the entire catchment under protection.

For larger scales, the correlation between spatial extent of PAs and its effect on aquatic ecology is less well studied. There is currently no scientifically derived minimum threshold that could be applied to all large basins globally, yet 100% upland protection is clearly unrealistic for the largest of rivers with millions of square kilometers in upstream area. While the CBD target of a minimum 17% coverage, derived through policy negotiations, could serve as an interim lower limit, considerably higher protection levels may be needed to achieve broad conservation goals (Woodley et al. 2012; Butchart et al. 2015).

In the absence of more robust research, we here propose a sliding upland protection target by defining an area-based threshold line below which upland protection is considered increasingly insufficient. We derive this threshold line (Figure 1) as a piecewise function by setting upper and lower boundary conditions, two inflection points, and a logarithmic transition in between:

(1) We propose that the entire catchments of headwater streams should be protected; and we define headwater catchments as those below 100 km² in size. However, to accommodate minor land cover changes that may be considered acceptable, and to cover small spatial uncertainties in defining the boundaries of catchments and PAs, we set a target of 95% to provide "sufficient" upland protection.

(2) We propose that even the largest basins worldwide (e.g., the Amazon with approximately 6 million km²) should reach at least 17% upland protection, as this value represents the minimum requirement for inland water protection according to the CBD.

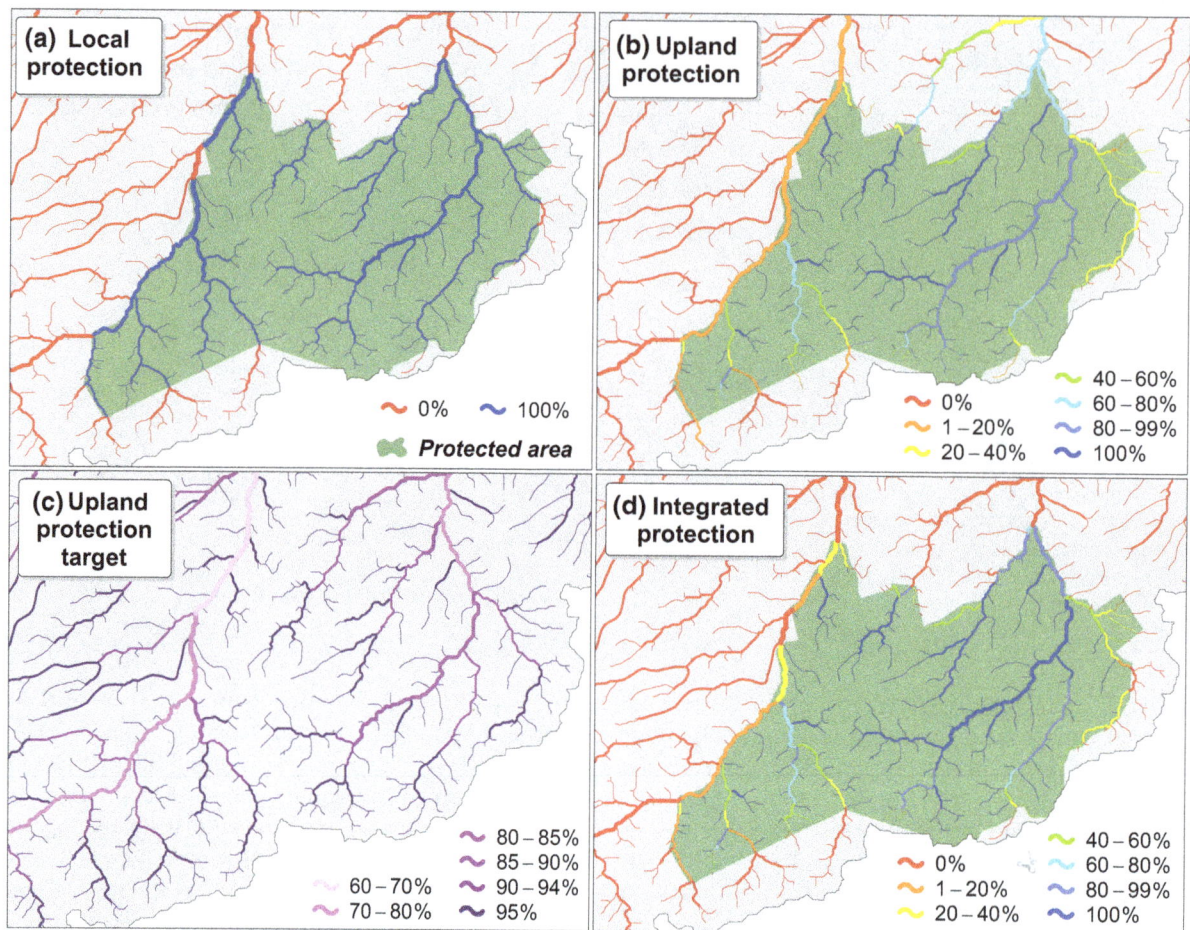

Figure 2 Conceptual approach of local versus integrated protection. Local protection (a) only distinguishes between 0% and 100% protection for each river reach. Upland protection (b) measures the percentage of protected area within the upstream catchment of each reach (0–100%). The upland protection target (c) assigns an area-based threshold of "sufficient protection" to each reach (17–95%). And finally, integrated protection (d) relates the upland protection of each reach to its protection target (0–100%). Reaches outside the protected area are considered unprotected (0%) in both local and integrated approaches. By design, integrated protection is equal or lower than local protection, and equal or higher than upland protection within the protected area.

(3) For a transition between these two boundaries, we propose a logarithmic decline in minimum upland protection targets. The proposed line is drawn in Figure 1 from 100% protection for a catchment size of 100 km^2 to 20% for a catchment size of 1 million km^2; thus, the protection target declines by 20% for each order of magnitude in catchment size, with the curve leveling off at 17%.

We consider basins below the target line to lack comprehensive protection because of the risk of significant impacts from modifications in the unprotected upstream catchment area. This risk grows proportionally with the deviation from the target line. We can quantify this deviation by calculating the "achieved target protection" of a river reach as the ratio (0–100%) between its actual

upland protection and the assigned upland protection target, which depends on the reach's individual catchment size. All reaches that meet or exceed the line of sufficient upland protection receive a value of 100%.

Integrated protection

Finally, we propose an "integrated protection" metric as a double-criteria index assuming that true protection of a river requires that it lies within a PA (local protection) and that its upstream catchment is under some degree of protection as well (achieved target protection). The index is calculated at a reach level by applying the achieved target protection ratio to each river reach inside PAs (see Figure 2). In this approach, full (100%) integrated protection of a river reach is only achieved if the reach is

(1) locally protected and (2) meets or exceeds its area-based target of sufficient upland protection. Reaches that fall inside PAs but have less than sufficient portions of their upstream catchment protected achieve only partial (0–100%) integrated protection. Reaches outside PAs are considered unprotected (0%), even if some of their upland catchment is protected.

Like other traditional gap assessments, these metrics make no assumptions about the effectiveness of protection afforded to the upstream catchment or to the river reach. As such, they may be considered to represent maximum potential protection rather than actual protection.

Calculation and comparison of protection levels for different regions and river size classes

Based on the definitions above, we determined all four indices (local protection, upland protection, achieved target protection, and integrated protection) for each river reach globally. We then summarized the two main indices of local and integrated protection for a variety of spatial units: globally, by continent, for a selection of large river basins, and for six river size classes based on orders of flow magnitude (defined by logarithmic scaling of the long-term average discharge). The average local and integrated protection levels of a spatial unit (e.g., a basin or streamflow size class) were calculated as the average protection ratios of all reaches constituting the spatial unit, weighted by their individual reach lengths.

As the local protection ratio of a river reach is binary (either 0% or 100%), the resulting average local protection of the spatial unit (in %) automatically represents the length of rivers (in %) that are inside PAs. The average integrated protection level, however, is more complex to interpret: an average integrated protection of 40% may indicate that 40% of all rivers (by length) are inside PAs and all of them achieved 100% of their upland protection target; or that 80% of rivers are inside PAs yet they only achieved 50% of their upland protection target; or any other combination of local and achieved target protection that leads to the same average protection level. Given this complexity, measures of "average integrated protection" should only be interpreted as a general index of global, continental, or basin-wide riverine protection.

Results

Globally, 16.0% of the length of rivers are within PAs or form their borders and are therefore considered locally protected. There are, however, wide geographic disparities in local protection (Table 1 and Figure 3a) ranging from very high in the Amazon Basin (44.2%), which alone contains 8.4% of global river length, to very low in the Euphrates-Tigris (1.4%).

In terms of upland protection, we found that 69.5% of rivers around the world (by length) have no PAs in their upstream catchments. While the remaining 30.5% of rivers have at least some kind of upland protection, this level varies between 0.1% and 100%. Only 10.9% of rivers (by length)—mostly smaller headwater streams with catchment areas of less than 100 km^2 and average flows below 1 m^3/second—achieve an upland protection of 95% or above. Including these headwater streams, a total of 11.5% of global river length meets or exceeds our defined target threshold of sufficient upland protection.

Combining local and achieved target protection, we found that 11.1% of global rivers (by length) are located within PAs and meet or exceed our defined protection target, i.e., the associated river reaches are under full integrated protection as defined by our proposed threshold line. When rolling up the results into different spatial units, the global average of integrated river protection is found to be 13.5% (Table 1). Results are substantially lower in many basins (Figure 3b), and within basins there can be high variation among river size classes (Table 1 and Figure 4). Small rivers show the highest averages of integrated protection, at around 14% globally. Medium to large rivers tend to be less well protected.

At the regional level, South America has by far the highest proportion of rivers under local and integrated protection (Table 1), yet the overwhelming influence of the Amazon on this result is apparent (Figures 3 and 4). The Middle East, Europe, and Asia, on the other hand, all have average levels of integrated protection below 10%.

Integrated river protection is, by definition, lower than local protection, and the discrepancy can reveal the appropriateness of PA design for inland water systems (Table 1). Europe, for instance, shows an especially high divergence between local (13.1%) and integrated (8.3%) protection levels, suggesting that many river reaches lie within PAs but lack sufficient headwater protection. At the basin scale, differences are even starker. For example, while local river protection in Australia's Murray-Darling Basin is 8.1%, integrated protection is much lower at only 3.5%. The spatial arrangement of PAs in the Mississippi Basin is similarly inadequate, with only 1.9% of integrated protection achieved basin-wide despite 5.6% of local protection.

When stratified by flow quantities, local protection of the world's rivers is roughly equally distributed among river size classes (Figure 4). Slightly lower levels are observed for larger rivers (>1,000 m^3/second), which may be due to higher degrees of anthropogenic pressures around them, making PA designation more challenging. For smaller streams, average integrated protection is

Table 1 Average local versus integrated protection levels (%) calculated globally, by continent, and for a selection of large river basins. Asia excludes European part of Russia; North America includes Central America and the Caribbean.

Spatial unit	Total protection		By streamflow size (m³/second)					
			0.1–1	1–10	10–100	100–1,000	1,000–10,000	> 10,000
Global	Local	16.0	15.5	16.8	16.9	16.7	15.2	11.6
	Integrated	**13.5**	**13.9**	**13.8**	**11.2**	**9.8**	**9.5**	**9.6**
Africa	Local	13.8	13.9	13.1	15.4	14.3	7.3	0.0
	Integrated	**11.2**	**12.3**	**9.6**	**8.1**	**7.2**	**4.9**	**0.0**
Asia	Local	10.8	11.0	10.7	10.6	8.3	7.5	7.1
	Integrated	**8.9**	**9.7**	**8.3**	**6.2**	**3.7**	**3.1**	**4.6**
Australia	Local	14.6	14.4	14.9	15.3	12.5	12.7	
	Integrated	**12.1**	**12.5**	**12.1**	**10.4**	**6.9**	**9.5**	
Europe	Local	13.1	12.2	14.3	15.0	17.6	18.8	
	Integrated	**8.3**	**8.7**	**8.1**	**6.1**	**5.9**	**8.9**	
Middle East	Local	9.2	9.8	7.6	6.0	7.3	0.0	
	Integrated	**7.6**	**8.6**	**6.0**	**1.8**	**0.6**	**0.0**	
North America	Local	13.5	12.9	14.5	15.0	14.8	15.3	9.2
	Integrated	**10.8**	**11.1**	**11.4**	**8.7**	**5.8**	**6.3**	**3.3**
South America	Local	29.3	28.8	30.4	29.5	30.5	27.3	17.6
	Integrated	**27.5**	**27.8**	**28.4**	**25.3**	**24.2**	**20.5**	**16.2**
Amazon	Local	44.2	44.7	44.1	44.8	43.9	33.5	18.4
	Integrated	**42.5**	**43.8**	**42.3**	**40.1**	**37.3**	**27.9**	**17.4**
Yukon	Local	33.2	33.2	34.2	36.1	19.5	29.0	
	Integrated	**30.2**	**31.3**	**30.2**	**27.1**	**15.9**	**23.0**	
Zambezi	Local	25.7	25.7	23.3	28.4	37.1	30.6	
	Integrated	**21.5**	**23.2**	**17.5**	**14.9**	**26.9**	**26.8**	
Mekong	Local	17.9	18.3	18.0	17.8	14.7	11.3	0.0
	Integrated	**15.8**	**17.1**	**15.6**	**12.4**	**7.9**	**8.8**	**0.0**
Danube	Local	14.9	13.4	15.6	16.5	28.9	31.4	
	Integrated	**9.2**	**9.3**	**8.8**	**7.7**	**10.2**	**18.1**	
Yangtze	Local	14.7	16.0	13.5	12.3	12.1	10.4	8.7
	Integrated	**12.6**	**14.5**	**10.7**	**8.3**	**8.3**	**4.5**	**8.7**
Colorado	Local	14.9	14.1	13.7	13.3	38.1		
	Integrated	**7.2**	**8.1**	**4.2**	**3.3**	**13.0**		
Congo	Local	11.4	11.6	10.7	13.3	11.5	0.7	0.0
	Integrated	**10.1**	**11.0**	**9.4**	**8.7**	**6.2**	**0.0**	**0.0**
Niger	Local	10.8	10.6	10.9	15.1	7.0	0.8	
	Integrated	**7.9**	**8.9**	**6.3**	**6.1**	**2.1**	**0.5**	
Amur	Local	10.1	9.5	10.3	15.3	9.3	6.1	1.5
	Integrated	**7.0**	**7.7**	**6.1**	**5.4**	**2.0**	**2.7**	**0.9**
Volga	Local	8.2	7.4	9.1	12.2	7.6	11.5	
	Integrated	**4.1**	**4.7**	**3.2**	**2.4**	**1.0**	**3.3**	
Murray-Darling	Local	8.1	6.0	8.4	11.6	35.8		
	Integrated	**3.5**	**3.4**	**3.1**	**2.5**	**10.1**		
Rio Grande	Local	6.1	5.1	7.5	7.4	23.3		
	Integrated	**3.3**	**3.4**	**3.2**	**1.4**	**5.8**		
Orange	Local	5.7	4.1	6.1	14.7	12.2		
	Integrated	**1.7**	**1.8**	**1.2**	**1.8**	**1.9**		
Mississippi	Local	5.6	4.6	5.7	9.3	13.9	15.2	1.2
	Integrated	**1.9**	**2.1**	**1.5**	**1.1**	**1.4**	**2.5**	**0.2**
Euphrates-Tigris	Local	1.4	1.3	0.8	2.7	4.7	0.0	
	Integrated	**0.9**	**1.1**	**0.6**	**0.3**	**0.4**	**0.0**	

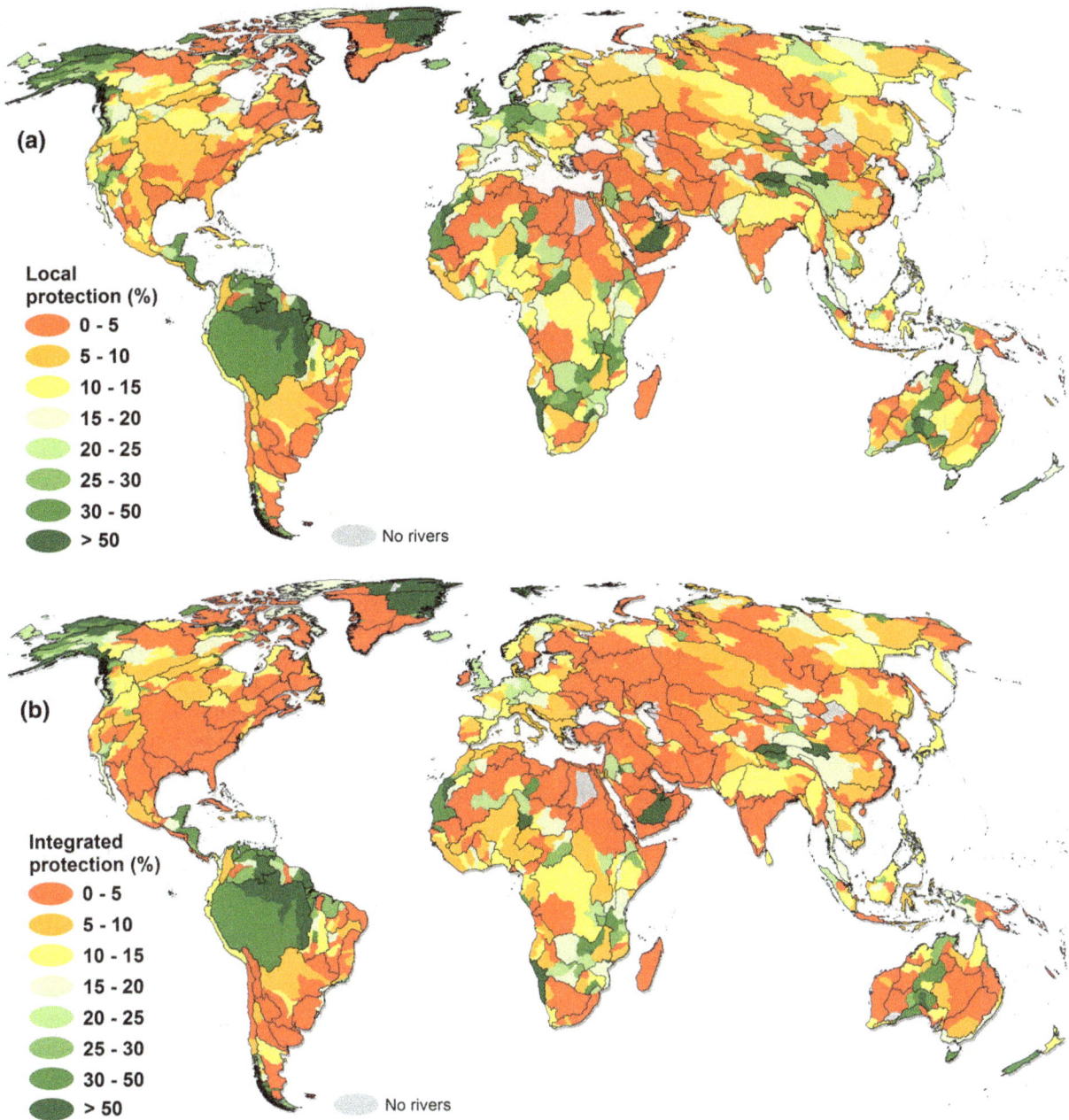

Figure 3 Global pattern of (a) local and (b) integrated river protection. High protection in river-rich areas is found in the Amazon Basin and in Alaska. Integrated protection levels are visibly lower in many places, e.g., in the Mississippi Basin and most parts of Europe. Figure shows a breakdown into subbasins of approximately 100,000 km² in average size, as well as black outlines for major basins.

similar to local protection levels, as local and upland protection are mostly coinciding in the associated smaller headwater catchments. Average integrated protection of the largest river class also tends to mirror local protection; this finding indicates that our chosen target of "sufficient upland protection" is often approached or exceeded and local protection represents the limiting factor. Mid-sized rivers (defined here as 10–10,000 m³/second), however,

deviate the most between local and integrated protection; they are thus of high priority for improving the spatial alignment of local and upland protection.

The example of the Mississippi Basin given in Figure 4 reveals big discrepancies between local and integrated protection across size classes; the smaller headwater streams are not well protected and this limitation is propagated downstream to affect larger rivers,

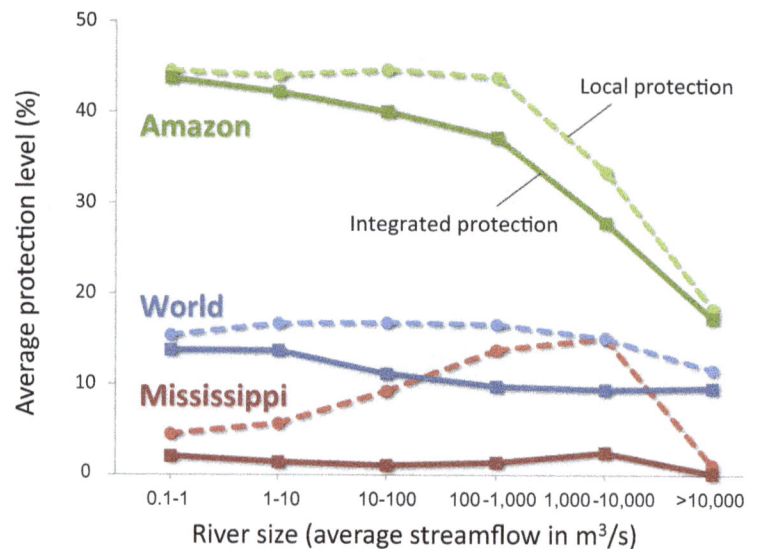

Figure 4 Local versus integrated protection for rivers of different streamflow sizes (flow values represent long-term averages). The global distribution shows that integrated protection is significantly lower than local protection, in particular for middle- to large-sized rivers.

which show increasing local protection while their integrated protection remains small. The Amazon Basin is an example of high protection, both locally and integrated, although the largest rivers remain less well protected.

To gain some insights into the sensitivity of the results regarding the chosen shape of the target line (Figure 1), we tested the effect of shifting the line by an entire order of magnitude to the left (i.e., making the target easier to achieve). As a result, the global average of integrated protection increased by only 0.3%, with the strongest increase of 2.3% found for larger rivers (1,000–10,000 m³/second). Based on this finding, we conclude that the general approach is fairly robust and not highly susceptible to the chosen inflection points of the target line. We recognize, however, that results for larger rivers are sensitive to the definition of the lower boundary (here set at the CBD target of 17%).

Discussion

This study represents the first comprehensive assessment of the extent of integrated river protection worldwide. Not only is local (i.e., traditional) protection accounted for, but each river segment is also evaluated in the context of its landscape position. Our results of integrated river protection suggest that even if all existing PAs were well managed to conserve riverine targets inside them, many of the world's rivers are poorly protected in their respective upland areas. Average integrated protection levels remain significantly below the CBD's 17% target when assessed globally, regionally, and by streamflow size class.

We have also produced traditional protection results (inside PAs) that can be used for comparisons with similar analyses for terrestrial and marine systems. We find that globally 16.0% of the length of river reaches with an average flow of at least 100 l/second are located within or along the border of PAs, suggesting that local river protection is higher than protection in both terrestrial (14.3%) and marine (3.2%) realms (our own calculations using the same PA coverage). The slight exceedance over the terrestrial value is indicative of global PA distribution being biased toward areas with higher river density as compared to deserts. When our new measure of integrated protection is applied, the global average of river protection drops to 13.5%, i.e., below terrestrial protection, due to locally protected rivers with insufficient upland protection. Only 11.1% of all rivers (by length) are under full integrated protection.

Parsed by region, the results tell a more differentiated story. Unsurprisingly, the patterns for rivers across geographic regions mirror those seen in terrestrial gap analyses, with hotspots of low protection in the Middle East, parts of Central and South Asia, North America, southern South America, northern Africa, and parts of Australia. Given that riverine protection gaps are as important to address as those for terrestrial and marine systems, new or expanded PAs in underprotected regions should be designed with a landscape view of the river network and associated species and ecosystems (Abell *et al.* 2011; Hermoso *et al.* 2015).

Globally, smaller headwater streams dominate (by length) the world's running waters. These streams provide important sources of water, sediments, and biota (Meyer *et al.* 2007; Clarke *et al.* 2008) and thus are vitally in need of comprehensive protection (Freeman *et al.* 2007). We found that smaller headwater streams in many basins have levels of integrated protection below the CBD's 17% target; this is a worrying gap, but also one that might be addressed more easily than similar gaps for larger systems. The lack of integrated protection of mid-sized to large rivers, on the other hand, raises concerns about the representativeness of different habitat types within PA networks (Linke *et al.* 2011).

Significant improvements in PA design could be achieved by designating as new PAs those unprotected headwaters that sit upstream of existing PAs. Alternately, where headwaters are already well protected, they may provide some degree of downstream protection even when no local PA exists there. In fact, these locations may offer a primary avenue for optimizing PA networks by adding downstream PAs for river corridors that already exceed the upland protection target. Our approach can highlight where strategic planning for upstream protection (Linke *et al.* 2007; Moilanen *et al.* 2008) and systematic approaches to catchment zoning (Abell *et al.* 2007; Nel *et al.* 2011) have the most potential to improve the condition of rivers and other inland waters.

While this study provides proof-of-concept of the suggested approach, both technical and methodological challenges remain. Our accounting of local and upland protection hinges on the quality of both the World Database on Protected Areas and the underpinning hydrographic data. We believe, however, that these technical issues are less problematic than the uncertainties introduced by the definitions and decisions required for the gap assessment. For example, it remains conceptually ambiguous whether to consider a river at the boundary of a PA as being locally protected or not.

We also acknowledge that there is a lack of existing evidence of successful river protection for testing the shape and inflection points of our proposed target line of "sufficient protection," yet we consider it a reasonable placeholder until such evidence emerges. We recognize that local geomorphological characteristics, hydrological connectivity, species distribution ranges, or varying runoff contribution can render some areas of a catchment more important than others, and optimized protection should focus on these critical landscape and ecological elements rather than relying on generalized percentage thresholds (Higgins 2003). As well, to the extent that PAs serve to mitigate riverine impacts originating outside PA boundaries, issues of size and configuration will be tied to the scale and intensity of those impacts. However, the required level of catchment protection remains a largely unresolved question (Gergel *et al.* 2002). Our sliding target line, based on increasing upland area, is intended to

integrate some of these scale issues, but as a global approach it can only cover broad patterns.

To provide stronger support for the validity of our approach and the chosen settings, it is of paramount importance to improve the monitoring of actual effects of PAs on ecosystem and hydrological integrity both locally and downstream. Depending on the outcomes, the method and settings should be adjusted accordingly. As for now, there is ample evidence that the proposed target line is already met or exceeded today in many river basins and throughout all streamflow size classes worldwide, demonstrating its general achievability.

As noted in our results, global figures mask regional and basin-level trends, with a significant bias due to the Amazon's very high protection levels (44.2% local protection and 42.5% integrated protection). A large proportion of the Amazon's PA network is composed of IUCN Category VI reserves, designed to "conserve ecosystems and habitats, together with associated cultural values and traditional natural resource management systems" (Dudley 2008). There is a history of debate around whether such reserves qualify as "true" PAs due to the range of activities allowed within them (Dudley *et al.* 2010). We included them in our analysis to be consistent with terrestrial accounting systems (Watson *et al.* 2014). Studies into how well each of the PA categories confers protection to inland waters both within their boundaries and downstream would be a valuable research direction.

In the meantime, protection results should be interpreted cautiously, given the focus of the gap assessment on extent of coverage and our inability to evaluate management effectiveness. Permitted developments in PAs of less strict protection may compromise the health of a river, yet a well-managed category VI PA may still provide better protection to inland waters than a poorly managed category III or IV PA. In general, we know that many PAs are not managed with riverine systems in mind, so the conservation picture is undoubtedly worse than our results suggest (Thieme *et al.* 2012). The location of threats relative to PAs may be a larger driver of the status of ecosystems within the PA than the proportion of upstream basin under protection. On the other hand, in the increasingly few remote areas of the world, such as many of the Amazon's headwaters, *de facto* protection of inland waters may occur without the presence of formal PAs at all (Joppa *et al.* 2008).

Conclusions

We have demonstrated that our proposed indicator of "integrated river protection" is a viable alternative to the traditional "within the fenceline" approach to gap assessments, in which systems are counted as protected or not based on whether they fall within PA boundaries. This binary approach is clearly inadequate for inland water systems, which sit at the lowest points in their landscapes and integrate hydrologically mediated impacts from their catchments. Because rivers are shaped by these complex processes, PAs alone will rarely ensure their conservation, but with effective design and management they can make important contributions. Identifying gaps in the extent of river protection is a first step, and we believe that our indicator can illuminate gaps from the small scale of individual subbasins to any aggregated ecological or political unit.

If our approach is adopted as an improvement over current indicators, or receives consideration by the CBD-mandated Biodiversity Indicators Partnership, we propose that a next-generation indicator be explored that incorporates other inland water types as well. To account for the effectiveness of protection, additional work could be pursued to combine the gap assessment with information on existing anthropogenic disturbances, ranging from dam construction to aquatic pollution. Some of this information is readily available at a global scale and could be used to assess how different types of upland activities (such as hydropower development, mining, or timber harvesting) may compromise the protection status of rivers. Regardless, we strongly believe that the proposed indicator method of measuring the extent of integrated protection can elevate the profile of inland waters within PA discussions and, consequently, take accounting of the protection of these systems to a higher level.

Acknowledgments

We would like to thank two anonymous reviewers for their contributions to improve this manuscript. BL was supported by Discovery Grant RGPIN/341992-2013 from the Natural Sciences and Engineering Research Council of Canada (NSERC); SL was supported by Grant DE130100565 from the Australian Research Council.

References

Abell, R., Allan, J.D. & Lehner, B. (2007). Unlocking the potential of protected areas for freshwaters. *Biol. Conserv.*, **134**, 48-63.

Abell, R., Thieme, M., Ricketts, T.H. *et al.* (2011). Concordance of freshwater and terrestrial biodiversity. *Conserv. Lett.*, **4**, 127-136.

Balian, E., Segers, H., Lévêque, C. & Martens, K. (2008). Freshwater animal diversity assessment: an overview of the results. *Hydrobiologia*, **595**, 627-637.

Butchart, S.H.M., Clarke, M., Smith, R.J. *et al.* (2015). Shortfalls and solutions for meeting national and global conservation area targets. *Conserv. Lett.*, **8**, 329-337.

Chape, S., Blyth, S., Fish, L., Fox, P. & Spalding, M. (2003). *United Nations List of Protected Areas*. IUCN and UNEP-WCMC, Gland, Switzerland and Cambridge, UK.

Clarke, A., Mac Nally, R., Bond, N. & Lake, P.S. (2008). Macroinvertebrate diversity in headwater streams: a review. *Freshwater Biol.*, **53**, 1707-1721.

Convention on Biological Diversity. (2010). Aichi Biodiversity Targets. Available from http://www.cbd.int/sp/targets. Accessed 1 March 2015.

Cuffney, T.F., Brightbill, R.A., May, J.T. & Waite, I.R. (2010). Responses of benthic macroinvertebrates to environmental changes associated with urbanization in nine metropolitan areas. *Ecol. Appl.*, **20**, 1384-1401.

Döll, P., Kaspar, F. & Lehner, B. (2003). A global hydrological model for deriving water availability indicators: model tuning and validation. *J. Hydrol.*, **270**, 105-134.

Dudgeon, D., Arthington, A.H., Gessner, M.O. *et al.* (2005). Freshwater biodiversity: importance, threats, status and conservation challenges. *Biol. Rev.*, **81**, 163-182.

Dudley, N. (2008). *Guidelines for applying protected area management categories*. IUCN, Cambridge, UK.

Dudley, N., Parrish, J.D., Redford, K.H. & Stolton, S. (2010). The revised IUCN protected area management categories: the debate and ways forward. *Oryx*, **44**, 485-490.

Finlayson, C.M. & D'Cruz, R. (2005). Inland water systems. Pages 551-583 in R. Hassan, R. Scholes, N. Ash, editors. *Ecosystems and human well-being: current state and trends, Volume I, Millennium Ecosystem Assessment*. Island Press, Washington, D.C.

Freeman, M.C., Pringle, C.M. & Jackson, C.R. (2007). Hydrologic connectivity and the contribution of stream headwaters to ecological integrity at regional scales. *J. Am. Water Res. Assoc.*, **43**, 5-14.

Geldmann, J., Barnes, M., Coad, L., Craigie, I.D., Hockings, M. & Burgess, N.D. (2013). Effectiveness of terrestrial protected areas in reducing habitat loss and population declines. *Biol. Conserv.*, **161**, 230-238.

Gergel, S.E., Turner, M.G., Miller, J.R., Melack, J.M. & Stanley, E.H. (2002). Landscape indicators of human impacts to riverine systems. *Aquat. Sci.*, **64**, 118-128.

Hermoso, V., Cattarino, L., Kennard, M.J., Watts, M. & Linke, S. (2015). Catchment zoning for freshwater conservation: refining plans to enhance action on the ground. *J. Appl. Ecol.*, **52**, 940-949.

Higgins, J.V. (2003). Maintaining the ebbs and flows of the landscape: conservation planning for freshwater ecosystems. Pages 291-318 in C. Groves, editor. *Drafting a conservation blueprint: a practitioner's guide to planning for biodiversity*. The Nature Conservancy and Island Press, Washington, D.C.

IUCN & UNEP-WCMC. (2014). The World Database on Protected Areas (WDPA). Available from http://www.protectedplanet.net. Accessed 20 October 2014.

Johnson, L.B. & Host, G.E. (2010). Recent developments in landscape approaches for the study of aquatic ecosystems. *J. North Am. Benthol. Soc.*, **29**, 41-66.

Joppa, L.N., Loarie, S.R. & Pimm, S.L. (2008). On the protection of "protected areas". *Proc. Natl. Acad. Sci. U. S. A.*, **105**, 6673-6678.

Lehner, B. & Grill, G. (2013). Global river hydrography and network routing: baseline data and new approaches to study the world's large river systems. *Hydrol. Process.*, **27**, 2171-2186.

Lehner, B., Verdin, K. & Jarvis, A. (2008). New global hydrography derived from spaceborne elevation data. *Eos*, **89**, 93-94.

Linke, S., Pressey, R.L., Bailey, R.C. & Norris, R.H. (2007). Management options for river conservation planning: condition and conservation re-visited. *Freshwater Biol.*, **52**, 918-938.

Linke, S., Turak, E. & Nel, J. (2011). Freshwater conservation planning: the case for systematic approaches. *Freshwater Biol.*, **56**, 6-20.

Meyer, J.L., Strayer, D.L., Wallace, J.B., Eggert, S.L., Helfman, G.S. & Leonard, N.E. (2007). The contribution of headwater streams to biodiversity in river networks. *J. Am. Water Res. Assoc.*, **43**, 86-103.

Moilanen, A., Leathwick, J. & Elith, J. (2008). A method for spatial freshwater conservation prioritization. *Freshwater Biol.*, **53**, 577-592.

Nel, J.L., Reyers, B., Roux, D.J., Impson, N.D. & Cowling, R.M. (2011). Designing a conservation area network that supports the representation and persistence of freshwater biodiversity. *Freshwater Biol.*, **56**, 106-124.

Nel, J.L., Roux, D.J., Maree, G. *et al.* (2007). Rivers in peril inside and outside protected areas: a systematic approach to conservation assessment of river ecosystems. *Divers. Distrib.*, **13**(3), 341-352.

Poff, N.L., Allan, J.D., Bain, M.B. *et al.* (1997). The natural flow regime: a paradigm for river conservation and restoration. *BioScience*, **47**, 769-784.

Schueler, T.R., Fraley-McNeal, L. & Cappiella, K. (2009). Is impervious cover still important? Review of recent research. *J. Hydrol. Eng.*, **14**, 309-315.

Sowa, S.P., Annis, G.M., Morey, M.E. & Diamond, D.D. (2007). A gap analysis and comprehensive conservation strategy for riverine ecosystems of Missouri. *Ecol. Monogr.*, **77**(3), 301-334.

Stein, J. & Nevill, J. (2011). Counting Australia's protected rivers. *Ecol. Manage. Restor.*, **12**(3), 200-206.

Thieme, M.L., Rudulph, J., Higgins, J. & Takats, J.A. (2012). Protected areas and freshwater conservation: a survey of protected area managers in the Tennessee and Cumberland River Basins, USA. *J. Environ. Manage.*, **109**, 189-199.

Vörösmarty, C.J., McIntyre, P.B., Gessner, M.O. *et al.* (2010). Global threats to human water security and river biodiversity. *Nature*, **467**(7315), 555-561.

Watson, J.E.M., Dudley, N., Segan, D.B. & Hockings, M. (2014). The performance and potential of protected areas. *Nature*, **515**, 67-73.

Woodley, S., Bertzky, B., Crawhill, N. *et al.* (2012). Meeting Aichi Target 11: what does success look like for protected area systems? *Parks*, **18**, 21-34.

2

Informing Strategic Efforts to Expand and Connect Protected Areas Using a Model of Ecological Flow, with Application to the Western United States

Brett G. Dickson[1,2], Christine M. Albano[1,3], Brad H. McRae[4], Jesse J. Anderson[1], David M. Theobald[1], Luke J. Zachmann[1,2], Thomas D. Sisk[2], & Michael P. Dombeck[5]

[1] Conservation Science Partners, Inc., 11050 Pioneer Trail, Suite 202, Truckee, CA 96161, USA
[2] Landscape Conservation Initiative, Northern Arizona University, Box 5694, Flagstaff, AZ 86011, USA
[3] John Muir Institute of the Environment, University of California - Davis, One Shields Ave., Davis, CA 95616, USA
[4] The Nature Conservancy, North America Region, 117 Mountain Ave, Suite 201, Fort Collins, CO 80524, USA
[5] College of Natural Resources, University of Wisconsin-Stevens Point, 800 Reserve St., Stevens Point, WI 54481, USA

Keywords
Centrality; connectivity; conservation planning; ecological flow; effective resistance; protected areas; Bureau of Land Management.

Correspondence
Brett G. Dickson, Conservation Science Partners, Inc., 11050 Pioneer Trail, Suite 202, Truckee, CA 96161, USA.
E-mail: brett@csp-inc.org

Editor
Richard Zabel

Abstract

Under rapid landscape change, there is a significant need to expand and connect protected areas (PAs) to prevent further loss of biodiversity and preserve ecological functions across broad geographies. We used a model of landscape resistance and electronic circuit theory to estimate patterns of ecological flow among existing PAs in the western United States. We applied these results to areas previously identified as having high conservation value to distinguish those best positioned to maintain and enhance ecological connectivity and integrity. We found that current flow centrality was highest and effective resistance lowest in areas that spanned the border between southern Oregon and Idaho, and in northern Arizona and central Utah. Compared to other federal jurisdictions, Bureau of Land Management lands contributed most to ecological connectivity, forming "connective tissue" among existing PAs. Our models and maps can inform new conservation strategies and critical land allocation decisions, within or among jurisdictions.

Introduction

Fundamental principles of systematic conservation planning (e.g., Margules & Pressey 2000) suggest that an ecologically functional system of protected areas (PAs) requires large and intact landscapes, should protect a variety of habitats, and critically, needs to be interconnected (Defries *et al.* 2007; Cumming *et al.* 2015). Globally, well-connected PA networks have the potential to enhance biodiversity within and beyond their boundaries, on land or in the sea (Brudvig *et al.* 2009; Foley *et al.* 2010). Yet, PAs have rarely been selected to conserve biodiversity and maintain ecosystem function (Pressey 1994; Watson *et al.* 2016), nor have they been selected to contribute to a well-connected ecological network. To address this gap, we developed a model of ecological flow among existing PAs to identify areas that are best positioned to maintain and enhance ecological connectivity and integrity across broad landscapes with multiple ownerships and applied it to the western United States.

Vast areas of currently unprotected public lands in the western United States have the potential to enhance the ecological effectiveness of the U.S. PA network. This is particularly true if the ecological significance and context of these lands are used to determine the location of new areas for conservation and protection (Dickson *et al.* 2014, Watson *et al.* 2016). Over 1.4 million km² (57%) of land in the 11 conterminous western states is owned by the American public and managed by the federal government. Although a sizable proportion of these lands remain largely undeveloped, the expansion of energy development, mining, timber harvesting, and other extractive land uses threaten to fragment these areas, reducing their ecological function (Hansen & Defries 2007). Furthermore, their degradation could undermine the PA network by reducing existing landscape connectivity and, in turn, lowering the conservation value of adjacent PAs (Berger *et al.* 2014).

Here we apply our model of ecological flow to the western United States to identify areas that can most enhance the ecological value of the U.S. PA network using two complementary estimates of how the areas contribute to ecological connectivity and integrity. These include (1) current flow centrality, which identifies areas important for keeping networks connected for ecological processes such as gene flow (McRae *et al.* 2008) and can be used to predict movement probabilities for animals at local or regional scales (e.g., McClure *et al.* 2016) and (2) effective resistance, which quantifies the isolation of sites or populations (McRae & Beier 2007). We use these estimates to assess and demonstrate the potential contributions of currently unprotected Bureau of Land Management (BLM) roadless lands with high conservation value (Dickson *et al.* 2014) to regional connectivity and PA integrity, where areas with relatively high current flow centrality contribute to maintenance of connectivity across the PA network, and areas of relatively low resistance to nearby PAs have the potential to enhance the integrity of existing PAs.

Methods

Our study area included the eleven western states in the contiguous United States: Arizona, California, Colorado, Idaho, Montana, Nevada, New Mexico, Oregon, Utah, Washington, and Wyoming (Figure 1). We defined the network of existing PAs within this extent using land management designations from the U.S. PA Database v1.3 (USGS 2012). We considered only those PAs that were designated within IUCN categories I-IV and that were $\geq$20.2 km^2 in size, as this is the federally mandated minimum size for wilderness areas in the United States (Wilderness Act 1964). We combined PA polygons when they were immediately adjacent and used geometric centroids (i.e., single pixels, constrained to polygon interiors) to represent each unique polygon in the connectivity analyses ($n = 1,043$ centroids).

As a first step to modeling landscape connectivity, we created a resistance surface (R) that combined data on human modification and slope using the equation $R = (H + 1)^{10} + s/4$, where H is the human modification score and s is percent slope (see also SI Methods). After Theobald (2013), we quantified the degree of human modification (H) of the western landscape circa 2010, with scores

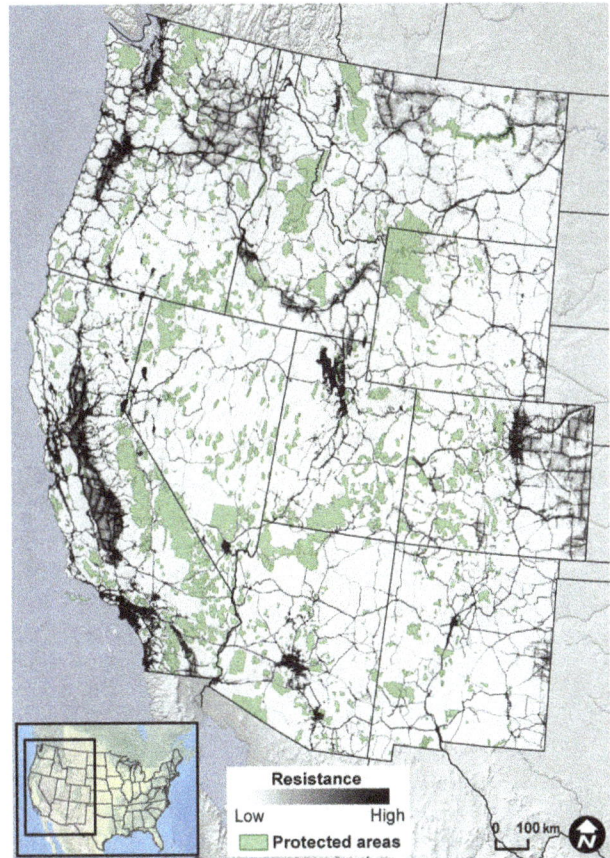

Figure 1 Distribution of protected areas and landscape resistance values across 11 states in the western United States. Protected areas are defined as those in IUCN categories I-IV and $\geq$ 20.2 km^2 in size.

ranging from 0.00 (unmodified) to 1.00 (completely converted) using multiple data layers including land cover, transportation, housing density, and oil and gas well density. To account for possible movement processes that avoided relatively large elevation changes or steep terrain (e.g., crossing over mountain ranges or through deep valleys), we added to H a penalty for areas with steep slopes, following Theobald *et al.* (2012). Finally, we used the National Hydrography Dataset Plus (USGS 2008) and assigned all rivers with annual mean flow >1,000 cubic feet per second a resistance of 1,000 to reflect their role as barriers to movement for many terrestrial organisms. Overall, this resulted in resistances value R ranging from 1.00 for unmodified lands to approximately 1024.00 for completely developed lands. We evaluated the sensitivity of our centrality results to different modeling choices for resistance surface parameterizations (SI Methods).

Next, we treated the PA centroids as focal nodes to be connected and calculated cumulative current flow among centroids using Circuitscape v4.0.5 (McRae *et al.* 2013), which implements a model of potential

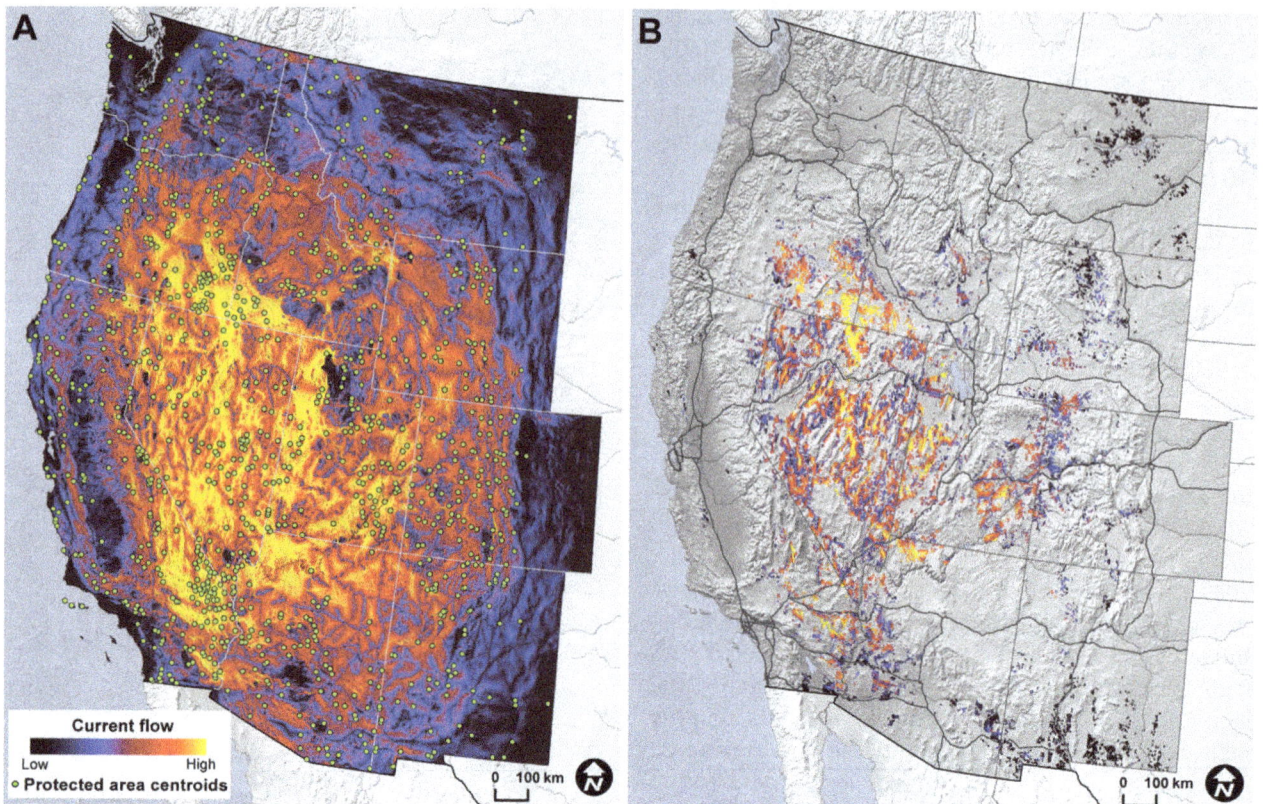

Figure 2 Current flow across the 11 western states (A) and within unprotected, roadless BLM lands (B). Protected area centroids were calculated as the geometric center of individual or immediately adjacent protected areas. Major interstate highways also are shown.

connectivity based on electrical circuit theory (McRae *et al.* 2008). Similarly to Theobald *et al.* (2012), our novel use of PA centroids (i.e., single pixels and not polygons) allowed us to more realistically estimate flow within each PA boundary as a continuous process, rather than treat each PA as a homogeneous patch with zero resistance, which is typical for connectivity studies. This approach also permitted a comprehensive comparison of current flow values among jurisdictions, including those that encompassed the PAs we analyzed. In addition, the use of centroids resulted in a substantial decrease in computation time. We used Circuitscape's all-to-one mode iterated across all centroids, connecting each to ground while injecting 1 Amp of current into the remaining centroids and allowing current to flow across the study area to the grounded centroid (McRae *et al.* 2013). Current densities were added across all iterations to produce a map of current flow centrality values for each intervening grid cell (pixel). This provided a measure of betweenness centrality (Newman 2005), which reflects the importance of each grid cell for maintaining connectivity among all centroid pairs. Current flow centrality includes contributions

from all paths between nodes, not just the shortest or least-cost paths, while still giving more weight to low-resistance paths (Newman 2005).

We estimated average current flow centrality across the entire extent of each of five federal jurisdictions—BLM, National Park Service (NPS), Department of Defense, US Fish and Wildlife Service, and US Forest Service (USFS)—and for each of 1,678 high conservation value areas by calculating the total current flow across it and dividing by its area. Our database of high conservation value areas (average size = 54.2 km^2, SD = 56.8, range = 20.2–959.5) was drawn from a systematic analysis of seven ecological indicators on contiguous areas of roadless BLM land (the "80/20 scale-dependent core" results and "roadless" definition described in detail by Dickson *et al.* 2014) (see also SI Methods). We considered high conservation value areas with relatively high centrality to be those places more likely to be important for maintaining connectivity among PAs. In doing so, we demonstrate the utility of our connectivity results in identifying areas of high conservation value that also had exceptionally high value for promoting ecological flows.

We also calculated the effective resistance between each high conservation value area and all PAs by setting all PAs to ground and iteratively injecting 1 Amp of current into each high conservation value area. Effective resistance reflects the degree to which a particular node (e.g., a PA) is connected to (or isolated from) others and is affected by proximity but also the resistance and number of connections (or pathways) between nodes (McRae *et al.* 2008). Because effective resistance is most influenced by nearby nodes, and because PA integrity would be most enhanced by nearby PAs, this analysis was restricted to a 250-km radius of each high conservation value area for computational efficiency. We considered high conservation value areas with relatively low effective resistance values to be those places more likely to enhance ecological integrity across the existing PA network.

As a final step, and to identify concentrations of regionally important areas with simultaneously high conservation and ecological connectivity values, we combined our centrality and effective resistance results. To do this, we simply summed each of the 1,678 high conservation value areas according to its ranked values for centrality (from low to high) and effective resistance (high to low).

Results

Patterns of current flow reflected a high overall contribution to ecological connectivity among PAs in the central portion of the western United States (Figure 2A), including large areas of unprotected, roadless BLM land (Dickson *et al.* 2014; Figure 2B). Within this extent, we observed the highest levels of current flow across southeastern Oregon, Nevada, and Utah. In Nevada, the north–south orientation of basin and range landforms appears to facilitate high current flow between the northwestern and southwestern states. Portions of southern Idaho and the desert regions of southern California and western Arizona also exhibit high levels of current flow, which may be driven by close proximity to existing PAs. Overall, we found that BLM lands contribute more to connectivity among PAs (per unit area) than lands managed by the other federal management agencies we analyzed (Table 1).

Among the best (95th percentile) high conservation value areas identified on unprotected roadless BLM lands by Dickson *et al.* (2014), flow centrality was greatest for those spanning southeastern Oregon and southwestern Idaho, central and eastern Nevada, southern Nevada, and northwestern Arizona (Figure 3A). Flow centrality was lowest in southern Arizona and southern New Mexico, where there are relatively few PAs in close proximity,

Table 1 Average current flow (betweenness) centrality for each of five federal jurisdictions in the western United States, ranked according to mean current density

Jurisdiction	Area (km^2)	Mean current density	SD
BLM	706,256	444.4	219.0
NPS	80,624	426.9	241.5
DOD	65,119	423.4	295.5
FWS	29,439	370.1	253.7
USFS	638,200	295.8	134.8

BLM, Bureau of Land Management; NPS, National Park Service; DOD, Department of Defense; FWS, Fish and Wildlife Service; USFS, United States Forest Service.

and where there are few large patches of unprotected, roadless BLM land. Effective resistance was lowest in high conservation value areas near or adjacent to PAs in southeastern Oregon and northern Nevada, but also in southeastern Utah and portions of northwestern Arizona and western Colorado (Figure 3B). Notably, areas of low effective resistance also were present in southern Arizona and southern New Mexico.

Based on combined (95th percentile) results for ranked centrality and effective resistance, areas of high conservation value that might simultaneously maintain and enhance connectivity were concentrated in the greater Owyhee Canyonlands of southeastern Oregon and southwestern Idaho, as well as southeastern Utah and the Mojave Desert of southern Nevada and northwestern Arizona (Figure 4). One of the largest (>940 km^2) high conservation value areas identified by Dickson *et al.* (2014) was found in southeastern Utah, overlapping the San Rafael Swell. This area was among over a dozen that neighbored multiple PAs and that would provide high connectivity value across the southwestern region.

Discussion

Around the world, intact but unprotected habitats are rapidly disappearing (e.g., Tilman *et al.* 2001; Butchart *et al.* 2010). These habitats may provide both movement pathways among PAs and sources for colonization of adjacent PAs (Goetz *et al.* 2009) similar to the "spillover" effect documented for marine PAs (Olds *et al.* 2012). Nevertheless, new PAs are rarely established because of their potential to improve connectivity among PAs or enhance the ecological integrity of existing PAs. A flow-based model of ecological connectivity, such as ours, can provide a flexible, yet theoretically grounded and robust method for identifying the location of new PAs wherever data on landscape (or seascape) resistance can be derived. Moreover, complementary estimates of centrality

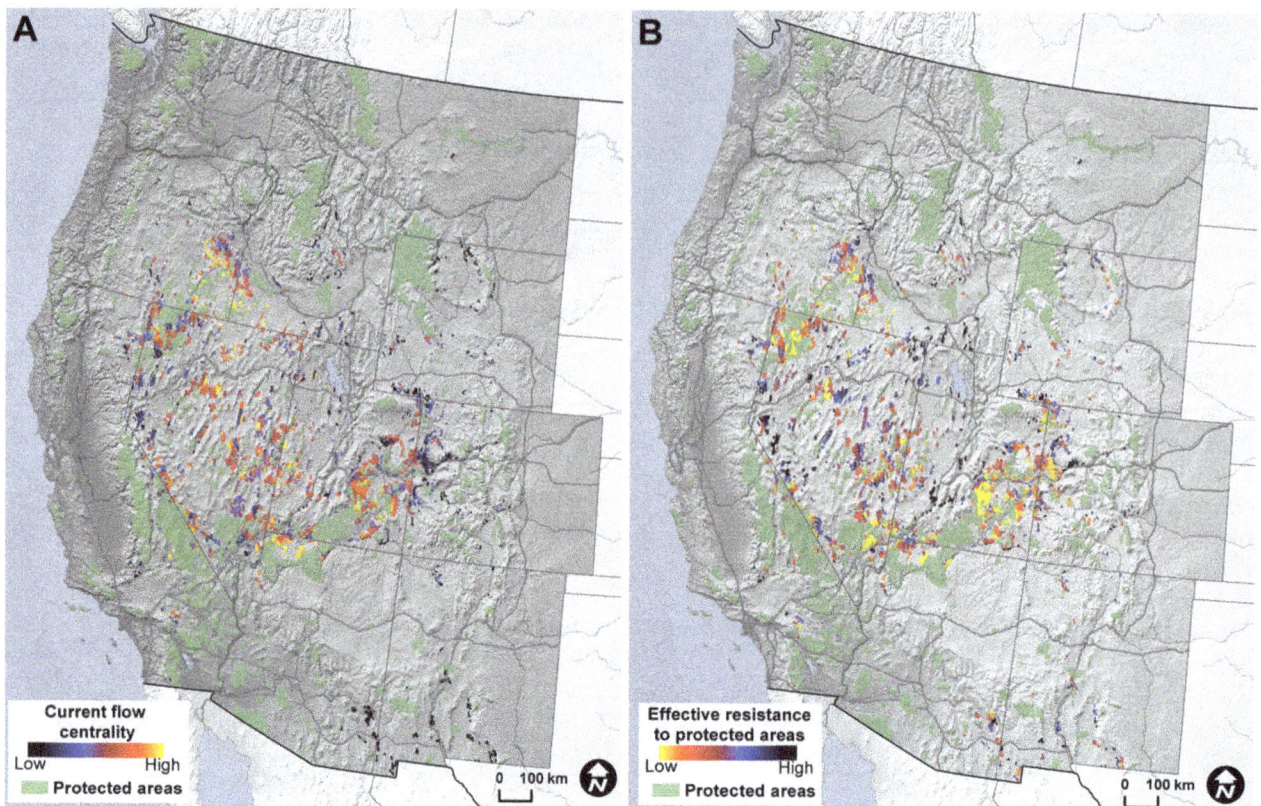

Figure 3 Current flow centrality (A) and effective resistance to protected areas (B) among high conservation value areas identified across unprotected, roadless BLM lands in the western United States. The 95th percentile of values is shown in yellow (scale is reversed for effective resistance). Major interstate highways also are shown.

and effective resistance can be combined to understand how new PAs might both facilitate and enhance ecological connectivity over administratively complex landscapes.

Our results suggest that intact, unprotected BLM lands disproportionately constitute the "connective tissue" among multiple PAs and jurisdictions. In fact, our cross-jurisdictional comparison revealed that new protections on BLM lands could do more to promote connectivity, on average, than new protections on other federal jurisdictions with many existing PAs. When compared to other federal jurisdictions, BLM lands were often proximate to existing PAs, presented fewer topographic (e.g., steep slopes) and hydrographic barriers to ecological flows, were relatively roadless, and generally encompassed lower levels of human modification (see also Dickson et al. 2014). In terms of connectivity, these unique qualities of BLM lands should be considered by efforts to delineate new PAs within this domain to increase the ecological effectiveness of the U.S. PA network as a whole.

Although our approach to modeling ecological connectivity was not conditioned on the movement parameters

of a particular focal species or taxonomic group, our results revealed patterns of regional connectivity overlapping with those observed by other studies centered on one or more species of conservation concern (see Krosby et al. 2015). For example, the central portion of our study area encompassed much of the highest priority habitat for Greater Sage-Grouse (*Centrocercus urophasianus*), in terms of potential connectivity among populations (Knick et al. 2013). Most of this habitat falls within BLM's jurisdiction. The areas of important ecological connectivity we identified on BLM lands also support long-distance migrations by herds of pronghorn antelope (*Antilocapra americana*) moving between Oregon and Nevada (Collins 2016) and through the upper Green River Basin of Wyoming (Seidler et al. 2014), as well as mule deer (*Odocoileus hemionus*) in Colorado (Lendrum et al. 2013) and Wyoming (Sawyer et al. 2012). Remaining, intact expanses of BLM land such as these should be the focus of efforts to maintain, restore, or establish key habitat linkages for wide-ranging species, and could be used to define new PAs that can secure the long term maintenance of connectivity. Concomitantly, thoughtfully

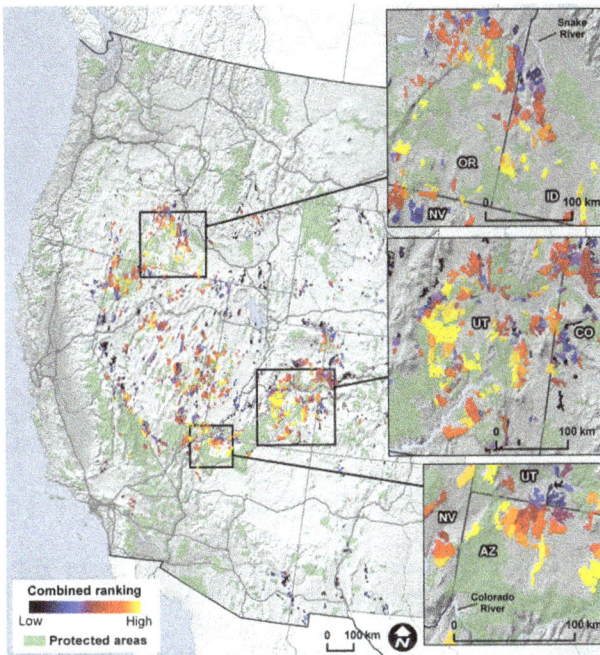

Figure 4 Result of a ranked combination of high conservation value areas with both high current flow centrality and low effective resistance across unprotected, roadless BLM lands in the western United States. The 95th percentile of the ranked combination values is shown in yellow. Insets detail concentrations of highly ranked conservation value areas in the greater Owyhee Canyonlands region (top), central Utah (middle), and northwestern Arizona (bottom). Major interstate highways also are shown.

planned linkages could enhance the movement of other species and ecological flows (Breckheimer *et al.* 2014).

Considering that BLM lands are the focus of increased energy production (USDOI 2011), our results are especially timely. For example, utility-scale energy development within this domain is expected to fragment habitats for wildlife, including key migration corridors (e.g., Beckmann *et al.* 2012). Because reliable spatial data and other scientific information on connectivity often are lacking, there remains uncertainty about the impacts of such developments on animal movement and dispersal (Northrup & Wittemyer 2013). As the demands for energy and other land-based resources in the western United States increase (Jones *et al.* 2015), so, too, does the need to balance these demands against the requirements of federal agencies to maintain large, ecologically functioning landscapes (Federal Land Policy and Management Act [Public Law 94–570, 90 Stat. 2743, *43 U.S.C. 1701*], U.S. Forest Service 2012 Planning Rule [77 FR 21162, Section 219.9]). Federal agencies are attempting to address this need through recent environmental initiatives, such as the strategy for improving the mitigation policies and practices of the Department

of the Interior (Clement *et al.* 2014) and the National Fish, Wildlife, and Plants Climate Adaptation Strategy (National Fish, Wildlife, and Plants Climate Adaptation Partnership 2012). Our West-wide results, which are the first to identify high-connectivity BLM lands that could serve as critical linkages between existing PAs across all public lands, have the potential to inform these efforts.

Our connectivity results can inform regional-scale planning and legislative or executive actions to expand existing systems of Wilderness Areas, Wilderness Study Areas, National Monuments, National Conservation Areas, and other lands in the National Conservation Lands (NCL) system. This system was established "to conserve, protect, and restore nationally significant landscapes that have outstanding cultural, ecological, and scientific values for the benefit of current and future generations" (Omnibus Public Land Management Act of 2009), and with the intent of maintaining or increasing ecological connectivity across NCL units (Secretarial Order 3308). Our methodology and the "wall-to-wall" nature of our results can help to illuminate the conservation context and significance of BLM and other public lands. These data also can be used to bolster and strategically site new land designations that more effectively protect—and truly network—biodiversity and ecological processes. Administrative designations, such as Areas of Critical Environmental Concern (ACEC) on BLM lands, which can be nominated on the basis of protecting natural processes, have the potential to secure the maintenance of regional connectivity among existing PAs. Although ACECs possess a lower level of protection than NCLs and often encompass small areas, they can provide critical stepping stones for the movement of organisms across western landscapes. For example, the Trapper's Point ACEC in west-central Wyoming was designated primarily to preserve a crucial route for ungulate migration amid oil and gas development (BLM 2008). Similarly, our results are being used to target the location of multiple new ACECs or NCL lands intended to facilitate connectivity in the Greater Yellowstone Ecosystem, in the sagebrush habitats of southwestern Colorado, and across the Owyhee Canyonlands, among other areas of the West. Designation of these areas also would help to mitigate the localized impacts of resource extraction, namely energy or minerals developments, that have the potential to impede ecological flows.

Conclusion

The success of conservation efforts in the United States—and other parts of the world—may hinge in part on the role that underappreciated ecological flows play in overall network effectiveness. Indeed, important land protection decisions are happening across the western United States, but are poorly informed regarding the acknowledged importance of potential ecological connectivity within and across public and private lands. Because BLM lands dominate much of the western United States, relatively roadless and intact landscapes in BLM's domain are likely key to the sustained or enhanced movement of organisms and the flow of fundamental ecological processes among PAs and across jurisdictional boundaries. New protections or special designations for select lands within this domain could help to build a significantly stronger and more ecologically effective network of PAs, one that promotes the environmental conditions that enhance landscape connectivity (Krosby et al. 2010) and the capacity for adaptation to future climate change (Dawson et al. 2011). Measures (and maps) of potential ecological connectivity, such as ours, should serve as a principal basis for decisions regarding new and critically needed conservation lands. Designations based on potential ecological connectivity would purposefully and strategically strengthen America's PA network.

Acknowledgments

Funding for this work was provided by the Pew Charitable Trusts. Additional support for C.M.A. was provided by the U.S. Department of the Interior Southwest Climate Science Center. The funders played no role in study design, in the collection, analysis, and interpretation of data, in the writing of the report, or in the decision to submit the article for publication.

Supporting Information

Table S1. Spearman rank correlation values of centrality results for high conservation value areas and the five resistance surfaces

References

Beckmann, J.P., Murray, K., Seidler, R.G. & Berger, J. (2012). Human-mediated shifts in animal habitat use: sequential changes in pronghorn use of a natural gas field in Greater Yellowstone. *Biol. Conserv.*, **147**, 222-233.

Berger, J., Cain, S.L., Cheng, E. *et al.* (2014). Optimism and challenge for science-based conservation of migratory species in and out of U.S. national parks. *Conserv. Biol.*, **28**, 4-12.

Breckheimer, I., Haddad, N.M., Morris, W.F. *et al.* (2014). Defining and evaluating the umbrella species concept for

conserving and restoring landscape connectivity. *Conserv. Biol.*, **28**, 1584-1593.

Brudvig, L., Damschen, E., Tewksbury, J., Haddad, N.M. & Levey, D. (2009). Landscape connectivity promotes plant biodiversity spillover into non-target habitats. *Proc. Natl. Acad. Sci.*, **106**, 9328-9332.

Bureau of Land Management (BLM) (2008). Pinedale Resource Management Plan [WWW Document]. http://www.blm.gov/wy/st/en/programs/Planning/rmps/pinedale/rod_armp.html. Accessed 13 May 2016.

Butchart, S.H.M., Walpole, M., Collen, B. *et al.* (2010). Global biodiversity: indicators of recent declines. *Science*, **328**, 1164-1168.

Clement, J., Belin, A., Bean, M., Boling, T. & Lyons, J. (2014). *A strategy for improving the mitigation policies and practices of the department of the interior*. A report to the Secretary of the Interior from the Energy and Climate Change Task Force, Washington, D.C.

Collins, G.H. (2016). Seasonal distribution and routes of pronghorn in the northern Great Basin. *West. North Am. Nat.*, **76**, 101-112.

Cumming, G., Allen, C., Ban, N. *et al.* (2015). Understanding protected area resilience: a multi-scale, social-ecological approach. *Ecol. Appl.*, **25**, 299-319.

Dawson, T.P., Jackson, S.T., House, J.I., Prentice, I.C. & Mace, G.M. (2011). Beyond predictions: biodiversity conservation in a changing climate. *Science*, **332**, 53-58.

Defries, R., Hansen, A.J., Turner, B., Redi, R. & Liu, J. (2007). Land use change around protected areas: management to balance human needs and ecological function. *Ecol. Appl.*, **17**, 1031-1038.

Dickson, B.G., Zachmann, L.J. & Albano, C.M. (2014). Systematic identification of potential conservation priority areas on roadless Bureau of Land Management lands in the western United States. *Biol. Conserv.*, **178**, 117-127.

Foley, M.M., Halpern, B.S., Micheli, F. *et al.* (2010). Guiding ecological principles for marine spatial planning. *Mar. Pol.*, **34**, 955-966.

Goetz, S.J., Jantz, P. & Jantz, C.A. (2009). Connectivity of core habitat in the Northeastern United States: parks and protected areas in a landscape context. *Rem. Sens. Environ.*, **113**, 1421–1429.

Hansen, A.J. & Defries, R. (2007). Ecological mechanisms linking protected areas to surrounding lands. *Ecol. Appl.*, **17**, 974-988.

Jones, N.F., Pejchar, L. & Kiesecker, J.M. (2015). The energy footprint: how oil, natural gas, and wind energy affect land for biodiversity and the flow of ecosystem services. *Bioscience*, **65**, 290-301.

Knick, S.T., Hanser, S.E. & Preston, K.L. (2013). Modeling ecological minimum requirements for distribution of greater sage-grouse leks: implications for population connectivity across their western range, U.S.A. *Ecol. Evol.*, **3**, 1539-1551.

Krosby, M., Breckheimer, I., Pierce, J. *et al.* (2015). Focal species and landscape "naturalness" corridor models offer complementary approaches for connectivity conservation planning. *Landsc. Ecol.*, **30**, 2121-2132.

Krosby, M., Tewksbury, J., Haddad, N.M. & Hoekstra, J. (2010). Ecological connectivity for a changing climate. *Conserv. Biol.*, **24**, 1686-1689.

Lendrum, P.E., Anderson Jr, C.R., Monteith, K.L., Jenks, J.A. & Bowyer, R.T. (2013). Migrating mule deer: effects of anthropogenically altered landscapes. *PLoS One*, **8**, e64548.

Margules, C.R. & Pressey, R.L. (2000). Systematic conservation planning. *Nature*, **405**, 243-253.

McClure, M.L., Hansen, A.J. & Inman, R.M. (2016). Connecting models to movements: testing connectivity model predictions against empirical migration and dispersal data. *Landsc. Ecol.*, **31**, 1419–1432.

McRae, B., Shah, V. & Mohapatra, T. (2013). *Circuitscape 4 User Guide.*

McRae, B.H. & Beier, P. (2007). Circuit theory predicts gene flow in plant and animal populations. *Proc. Proc. Natl. Acad. Sci. United States Am.*, **104**, 19885-19890.

McRae, B.H., Dickson, B.G., Keitt, T. & Shah, V. (2008). Using circuit theory to model connectivity in ecology, evolution, and conservation. *Ecology*, **89**, 2712-2724.

National Fish, Wildlife, and Plants Climate Adaptation Partnership. (2012). *National fish, wildlife and plants climate adaptation strategy.* Association of Fish and Wildlife Agencies, Council on Environmental Quality, Great Lakes Indian Fish and Wildlife Commission, National Oceanic and Atmospheric Administration, and U.S. Fish and Wildlife Service, Washington, DC.

Newman, M. (2005). A measure of betweenness centrality based on random walks. *Soc. Networks*, **27**, 39-54.

Northrup, J.M. & Wittemyer, G. (2013). Characterising the impacts of emerging energy development on wildlife, with an eye towards mitigation. *Ecol. Lett.*, **16**, 112-125.

Olds, A.D., Connolly, R.M., Pitt, K.A. & Maxwell, P.S. (2012). Habitat connectivity improves reserve performance. *Conserv. Lett.*, **5**, 56-63.

Pressey, R.L. (1994). Ad hoc reservations: forward or backward steps in developing representative reserve systems? *Conserv. Biol.*, **8**, 662-668.

Sawyer, H., Kauffman, M.J., Middleton, A.D., Morrison, T.A., Nielson, M. & Wyckoff, T.B. (2012). A framework for understanding semi-permeable barrier effects on migratory ungulates. *J. Appl. Ecol.*, **50**, 68-78.

Seidler, R.G., Long, R.A., Berger, J., Bergen, S. & Beckmann, J.P. (2014). Identifying impediments to long-distance mammal migrations. *Conserv. Biol.*, **29**, 99-109.

Theobald, D.M. (2013). A general model to quantify ecological integrity for landscape assessments and US application. *Landsc. Ecol.*, **28**, 1859-1874.

Theobald, D.M., Reed, S.E., Fields, K. & Soulé, M. (2012). Connecting natural landscapes using a landscape permeability model to prioritize conservation activities in the United States. *Conserv. Lett.*, **5**, 123-133.

Tilman, D., Fargione, J., Wolff, B. *et al.* (2001). Forecasting agriculturally driven global environmental change. *Science*, **292**, 281-284.

U.S. Department of Interior (USDOI) (2011). Strategic Plan for Fiscal Years 2011-2016. [WWW Document]. https://www.doi.gov/sites/doi.gov/files/migrated/bpp/upload/DOI_FY2011-FY2016_StrategicPlan.pdf. Accessed 1 May 2015.

U.S. Geological Survey (USGS). (2008). National Hydrography Dataset [WWW Document]. http://nhd.usgs.gov/. Accessed 1 June 2014.

U.S. Geological Survey (USGS) (2012). Protected Areas Database of the United States (PADUS) version 1.3 [WWW Document]. http://gapanalysis.usgs.gov/PADUS. Accessed 10 June 2015.

Watson, J., Darling, E., Venter, O. *et al.* (2016). Bolder science needed now for protected areas. *Conserv. Biol.*, **30**, 243-248.

Pokémon Go: Benefits, Costs, and Lessons for the Conservation Movement

Leejiah J. Dorward[1], John C. Mittermeier[2], Chris Sandbrook[3,4], & Fiona Spooner[5]

[1] Department of Zoology, University of Oxford Tinbergen Building, South Parks Road, Oxford OX1 3PS, UK
[2] School of Geography and the Environment, University of Oxford, South Parks Road, Oxford, OX1 3QY, UK
[3] United Nations Environment Programme World Conservation Monitoring Centre, 219 Huntingdon Road, Cambridge CB3 0DL, UK
[4] Department of Geography, University of Cambridge, Downing Place, Cambridge CB2 3EN, UK
[5] Department of Genetics, Evolution and Environment, Centre for Biodiversity and Environment Research, University College London, Gower Street, London, WC1E 6BT, UK

Keywords

Augmented reality; biodiversity; citizen science; conservation; natural history; nature deficit disorder; Pokémon; smartphone; technology; video games.

Correspondence

Leejiah J. Dorward, Department of Zoology, University of Oxford, Tinbergen Building, South Parks Road, Oxford OX1 3PS, UK.
E-mail: leejiah.dorward@zoo.ox.ac.uk

Editor

Edward Game

Abstract

Pokémon Go, an augmented reality (AR) smartphone game, replicates many aspects of real-world wildlife watching and natural history by allowing players to find, capture, and collect Pokémon, which are effectively virtual animals. In this article, we consider how the unprecedented success of Pokémon Go as a smartphone game might create opportunities and challenges for the conservation movement. By encouraging players to go outside and consider various aspects of virtual species' biology, the game could increase awareness and engagement with real-world nature. However, interacting with Pokémon could alternatively encourage exploitation of wildlife or replace players' desire to interact with real-world nature. We suggest a number of ways in which Pokémon Go could be adapted to increase its conservation impact, and how new conservation-orientated AR games could be created. We conclude that Pokémon Go sets a precedent for well-implemented AR games from which the conservation movement could borrow a number of ideas.

Introduction

On 6th July 2016, a San Francisco-based software company launched a large-scale citizen science project in New Zealand, Australia and the United States. Building on a series of scientific programmes started in Japan, the project aimed to improve our understanding of the distribution and abundance of over 150 species by having users enter data with a smartphone app. The launch was a huge success. In the first week there were an estimated 21 million active users in the United States (Allan 2016) and the project's app became the most downloaded in the Apple App Store's history (BBC 2016). Within days the app had surpassed Twitter in its number of daily active users (Allan 2016) and was beating Facebook, Twitter, Instagram and Snapchat in

terms of daily user engagement times (Nelson 2016). Furthermore, anecdotal evidence suggested high levels of behavioural change amongst users, with people making significant adjustments to their daily routines and to the amount of time spent outside in order to increase encounter rates with the target species (Armanet 2016; Butcher 2016). Though this sounds like one of the most successful citizen science initiatives in history, it is, unfortunately, an illusion. The "citizen science" is part of the game "Pokémon Go" and while the user and usage statistics are real, the research is fictional and the species being studied are "Pokémon"—fictional creatures from a series of television programmes and games.

Developed by Niantic, a Google spin-off company, Pokémon Go is the latest output of the hugely successful Pokémon franchise. Using a smartphone's GPS together

Figure 1 Screenshots of gameplay showing the implementation of augmented reality. Panel A shows how the player's real-world location is displayed. The tall white and blue pillar and turquoise squares show the location of a gym, and Pokéstops, respectively. The darker land above the gym shows the location of an urban park where grass Pokémon are more likely to be found. A Pokémon, a Rattata, is to the right of the player. Panel B shows a Pokémon, a Zubat, superimposed into the real world about to be caught. Images courtesy of Niantic, Inc.

with Google Maps the game provides users with an augmented reality (AR) experience where they encounter, catch, and collect virtual species of Pokémon while exploring the real world (Figure 1). Once caught, species are catalogued (by being registered to a "Pokédex") and added to a player's personal collection. The game also designates sites of cultural interest, such as monuments or notable buildings as "gyms" (where players fight their Pokémon) or "Pokéstops" (where players collect items that help them catch and train Pokémon). As the game develops, players are able to "evolve" their Pokémon into more powerful forms and fight Pokémon belonging to other players.

Satoshi Tajiri, who designed the first Pokémon game in 1996, wanted to create a preindustrial play-world for urban children, one that replicated his childhood experiences of collecting insects and crayfish (Allison 2003). Many Pokémon are based on real species; Caterpie, for example, strongly resembles the caterpillar of Eastern Tiger Swallowtail (*Papilio glaucus*, Linnaeus 1758), while the famous Pikachu is based on a Pika (*Ochotonidae* sp.). Also, like real-world species, characters in Pokémon Go are linked to different environments and vary in abundance. In these respects, searching for and collecting Pokémon is a hugely popular, virtual replication of types of natural history observation, such as birdwatching.

In this article, we consider how the unprecedented success of Pokémon Go as a smartphone game might create

opportunities and challenges for the conservation movement. Does its merger of virtual nature and the real environment offer an opportunity to use AR games to achieve conservation success? Or is this merely evidence of how little has changed since Balmford *et al.* (2002, p. 2367) concluded "conservationists are doing less well than the creators of Pokémon at inspiring interest in their subjects"? The potential for games to produce conservation outcomes has been explored previously (Sandbrook *et al.* 2015; Fletcher 2016a). However, with its novel use of AR, colossal popularity, and natural history parallels, Pokémon Go may represent a step-change in the potential relevance and impact of digital games for conservation. We begin by assessing the potential positive and negative impacts Pokémon Go could have for conservation. We then assess what lessons the conservation movement can take from the game, before concluding with recommendations for the future.

Pokémon Go as a conservation opportunity

As Satoshi Tajiri noticed during the 1990s, rapidly growing urban areas offer limited opportunities to connect directly with nature (Allison 2003). This reflects a widespread concern amongst conservationists that people, and particularly young people, have become disconnected from nature through urban living and are therefore less likely to value wildlife and wild places (Balmford *et al.* 2002; Balmford & Cowling 2006; Pergams & Zaradic 2006), a concept widely popularised as Nature Deficit Disorder (Louv 2005). Concern has also been expressed that interest in natural history is fading (Tewksbury *et al.* 2014) and that skilled natural historians and taxonomists are in increasingly short supply (Tancoigne & Dubois 2013). Two aspects of Pokémon Go in its current form have direct implications for addressing these urgent conservation problems.

First, and perhaps most obviously, the game encourages people to get outside. Successful players must explore new areas, visit a variety of environments, and cover lots of ground, preferably on foot. Special "eggs" that players collect, will only "hatch" after a player has walked a certain distance (2, 5, or 10 km), so there is a direct correlation between distance covered and success. There is growing evidence from social and traditional media indicating that Pokémon Go is already driving huge numbers of people outdoors and increasing the time they spend there (Armanet 2016; Butcher 2016). For example, there are reports of "hundreds if not thousands" of young people playing Pokémon Go at the National Mall and Memorial Parks in Washington, D.C., sites which normally attract older generations (Carlton 2016). The location of specific Pokéstops and gyms can play an important role in motivating people to visit sites they might not otherwise be aware of (Butcher 2016; Streitfeld 2016). Though these are primarily sites that have no connection to natural history, many are in municipal parks, nature reserves, and national parks (Figure 1) (Carlton 2016; Zachos 2016).

Pokémon Go makes no explicit attempt to connect people to nonvirtual wildlife or conservation issues, and spending time outside does not always translate into engagement with nature. Nonetheless, there is evidence that people are discovering nonvirtual wildlife while playing Pokémon Go (Brulliard 2016). This type of experience is widespread and has led to the Twitter hashtag #Pokeblitz which helps people to identify "real" species found and photographed while playing (Brulliard 2016). Anecdotally, playing Pokemon Go has led the authors of this article to encounter a European Hedgehog (*Erinaceus europaeus*, Linnaeus 1758) and Tawny Owl (*Strix aluco*, Linnaeus 1758) in areas they had not previously seen them, and find a Madagascar Pond-heron (*Ardeola idae*, Hartlaub 1860) in the wild for the first time.

Second, Pokémon Go exposes users first hand to basic natural history concepts such as species habitat preferences and variations in abundance. Niantic has not released specifics as to how it assigns Pokémon to particular locations, but they do use a number of spatial environmental variables (local climate, vegetation type, distance to water, soil or rock type and land-use classifications such as zoos or parks) to place Pokémon in certain environments (Bogle 2016). For example, "grass Pokémon" tend to occur in parks while water-related types are more likely close to water bodies. There are also four regional species that are continent restricted: Tauros to the Americas, Mr Mime to Western Europe, Farfetch'd to Asia, and the marsupial like Kangaskhan to Australasia. This differentiation captures a fundamental aspect of natural history observation; that exploring new habitats and continents will lead to encounters with different species.

Varying abundance of Pokémon species also exposes players to species accumulation curves; playing Pokémon Go in central London will result in many more Pidgeys and Rattatas than Squirtles, just as it will result in many more observations of pigeons and rats than of turtles. The allure of rarity has been a driving motivation for generations of natural historians to spend long hours outside and explore remote areas, and this clearly applies to the search for unusual Pokémon as well. The news that hundreds of people recently congregated near New York's Central Park to try to find a rare Vaporeon (Worley 2016), for example, will sound familiar to birdwatchers used to similar congregations to see a rare species.

A number of conservation and nature organizations are already trying to make the most of Pokémon Go. A recent editorial in the journal Nature encouraged Pokémon Go players to make a contribution to real-world taxonomy by photographing and identifying real species during their Pokémon hunts ("Gotta name them all" 2016), and the U.S. Fish & Wildlife Service has produced a blog comparing Pokémon to the real species that occur at National Wildlife Refuges (Brigida 2016). It has even been calculated that if Pokémon Go players were identifying real instead of virtual animals, they could collect as much data in 6 days as has been collected in 400 years of natural history effort (August 2016). National Park Service Rangers at the National Mall are offering "Catch the Mall" Pokémon hunts; guided walks where rangers explain the important cultural sites the group pass while catching Pokémon (Zachos 2016). Similar ideas could easily be used in other parks and reserves with expert-led Pokémon tours where both Pokémon and real species are pointed out and discussed. Interactions such as this could leave a lasting legacy if people fall in love with outdoor experiences, become more aware of their local environment, or develop outdoor-orientated habits.

"Pokémon No": potential downsides of Pokémon for conservation

Various commentators have pointed out some of the less positive aspects of the Pokémon Go phenomenon. These include concerns about the game being played in inappropriate ways, such as while driving a car, or places, such as at the Fukushima evacuation zones in Japan, or due to players not paying attention to their surroundings while playing and becoming lost in cave systems, stranded by tides, or being robbed as a result (Khomami 2016). Beyond these more general concerns, there are a number of ways in which Pokémon Go might have negative implications for conservation; here we highlight four notable examples.

First, by drawing players outside, Pokémon Go may create direct negative environmental impacts, such as erosion caused by gamers' footfall. This could be particularly damaging if large numbers are drawn to search for rare Pokémon species in particularly sensitive habitats. However the urban bias within the game as it is currently designed suggests that this is unlikely.

Second, by promoting the idea of "catching" creatures that are subsequently used to fight against each other, the game may create or reinforce utilitarian and exploitative relations between human and nonhuman nature, rather than the message of respect preferred by conservationists.

It is not difficult to imagine Pokémon Go players trying to catch real animals and fight them against each other, inspired by the game. There have already been examples of real animals being "caught" in "Pokéballs" on social media (Desejosdehomem 2016).

Third, the brightly colored, exciting, and easily accessible Pokémon species may distract people from real species and the problems they face (Sandbrook *et al.* 2015). Who cares about critically endangered tigers in a faraway land when there may be a Vaporeon in the nearby park? While there is clearly potential to link an interest in Pokémon to natural history, it should not be assumed that the one will automatically flow from the other. In the aforementioned Washington D.C., example, the director of the National Park Service was compelled to issue a warning to players stating that people need to be wary of stumbling into wildlife around them whilst focusing on their phone screens (Carlton 2016), suggesting that cognitive "engagement" with the real world and its wildlife was limited. Indeed, if it is the ostentatiously fictitious nature of Pokémon that explains their appeal to an audience seeking escape from the perceived mundanity of the nonvirtual world, it could be very challenging to inspire interest in real-world wildlife through the game.

Finally, it has been argued that conservation efforts to reconnect people with nature can have the opposite of the intended effect, because constructing the problem as one of dissociation with nature reinforces the idea that humans occupy a distinct category from nonhuman nature (Fletcher 2016b). On a related note, Schultz (2000) argues that direct experience of nature augments the sense of being part of nature and therefore caring about its conservation, whereas indirect learning (such as through visiting a zoo) fosters an egoistic attitude to nature which is less conducive to supporting conservation. In both cases, it is possible to imagine that experiencing "nature" on screen through playing Pokémon might undermine rather than augment positive and caring relations between people and real nonhuman wildlife.

What can conservation learn from Pokémon Go?

The spectacular success of Pokémon Go provides significant lessons for conservation. Importantly, it suggests that conservation is continuing to lag behind Pokémon in efforts to inspire interest in its portfolio of species, a situation first identified by Balmford *et al.* (2002). We see two possible explanations for this situation. First, Pokémon Go is extremely user-friendly, and has none of the bar-

riers to entry present in many types of natural history observation. It requires commonly available equipment (smartphones are widely owned and the game is free to download), no special knowledge (gameplay is very simple and the app locates and identifies Pokémon), no specific location (a short walk in any town or village is likely to produce interesting Pokémon), and rare "species" can be found in easily accessible and densely populated areas. By comparison real-world natural history activities such as birdwatching require specialist equipment (binoculars and field guides at a minimum), knowledge, and skills, access to certain habitats and locations, and a willingness to travel out of towns and cities for increased chances of seeing rare species. Studying taxonomic groups such as insects, plants and mammals often present steeper challenges. Finding ways to break down these barriers to engagement with real-world biodiversity is a priority for conservation.

Second, the Pokémon creatures encountered are not only species but also characters with specific story lines and histories from the Pokémon universe. Modern natural history study, in contrast, tends to frame itself entirely in scientific terms, avoiding anthropomorphizing its subjects. This overemphasis on a scientific framing may miss important opportunities for engagement based on affective relations with nonhuman nature (Lorimer 2015). Publicity surrounding the death of Cecil the lion, for example, highlights how easily individual animals can become anthropomorphized and the wealth of public interest that they often capture when this happens.

Conservation could potentially use digital games to address both of these issues, although with the risks and caveats outlined above. Most directly, there is clear potential to modify Pokémon Go itself to increase conservation content and impact above and beyond simply bringing gamers into closer physical proximity to nonhuman wildlife as a by-product of the game. Pokémon Go could be adapted to enhance conservation benefits by: (a) making Pokémon biology and ecology more realistic (e.g., stronger links between Pokémon species habitat requirements and real-world habitats to encourage learning about ecology); (b) adding real species to the Pokémon Go universe to expose those species to a huge number of users, and creating opportunities to raise awareness about them (e.g., the Zoological Society of London's endangered and unusual "EDGE" species); (c) deliberately placing Pokémon in more remote natural settings rather than urban areas to draw people to experience nonurban nature; or (d) adding a mechanism for users to catalogue real species, building on the popularity of the "Pokeblitz" concept (e.g., newly developed websites such as Pokemapper [www.pokemapper.co] and Poke Radar [www.pokeradar.io] already map the "distributions" of different Pokémon in ways that are striking similar to citizen science projects such as eBird [www.ebird.org], and iNaturalist [www.inaturalist.org]).

Less directly, lessons from Pokémon Go could be applied to conservation through the development of new conservation-focused AR games. Following the model of Pokémon Go, games that encourage users to look for real species could provide a powerful tool for education and engagement. AR could also be used in zoos and protected areas to provide visitors with information about species and their habitats. It has been argued that such virtual engagement with conservation issues through games can have a greater affective influence on gamers than first-hand experience not mediated through screens (Fletcher 2016a). Though these ideas are potentially promising, it is important to note that Pokémon Go is specifically designed to be entertaining and builds off a well-established brand and nostalgia of people who grew up with the Pokémon franchise, benefits that would likely not be applicable to a conservation-focused app. Given the cost and difficulty of developing new games from scratch, seeking to modify a successful, existing product such as Pokémon Go may be the best way for conservation to benefit from AR games (Sandbrook *et al.* 2015).

Conclusion

Pokémon Go demonstrates that cleverly implemented AR games can reach millions of people and trigger substantial levels of behavioral change. In its basic features, the game has strong parallels to natural history observation and encourages outdoor recreation, both of which can help to establish interest in conservation and build conservation ethics (Kellert 1985; McFarlane & Boxall 1996) but are widely viewed as declining (Pergams & Zaradic 2006; Tewksbury *et al.* 2014). Though there are potential pitfalls that must be carefully considered, we see this game as an exciting opportunity to build interest in natural history observation and learning. There are ways to do this within the framework of Pokémon Go itself, through the development of related AR games, or by simply taking some of the lessons of Pokémon Go's appeal and applying them to natural history education in general.

Acknowledgments

We would like to thank Paul Jepson for his helpful discussion on Pokémon Go, Niantic, Inc. for permission to use screenshots from the game and two anonymous reviewers for their very helpful and constructive comments.

References

Allan, R. (2016). Pokémon GO is now the biggest mobile game in U.S. history. Available from https://www.surveymonkey.com/business/intelligence/pokemon-go-biggest-mobile-game-ever/. Accessed 31 July 2016

Allison, A. (2003). Portable monsters and commodity cuteness: Pokémon as Japan's new global power. *Postcol. Stud.*, **6**, 381-395.

Armanet, J. (2016). Could Pokémon Go improve people's health? Available from https://www.theguardian.com/healthcare-network/2016/jul/27/pokemon-go-improve-health-walking. Accessed 31 July 2016

August, T. (2016). Pokémon-Go players could capture 400 years of wildlife sightings in 6 days. Available from http://www.ceh.ac.uk/news-and-media/blogs/pok%C3%A9mon-go-players-could-capture-400-years-wildlife-sightings-6-days. Accessed 6 August 2016

Balmford, A., Clegg, L., Coulson, T., Taylor, J. & Street, D. (2002). Why conservationists should heed Pokémon. *Science*, **295**, 2367

Balmford, A. & Cowling, R.M. (2006). Fusion or failure? The future of conservation biology. *Conserv. Biol.*, **20**, 692-695.

BBC. (2016). Should you believe those Pokémon Go download numbers? Available from http://www.bbc.co.uk/news/magazine-36868076. Accessed 31 July 2016

Bogle, A. (2016). The story behind "Pokémon Go"s' impressive mapping. Available from http://mashable.com/2016/07/10/john-hanke-pokemon-go/#qIS7o8y1CmqN. Accessed 31 July 2016

Brigida, D. (2016). The Pokémon around us. Available from https://www.fws.gov/news/blog/index.cfm/2016/7/14/The-Pokemon-Around-Us. Accessed 1 August 2016

Brulliard, K. (2016). If you must play Pokémon Go, "catch" some real animals while you're at it. Available from https://www.washingtonpost.com/news/animalia/wp/2016/07/13/if-you-must-play-pokemon-go-catch-some-real-animals-while-youre-at-it/?tid=a_inl. Accessed 1 August 2016

Butcher, A. (2016). Pokémon Go see the world in its splendor. Available from http://www.nytimes.com/2016/07/17/opinion/sunday/pokemon-go-see-the-world-in-its-splendor.html. Accessed 31 July 2016

Carlton, J. (2016). "Pokémon Go" gives boost to national parks. Available from http://www.wsj.com/articles/pokemon-go-gives-boost-to-national-parks-1468447301. Accessed 1 August 2016

Desejosdehomem. (2016). Pokémon da vida real. Available from https://www.instagram.com/p/BI8aH1ig2I8/. Accessed 12 August 2016

Fletcher, R. (2016a). Gaming conservation: Nature 2.0 confronts nature-deficit disorder. *Geoforum*, doi:10.1016/j.geoforum.2016.02.009

Fletcher, R. (2016b). Connection with nature is an oxymoron: a political ecology of "nature-deficit disorder." *J. Environ. Educ.*, **0**, 1-8.

Gotta name them all: how Pokémon can transform taxonomy. (2016). Available from http://www.nature.com/news/gotta-name-them-all-how-pok%C3%A9mon-can-transform-taxonomy-1.20275. Accessed 1 August 2016

Kellert, S.R. (1985). Birdwatching in American society. *Leisure Sci.*, **7**, 343-360.

Khomami, N. (2016). Pokémon Go: London players robbed of phones at gunpoint. Available from https://www.theguardian.com/uk-news/2016/jul/30/three-pokemon-go-players-robbed-of-phones-at-gunpoint-in-london. Accessed 1 August 2016

Lorimer, J. (2015). *Wildlife in the Anthropocene*. University of Minnesota Press, Minnesota.

Louv, R. (2005). *Last child in the wood*. Algonquin Books, New York, NY, USA.

McFarlane, B.L. & Boxall, P.C. (1996). Participation in wildlife conservation by birdwatchers. *Human Dimen. Wildl.*, **1**, 1–14.

Nelson, R. (2016). Mobile users are spending more time in Pokémon Go than Facebook. Available from https://sensortower.com/blog/pokemon-go-usage-data. Accessed 31 July 2016

Pergams, O.R.W. & Zaradic, P.A. (2006). Is love of nature in the US becoming love of electronic media? 16-year downtrend in national park visits explained by watching movies, playing video games, internet use, and oil prices. *J. Environ. Manage.*, **80**, 387-393.

Sandbrook, C., Adams, W.M. & Monteferri, B. (2015). Digital games and biodiversity conservation. *Conserv. Lett.*, **8**, 118-124.

Schultz, P.W. (2000). Empathizing with nature: the effects of perspective taking on concern for environmental issues. *J. Soc. Issues*, **56**, 391-406.

Streitfeld, D. (2016). Chasing Pokémon, a baby step toward virtual reality. Available from http://www.nytimes.com/2016/07/22/technology/personaltech/chasing-pokemon-a-baby-step-toward-virtual-reality.html. Accessed 1 August 2016

Tancoigne, E. & Dubois, A. (2013). Taxonomy: no decline, but inertia. *Cladistics*, **29**, 567-570.

Tewksbury, J.J., Anderson, J.G.T., Bakker, J.D., *et al.* (2014). Natural history's place in science and society. *Bioscience*, **64**, 300-310.

Worley, W. (2016). Pokémon Go: video shows moment rare Vaporeon appears in Central Park and all hell breaks loose. Available from http://www.independent.co.uk/news/world/americas/pokemon-go-video-central-park-vaporeon-rare-running-new-york-a7140801.html. Accessed 1 August 2016

Zachos, E. (2016). Can Pokémon Go get players into national parks? Available from http://news.nationalgeographic.com/2016/07/pokemon-go-national-parks-level-up/. Accessed 1 August 2016

4

Embedding Evidence on Conservation Interventions Within a Context of Multilevel Governance

Johan Ekroos[1], Julia Leventon[2], Joern Fischer[3], Jens Newig[2], & Henrik G. Smith[1,4]

[1] Centre for Environmental and Climate Research, Lund University, Ecology Building, Lund, Sweden
[2] Research Group Governance and Sustainability, Faculty of Sustainability, Leuphana Universität Lüneburg, Lüneburg, Germany
[3] Institute of Ecology, Faculty of Sustainability, Leuphana Universität Lüneburg, Lüneburg, Germany
[4] Department of Biology, Lund University, Ecology Building, Lund, Sweden

Keywords

Biodiversity conservation; ecology; environmental decision-making; evidence-informed conservation; general principles; science-policy interface.

Correspondence

Johan Ekroos, CEC, Lund University, Sölvegatan 37, S-223 62 Lund, Sweden.
E-mail: johan.ekroos@cec.lu.se, jeekroos@gmail.com

Editor

David Lindenmayer

Abstract

We outline a conceptual strategy for implementing conservation interventions in a multiscale, multiactor, and multilevel governance world. Using farmland as an example, we argue that conservation interventions should be implemented within a multiscale framework of guiding ecological principles. In this context, findings from multilevel governance research can inform a nuanced understanding of the role of evidence in conservation governance and decision-making. We propose that principles of evidence-based conservation can be used to refine guiding ecological principles across scales, thereby creating a comprehensive evidence base that underpins decision-making. This evolving evidence base, in turn, should be operationalized by considering the fit of ecologically relevant scales to governance levels, paying explicit attention to issues such as democratic legitimacy and interplay with existing governance structures. We outline two specific steps for meeting this challenge. Drawing on a strategic combination of conservation interventions, guiding ecological principles, and insights from multilevel governance research promises to improve both the effectiveness and legitimacy of conservation action.

Introduction

Evidence-based conservation is rapidly gaining currency in both scientific and policy circles. It arose from the notion that much conservation effort was not based on solid facts but instead on beliefs or even myths (Pullin & Knight 2001). The origins of the evidence-based conservation approach can be traced back to medicine, where the Cochrane collaboration was established in 1993 to systematically review the evidence for different medical interventions (Bero & Rennie 1995). Seeing the value in a similar approach for biodiversity conservation, evidence-based conservation was first suggested in the early 2000s (Sutherland *et al.* 2004). Since then, efforts to facilitate evidence-based conservation via systematic reviews, systematic mapping, and conservation evidence synopses have increased substantially.

Despite its success in the academic community, evidence-based conservation has also been met with criticism (Adams & Sandbrook 2013, but see Haddaway & Pullin 2013), including its ability to link to policy-making (Greenhalgh & Russell 2009). A major issue is that existing evidence on conservation is heavily biased toward small-scale interventions (Fazey *et al.* 2005). As a result, several small-scale conservation interventions have been promoted by conservationists and implemented by landowners, but it remains unclear how these multiple small-scale interventions fit together, and how they intersect with other considerations, to create effective conservation governance (Pelosi *et al.* 2010). Within the field of evidence-based conservation, better integrating evidence into decision support systems has recently been identified as a key challenge (Dicks, Walsh *et al.* 2014). However, greater attention also needs to be paid to the process of decision-making. This is because real-world conservation cannot follow a linear logic of knowledge transfer, but instead takes place within complex and sometimes messy social contexts, where issues of vested interests, power and democratic considerations influence conservation decisions (Adams & Sandbrook 2013). Such issues play out over governance systems comprising multiple jurisdictional levels that involve a variety of actors, their preferences and powers (Bache & Flinders 2004). Notably, governance issues are beginning to be considered in evidence-based conservation (Macura *et al.* 2013), but multilevel governance (especially as it applies to conservation in farmland) has not been addressed in this context to date.

In this article, we argue that applying evidence-based conservation interventions will be most effective by: (1) embedding such interventions within a multiscale guiding framework of ecological principles and (2) drawing on concepts and understandings from multilevel governance research to be more cognisant of the governance context in which conservation ultimately takes place. We focus primarily on European farmland because here, conservation interventions are particularly dominated by data collected at small scales (Kleijn *et al.* 2011). However, many of the issues raised will be equally relevant in other contexts. We first discuss the existing evidence base for farmland conservation interventions, and reflect on the challenges posed by most evidence pertaining to small spatial scales. The subsequent two sections highlight how these challenges could be overcome by explicitly embedding local-scale interventions within a broader multiscale context, including an appreciation of relevant institutions and actors, as well as their (power) relationships. Our goal is to guide scientists who wish to have a policy impact toward more nuanced thinking on the complex sociopolitical context influencing conservation decisions. This, in turn, may help to overcome some difficulties in implementing effective conservation interventions in farmland.

Generating a systematic understanding of conservation interventions

Evidence-based conservation heavily relies on systematic reviews of existing research (Dicks, Walsh *et al.* 2014). Following the identification of a sufficiently specific question of relevance to stakeholders, a review protocol is developed, and evidence is collected, assessed for its quality, and formally evaluated (Pullin & Stewart 2006). A key objective of systematic reviews is to avoid intentional or unintentional bias in the evaluation of evidence, which may easily enter more qualitative evaluations (Roberts *et al.* 2006). A less formal approach is to summarize evidence in the form of systematic mapping, which seeks to gather an unbiased database of research within an area, without the goal of answering a specific question. Finally, evidence can be gathered in comprehensive synopses (Dicks, Hodge *et al.* 2014). In contrast to system-

atic reviews, which focus on specific questions, synopses address broad topics such as farmland conservation interventions in general. While they differ in the level of detail they address, all of these comparisons allow for evaluating whether potential conservation interventions actually benefit biodiversity.

Despite clear benefits of systematically collating an evidence base, some factors limit our understanding of how individual pieces of evidence scale up into a comprehensive approach to conservation. This is particularly problematic in situations where most of the existing research has focused on local, small-scale interventions (e.g., in European farmland). First, although systematic reviews actively address context-dependent effects of various interventions, the necessary evidence for transferring an intervention to a new context may be lacking. For example, interventions designed from an understanding of Western European farming systems may be detrimental to promoting biodiversity in an Eastern European context (Sutcliffe *et al.* 2013). More critically, many researchers appear to assume that generating a stronger evidence base is a primary need to drive better conservation outcomes, thereby (often inadvertently) framing governance as a technocratic process. However, evidence about specific conservation interventions per se will constitute only one factor influencing policy decisions. How policy responds to evidence is shaped by the politics of policy-making, power, and the nature of the evidence itself (Juntti *et al.* 2009). Moreover, different institutional arrangements facilitate the integration of forms of knowledge in different ways, and offer variable opportunities for mediating actors' conflicting views on conservation. Importantly, integration of knowledge rarely happens at just the national or EU level, meaning that politics and polity influence the integration of knowledge at numerous points throughout a governance system (Leventon & Antypas 2012).

None of these challenges fundamentally undermines the general value of using the best available evidence for any given intervention, that is, the general notion of evidence-based conservation is without doubt better than haphazard approaches to conservation. However, the practical value of small-scale interventions could be greatly improved if they were more explicitly embedded within an understanding of regional ecological contexts, including explicit recognition of key principles of multilevel governance.

Embedding local interventions within a multiscale ecological context

One of the criticisms of evidence-based conservation has been that it may be overly reductionist (Adams & Sandbrook 2013, but see Pullin *et al.* 2009). Particularly in farmland, there can be a risk that complex problems are being subdivided into a series of specific local interventions, which are then tested for their effectiveness (e.g., Scheper *et al.* 2013). Thus, the best evidence at hand may not be relevant for those spatial scales controlling population-level responses, such as landscape complementation/supplementation, or meta-population processes (Smith *et al.* 2014). The lack of availability of larger-scale evidence can thus lead to an unfortunate mismatch in knowledge of the relative effects of local-scale interventions versus regional-scale ecological processes, and thus also to incomplete knowledge about where local conservation interventions would be most effective.

The above mismatch could be bridged if the implementation of specific interventions was explicitly considered in a context of general, multiscale guiding principles for biodiversity conservation. Guiding ecological principles have been based on conceptual or mechanism-informed models (e.g., Lindenmayer & Franklin 2002; Hanski 2011). For farmland systems, one of the best known general principles is the notion of promoting habitat heterogeneity at multiple spatial scales (Benton *et al.* 2003); other widely agreed upon principles are to expand the amount of native vegetation in a given landscape or buffer sensitive areas (Fischer *et al.* 2006).

Integrating specific intervention-guided conservation with a deeper understanding of moderating regional ecological contexts requires combining intervention-driven conservation thinking with "holistic" conservation thinking. As an example, incentives intended to benefit farmland biodiversity may increase one resource, such as food availability, but fail to provide other key resources such as nesting sites or overwintering habitats (Kleijn *et al.* 2011). Thus, it is not only important to know whether individual interventions benefit a particular outcome locally (e.g., species richness), but also how locally implemented interventions contribute to larger-scale ecological processes such as landscape complementation, spillover or meta-population dynamics (Smith *et al.* 2014). Embedding evidence on conservation interventions within general ecological principles across a range of spatial scales would add an improved understanding of the context to specific interventions.

While integration is essential for effective decision support systems (Dicks, Walsh *et al.* 2014), it is challenging because of geographical variation in environmental conditions, as well as in habitat specificity and mobility of organisms. Therefore, there may always be exceptions where a given general principle may not be desirable for biodiversity conservation. For example, despite the general value of connectivity, there may be instances in which wildlife corridors are undesirable because they exacerbate existing problems or cause new ones (Simberloff

et al. 1992). Evidence-based conservation thus retains an important role when embedded within a regional ecological context, in that evidence should be generated to validate, refute and refine an emerging set of guiding principles across multiple spatial scales.

Considering a context of multilevel governance

Embedding specific conservation interventions within the context of multiscale ecological principles could help alleviate the problem that a focus on local conservation interventions is unable to deal with multiscale phenomena. However, it does not yet address another main criticism raised in the past, namely that existing work on evidence-based practice in general has been overly technocratic in its conception of real-world policy implementation and governance (Greenhalgh & Russell 2009, but see Pullin *et al.* 2009). In this context, drawing on insights from multilevel governance research could help to understand key challenges of implementing evidence-based interventions in practice. Such insights can be applied to both specific interventions, as well as to a more general, multiscale approach to conservation that is based on guiding ecological principles.

Multilevel governance research brings together multiple forms of knowledge with the issues of politics, power, democratic legitimacy, accountability, and actor capacity for conservation governance. Multilevel governance systems have evolved, among others, in the United States and the EU as layered systems of decision–making, whereby each level (e.g., supranational, national, regional, local) has considerable autonomy (Bache & Flinders 2004)—although lower levels of government are typically bound by the rules set at higher levels. That said, multilevel governance is more than just multiple levels of government, and central themes in multilevel governance research include actors, their relationships and interactions within and between levels, as well as the way they make decisions (e.g., Hooghe & Marks 2003).

A multilevel governance system presents multiple science-policy interfaces at which decision-making engages with ecological evidence. Broadly, two types of multilevel governance have emerged (Hooghe & Marks 2003). Type I refers to nested systems of all-purpose jurisdictions, such that different sectors (e.g., biodiversity, water, energy) are integrated at each level of decision-making. In this form, decisions are taken at multiple jurisdictional levels. By contrast, Type II multilevel governance refers to more fluid, potentially overlapping levels of decision-making, which are functionally specific, that is, concerning either biodiversity or water, or another function. This, in turn, allows matching the scale of decision-making with the scales most relevant to a particular issue (e.g., watersheds or landscapes, etc.). For example, under the EU's Water Framework Directive, the river basin district is introduced as a management level that crosses boundaries of existing administrative jurisdictions. Multiple levels of administrative units within the river basin must collaborate to plan and implement water management (Newig & Koontz 2014). Moreover, decision spaces are also characterised by the actors engaged in decision-making. In some systems, central government units are primarily responsible for decision-making, implementation and enforcement, whereas in more participatory systems, a broader range of civil society is involved. Thus, science-policy interfaces can occur at a range of decision-making levels.

By understanding the principles of multilevel governance, we can think critically about the scale of knowledge needed for decision-making. For example, the EU embodies a principle of subsidiarity, whereby decisions should be made at the most locally appropriate level. Environmental federalism (Oates 2002) and institutionalism (Young 2002) advise that for environmental decision-making, this is the level that most closely matches the scale of the environmental issue. If decisions are taken on too local a level, jurisdictions will not cover the full area over which the ecological process plays out, so common good dilemmas might prevail, and rules and actions will be fragmented (Cumming et al. 2006). Similarly, if decisions are made on too high a level, decision-makers risk lacking the relevant knowledge or using knowledge that is too general to (appropriately) inform the management of very different areas. Although such spatial "fit" to ecological processes is important, it is not the only consideration in multilevel governance. For example, more local decisions can have higher democratic legitimacy because they are closer to citizens, but may be more development-oriented than decisions made at higher levels (Newig & Fritsch 2009). In practice, multilevel governance systems in the EU tend to have multiple levels of decision-making over a single issue, such that higher levels set agendas and goals, and lower levels translate these into practice in context-appropriate ways.

Multilevel governance provides a lens for understanding decision-making challenges that differs from dominant understandings in the conservation literature. Notably, the ecologically appropriate level of decision-making will vary for different species (e.g., grassland plants vs. large-ranging mammals), or different aspects of biodiversity (e.g., rare species vs. ecosystem service providers; see Kleijn et al. 2011). An effective response to the wide variety in governance and ecological systems therefore calls for the creation of new decision-making forums that engage diverse constellations of actors and knowledge across spatial and temporal scales, in ways relevant to specific decisions (Paavola et al. 2009). This in turn raises issues of democratic legitimacy and accountability, because for citizens it may become difficult to assume democratic responsibilities when being part of overlapping sites of decision-making (Peters & Pierre 2005). It also increases governance complexity and the likelihood of negative interplay, where actions taken in one policy arena hinder those in another (Moss 2003; Paavola et al. 2009).

Conservation interventions in a complex, multiscale world

In practice, small-scale conservation interventions would be substantially enhanced if they took more careful account of the governance context—in many instances, governance arrangements will be just as important in shaping the success of a particular intervention as the ecological science underpinning the intervention. While locally specific conservation interventions can be informed by solid empirical evidence, this, on its own, is necessary but insufficient to inform effective governance of entire landscapes or regions. We therefore highlight two challenges in accounting for the governance context in evidence-informed decision–making, and two practical suggestions for how conservation decision-making can address these challenges.

In terms of challenges, first we contend that a management approach based on a combination of context-dependent, local conservation interventions, and general ecological principles should underpin decision-making at all spatial scales, by drawing on, and seeking out, relevant evidence at multiple scales to fill evidence gaps. Second, we argue that this approach should be embedded within a multilevel governance context (Figure 1). Doing so implies that decisions will need to be taken at multiple governance levels, with some match between ecological scale and governance level. These governance levels should be informed by considering appropriate democratic legitimacy, fit to ecological scale, and fit to existing governance structures.

To meet these challenges, the first step, according to our nested framework, is to implement conservation interventions in the context of nested, regional-scale conservation plans. Such plans will draw on general models, experience (including traditional knowledge), and large-scale data sets. They will be refined at increasingly local levels, drawing on empirical evidence relevant to local-scale interventions in particular contexts, and accounting for prioritizations concerning particularly important habitat types (in Europe, e.g., Natura 2000 sites). Broader-scale ecological principles will need to be communicated primarily to higher levels of governance (e.g., on the need

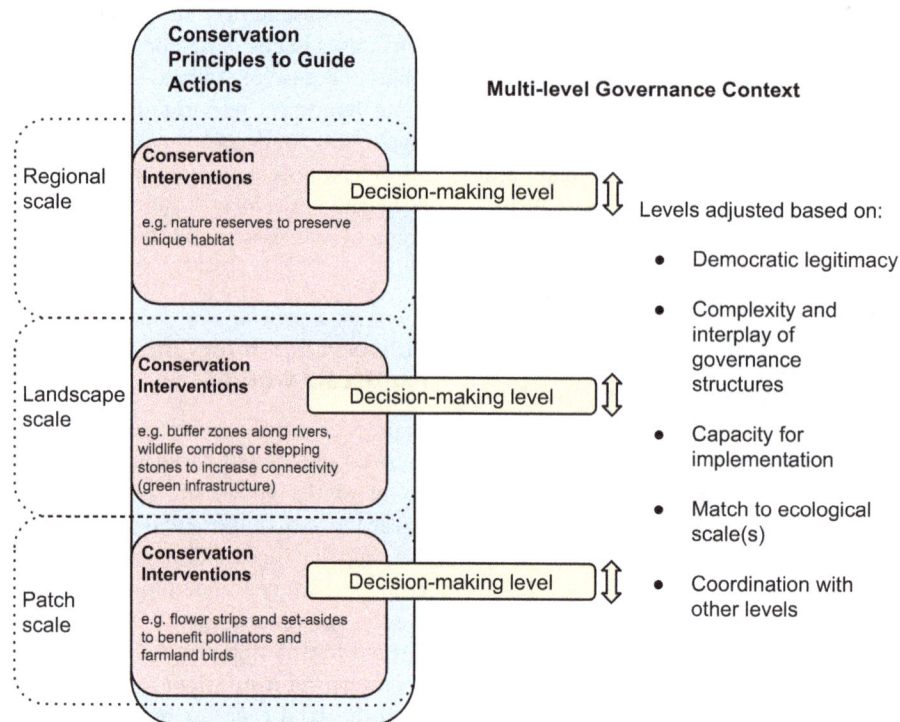

Figure 1 Schematic showing how specific conservation interventions should be considered within a context of general ecological principles (i.e., conceptual or mechanistic models), which in turn plays out within the context of a multilevel governance system to facilitate the fit between decision-making and ecological scales. The body of conservation evidence and principles (blue, central box) should be drawn on at a range of ecological scales in order to design scale-specific conservation interventions (pink boxes) to best match landscape contexts where individual interventions would be most efficient. These interventions should in turn be designed and adopted in coordination with aims set at a range of decision-making levels (yellow boxes). Decision-making levels should be adjusted to match ecological scales, but also to consider the implications to a range of context-dependent concerns highlighted on the figure.

for functional green infrastructure at regional levels). In contrast, local studies will help inform individual farmers and extension services to shape the way in which targets are met through on-ground actions (e.g., on how local practices can best promote a functional green infrastructure across larger spatial scales). This is shown by the regional- and landscape-scale interactions between conservation interventions and decision-making levels in Figure 1. Such an approach will ensure that local interventions are not piecemeal, but work together as part of a multilevel biodiversity strategy.

The second step will be to ensure that scientists, policy makers, and practitioners participate in the cocreation of policy-relevant science, going beyond identifying stakeholder-relevant questions for systematic reviews. From the outset, scientists and decision-makers should jointly consider how administrative and ecological scales fit in order to balance democratic legitimacy and ecological efficacy. Plans therefore will not be forced into existing, mismatched jurisdictional units, but instead

will be relevant to biodiversity conservation, while also being mindful of the complexity this creates (e.g., through interplay with other environmental issues, such as water management). By being clear as to the types and scales of knowledge needed, and the limitations of existing knowledge to inform policy, decision-makers will also play a role in highlighting knowledge gaps. We thus frame decision-makers as actively participating stakeholders in shaping what evidence base is needed for conservation, rather than framing conservation policy as something that must respond to the agenda of scientists who produce evidence. As a consequence, there is a strong need to develop practical solutions, based on a joint effort by researchers, decision-makers and land-use planners, on how to integrate evidence-based practices and general ecological principles within a multilevel governance framework. Through embedding locally implemented conservation interventions within a broader context, we are confident they would gain both in legitimacy and effectiveness.

Acknowledgments

We thank Mark Brady and Felix Schläpfer for constructive discussions on the economic side of multilevel governance and Tamara Schaal for editorial input. David Lindenmayer and four reviewers provided valuable feedback to an earlier draft of this article. JE was supported by the strategic research environment BECC and Formas, while HGS was supported by the strong research area SAPES (through Formas). The ERA-NET project MULTAGRI provided support to HGS, JE, JF, JN, and JL.

References

Adams, W.M. & Sandbrook, C. (2013). Conservation, evidence and policy. *Oryx*, **47**, 329-335.

Bache, I. & Flinders, M.V., editors. (2004). *Multi-level governance*. Oxford University Press, Oxford.

Benton, T.G., Vickery, J.A. & Wilson, J.D. (2003). Farmland biodiversity: is habitat heterogeneity the key? *Trends Ecol. Evol.*, **18**, 182-188.

Bero, L. & Rennie, D. (1995). The Cochrane Collaboration: preparing, maintaining, and disseminating systematic reviews of the effects of health care. *JAMA*, **274**, 1935-1938.

Cumming, G.S., Cumming, D.H. & Redman, C.L. (2006). Scale mismatches in social-ecological systems: causes, consequences, and solutions. *Ecol. Soc.*, **11**, 14.

Dicks, L.V., Hodge, I., Randall, N.P., *et al.* (2014). A transparent process for "evidence-informed" policy making. *Conserv. Lett.*, **7**, 119-125.

Dicks, L.V., Walsh, J.C. & Sutherland, W.J. (2014). Organising evidence for environmental management decisions: a '4S' hierarchy. *Trends Ecol. Evol.*, **29**, 607-613.

Fazey, I., Fischer, J. & Lindenmayer, D.B. (2005). What do conservation biologists publish? *Biol. Conserv.*, **124**, 63-73.

Fischer, J., Lindenmayer, D.B. & Manning, A.D. (2006). Biodiversity, ecosystem function, and resilience: ten guiding principles for commodity production landscapes. *Front. Ecol. Environ.*, **4**, 80-86.

Greenhalgh, T. & Russell, J. (2009). Evidence-based policymaking: a critique. *Perspect. Biol. Med.*, **52**, 304-318.

Haddaway, N. & Pullin, A.S. (2014). Evidence-based conservation and evidence-informed policy: a response to Adams & Sandbrook. *Oryx*, **47**, 336-338.

Hanski, I. (2011). Habitat loss, the dynamics of biodiversity, and a perspective on conservation. *AMBIO*, **40**, 248-255.

Hooghe, L. & Marks, G. (2003). Unraveling the central state, but how? Types of multi-level governance. *Am. Polit. Sci. Rev.*, **97**, 233-243.

Juntti, M., Russel, D. & Turnpenny, J. (2009). Evidence, politics and power in public policy for the environment. *Environ. Sci. Policy*, **12**, 207-215.

Kleijn, D., Rundlöf, M., Scheper, J., Smith, H.G. & Tscharntke, T. (2011). Does conservation on farmland contribute to halting the biodiversity decline? *Trends Ecol. Evol.*, **26**, 474-481.

Leventon, J. & Antypas, A. (2012). Multi-level governance, multi-level deficits: the case of drinking water management in Hungary. *Env. Pol. Gov.*, **22**, 253-267.

Lindenmayer, D. & Franklin, J.F. (2002). *Conserving forest biodiversity. A comprehensive multiscaled approach*. Island Press, Washington.

Macura, B., Secco, L. & Pullin, A.S. (2013). Does the effectiveness of forest protected areas differ conditionally on their type of governance? *Environ. Evidence*, **2**, 14.

Moss, T. (2003). Solving problems of 'fit' at the expense of problems of 'interplay'? The spatial reorganisation of water management following the EU water framework directive. Pages 85-121 in H. Breit, A. Engels, T. Moss, M. Troja, editors. *How institutions change*. Leske + Budrich, Opladen, Leverkusen.

Newig, J. & Fritsch, O. (2009). Environmental governance: participatory, multi-level – and effective? *Env. Pol. Gov.*, **19**, 197-214.

Newig, J. & Koontz, T.M. (2014). Multi-level governance, policy implementation and participation: the EU's mandated participatory planning approach to implementing environmental policy. *J. Eur. Publ. Pol.*, **21**, 248-267.

Oates, W.E. (2002). A reconsideration of environmental federalism. Pages 125-156 in W.E. Oates, editor. *Environmental policy and fiscal federalism. Selected essays of Wallace E. Oates*. Edward Elgar Publishing, Cheltenham, UK.

Paavola, J., Gouldson, A. & Kluvánková-Oravská, T. (2009). Interplay of actors, scales, frameworks and regimes in the governance of biodiversity. *Env. Pol. Gov.*, **19**, 148-158.

Pelosi, C., Goulard, M. & Balent, G. (2010). The spatial scale mismatch between ecological processes and agricultural management: do difficulties come from underlying theoretical frameworks? *Agr. Ecosyst. Environ.*, **139**, 455-462.

Peters, B.G. & Pierre, J. (2005). Multi-level governance and democracy: a Faustian bargain? Pages 75-89 in I. Bache, M. Flinders, editors. *Multi-level governance*. Oxford University Press, Oxford.

Pullin, A.S. & Knight, T.M. (2001). Effectiveness in conservation practice: pointers from medicine and public health. *Conserv. Biol.*, **15**, 50-54.

Pullin, A.S. & Stewart, G.B. (2006). Guidelines for systematic review in conservation and environmental management. *Conserv. Biol.*, **20**, 1647-1656.

Roberts, P.D., Stewart, G.B. & Pullin, A.S. (2006). Are review articles a reliable source of evidence to support conservation and environmental management? A comparison with medicine. *Biol. Conserv.*, **132**, 409-423.

Scheper, J., Holzschuh, A., Kuussaari, M., *et al.* (2013). Environmental factors driving the effectiveness of

European agri-environmental measures in mitigating pollinator loss – a meta-analysis. *Ecol. Lett.*, **16**, 912-920.

Simberloff, D., Farr, J.A., Cox, J. & Mehlman, D.W. (1992). Movement corridors: conservation bargains or poor investments? *Conserv. Biol.*, **6**, 493-504.

Smith, H.G., Birkhofer, K., Clough, Y., Ekroos, J., Olsson, O. & Rundlöf, M. (2014). Beyond dispersal: the role of animal movement in modern agricultural landscapes. Pages 51-70 in L.-A. Hansson, S. Åkesson, editors. *Animal movement across scales*. Oxford University Press, Oxford.

Sutcliffe, L., Paulini, I., Jones, G., Marggraf, R. & Page, N. (2013). Pastoral commons use in Romania and the role of the Common Agricultural Policy. *Int. J. Commons*, **7**, 58-72.

Sutherland, W.J., Pullin, A.S., Dolman, P.M. & Knight, T.M. (2004). The need for evidence-based conservation. *Trends Ecol. Evol.*, **19**, 305-308.

Young, O.R. (2002). The institutional dimensions of environmental change: fit, interplay, and scale. MIT Press, Cambridge, MA.

5

Exploring the Permanence of Conservation Covenants

Mathew J. Hardy[1], James A. Fitzsimons[2,3], Sarah A. Bekessy[1], & Ascelin Gordon[1]

[1] School of Global, Urban and Social Studies, RMIT University, Melbourne, VIC 3001, Australia
[2] The Nature Conservancy, Carlton, VIC 3053, Australia
[3] School of Life and Environmental Sciences, Deakin University, Burwood, VIC 3125, Australia

Keywords
Covenants; breach; release; private land conservation; private protected areas, easements; biodiversity.

Correspondence
Mathew J. Hardy, School of Global, Urban and Social Studies, RMIT University, GPO Box 2476, Melbourne, VIC 3001, Australia.
E-mail: mat.hardy@rmit.edu.au

Editor
Derek Armitage

Abstract

Conservation on private land is a growing part of international efforts to stem the decline of biodiversity. In many countries, private land conservation policy often supports in perpetuity covenants and easements, which are legally binding agreements used to protect biodiversity on private land by restricting activities that may negatively impact ecological values. With a view to understand the long-term security of these mechanisms, we examined release and breach data from all 13 major covenanting programs across Australia. We report that out of 6,818 multi-party covenants, only 8 had been released, contrasting with approximately 130 of 673 single-party covenants. Breach data was limited, with a minimum of 71 known cases where covenant obligations had not been met. With a focus on private land conservation policy, we use the results from this case study to argue that multi-party covenants appear an enduring conservation mechanism, highlight the important role that effective monitoring and reporting of the permanency of these agreements plays in contributing to their long-term effectiveness, and provide recommendations for organizations seeking to improve their monitoring programs. The collection of breach and release data is important for the continuing improvement of conservation policies and practices for private land.

Introduction

It is widely recognized that stemming the decline of biodiversity requires a greater focus on conservation efforts targeting private land. With private land covering a large part of the terrestrial landmass and supporting important biodiversity, its significance for conservation is gaining prominence in many countries, including Australia, Canada, the USA, New Zealand, Chile, and South Africa (Langholz & Lassoie 2001; Ewing 2008; Fishburn et al. 2009; von Hase et al. 2010). The approaches used by policy-makers to conserve biodiversity on private land vary considerably, from voluntary to incentives-based schemes to regulation. A number of studies have recently evaluated these various approaches, including the effectiveness of incentive-based programs to protect biodiversity (von Hase et al. 2010), the ability of voluntary stewardship programs to conserve habitat (Platt & Ahern 1995), and the extent to which conservation easement programs contribute to reducing development pressure and maintaining biodiversity (Pocewicz et al. 2011). Studies have also looked at the degree to which private land conservation aligns with strategic conservation goals (Kiesecker et al. 2007; Adams et al. 2014). Yet important questions still remain about the effectiveness and long-term consequences of private land conservation mechanisms (Merenlender et al. 2004).

Of growing importance in private land conservation policy is the establishment of Private Protected Areas (PPAs) - a protected area, as defined by the IUCN (Dudley 2008), under private governance (Stolton et al. 2014). PPAs are established in different ways in different countries, and the mechanisms used to protect biodiversity through legal or other effective means also vary. Here, we investigate two components central to private land conservation policy; the permanence (duration) and security (resistance to removal) of conservation agreements with landholders, focusing on conservation covenants as one form of PPA. We focus on examining these issues in Australia, which has a large number of individual conservation covenants (Stolton *et al.* 2014; Fitzsimons 2015). We first provide background information on our case study and the challenges around permanence and security for policy-makers, before presenting our results and using them as context to highlight the central role that monitoring and reporting of covenant releases and breaches plays in ensuring the long-term effectiveness of these agreements.

Private land conservation in Australia

As in many countries, conservation policy in Australia has historically focused on public land (Figgis 2004). Although public protected areas cover more than 65 million ha across 8.5% of the continent (DotE 2014), private and leasehold land covers over 62% of Australia's land area (AUSLIG 1993), and contains significant biodiversity value (Fitzsimons & Wescott 2001). Many of Australia's threatened species occur entirely outside of public protected areas (Watson *et al.* 2011), as do some of the most threatened ecosystems (Figgis *et al.* 2005; Taylor *et al.* 2011). Although the long-term security of private land conservation mechanisms is not yet clear, with the continuing loss of biodiversity, and broad acceptance that the public conservation estate is insufficient on its own, private land conservation policies and programs are increasingly important (Gordon *et al.* 2011).

Conservation covenants are an important component of Australia's private land conservation policy mix. Similar to conservation easements in North America, conservation covenants are mostly voluntary, legally binding agreements between an authorized organization and a landholder (Todd 1997). They can apply to all or part of a property and are registered on the property title (Figgis 2004), usually running in perpetuity. The vast majority are established primarily to protect land with high nature conservation value, where the landholder retains ownership but has a reduced 'bundle of rights', in effect giving up development and land-use rights incompatible with conservation (Iftekhar *et al.* 2014). Whilst covenants can be tailored to individual properties (Adams & Moon 2013) each covenant contains a standard set of obligations which remain relatively fixed over the term of the agreement, with limited, site-specific management requirements determined during establishment (Figgis *et al.* 2005). All Australian covenants are backed by specific enabling legislation (Fitzsimons 2015), specifying the body authorized to administer the covenant, typically a statutory authority.

Since the creation of the first conservation covenant in Australia, a Wildlife Refuge in New South Wales in 1951 (DECCW 2010), the number of covenants has grown considerably to 7,491 in 2014 (Figure 1). This in-

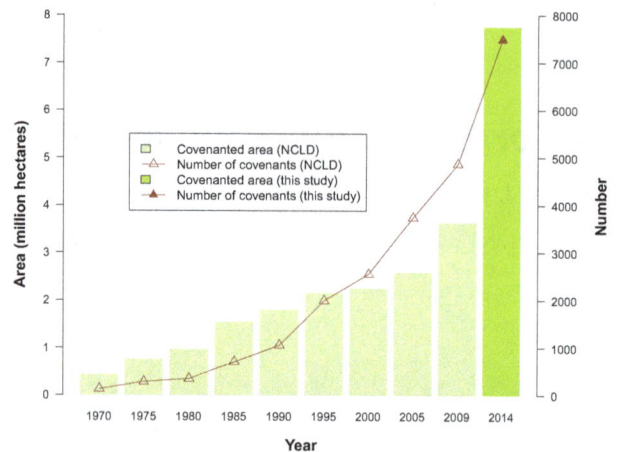

Figure 1 Cumulative trend in the number and area of covenanted properties in Australia. Columns represent covenanted area, and the triangles represent the number of covenants. Lighter green columns and hollow triangles indicate National Conservation Lands Database (DSEWPaC 2011) data, the darker green column and the filled triangle represent data collected for this study.

cludes 4,894 covenants likely to meet the private protected area criteria in Australia, which require the area to be valuable, secure through statutory provisions, well-managed for conservation, and clearly defined (see Fitzsimons 2015). With the number of covenants set to grow further, it is important to evaluate their permanence as a conservation mechanism.

Permanence and security

From a conservation policy perspective, the permanence and security of agreements with private landholders are central issues. Whilst permanence can relate to a number of ecological and social factors in conservation, here we focus on 'permanence' as the length of time that a conservation agreement (e.g., a covenant or easement) remains in place to protect conservation values (Fitzsimons 2006). An agreement's permanence can have substantial implications for the persistence of conservation values (Jones *et al.* 2005), and is of particular importance on private land, where landholders and land uses can change frequently, especially amid pressure from mining, agriculture, and other types of development (Cox & Underwood 2011; Pocewicz *et al.* 2011; Adams & Moon 2013). With covenant restrictions typically associated with the property title and lasting in-perpetuity, they are commonly considered the most permanent private land conservation mechanism in Australia. Thus they are formally able be classified as protected areas and can contribute to Australia's international protection targets (Fitzsimons 2006, 2015).

Related to permanence is an agreement's strength (its "security"), which refers to the level of authority required to establish, alter and/or terminate or extinguish ("release") that agreement (Fitzsimons 2006). Although security provisions vary between programs, all covenants in Australia are backed by legislation (Fitzsimons 2015), with release usually requiring approval from multiple parties including a government Minister. The exception is the Wildlife Refuge program, which is only available in the state of New South Wales and is unique amongst Australian covenants for only requiring approval for release from a single party (e.g., the landholder) (Figgis 2004).

Threats to permanence

Although protected area downgrading, downsizing and degazettement (PADDD) is a known policy issue and has been noted as a threat to public reserves (Mascia & Pailler 2011), some see covenants as less secure than public protected areas (e.g., Centre for Environmental Management 1999). Of particular importance here are mineral exploration and extraction rights, which have been identified as an emerging threat to the natural values on covenanted land (Adams & Moon 2013; Root-Bernstein *et al.* 2013), although covenants in Australia do not have the legal ability to prevent such activities as mineral rights rest with governments, not landholders. Changing property ownership, market conditions and government policy have also been noted as threats to the permanence of private land conservation more generally (Figgis *et al.* 2005; Jones *et al.* 2005). In Australia, concerns over covenant permanence also relate to their relatively recent adoption (most covenants have been established since the 1990s (Fitzsimons & Carr 2014; Figure 1), compared to public protected areas, which saw considerable expansion in the mid to late 20th century and some 274 (3.6%) of which are over 100 years old (DotE 2014).

Beyond these broader issues, a particular challenge for private land conservation policy globally is the identification and enforcement of "breaches," which are instances of landholders failing to meet their obligations or violating the conditions of their agreement in some way (Owley 2011). Breaches can vary in severity, and in extreme cases could lead to a release of the covenant. It is possible that the reasons behind breaches are similar to releases, and understanding these could allow for early and targeted intervention to prevent release. However, identifying breaches can be difficult for administering bodies, with the need to account for the agreement's flexibility (Rissman 2010), variability in permitted land uses (Rissman *et al.* 2007), changing ecological and

social conditions (Rissman 2014), and financial and practical limitations on their capacity to monitor covenanted land (Kiesecker 2007; Korngold 2007; Fitzsimons & Carr 2014). Moreover, how administering bodies respond to breaches is important for ensuring the effectiveness of these agreements, faced with the costs of pursuing legal action (Rissman & Butsic 2011) or the consequences of modifying the boundaries and/or obligations of these agreements ("amendments") through time (McLaughlin 2007; Jay 2013).

Little information exists on the permanence and security of PPAs in Australia. Here, for the first time, we collate and examine the available data on covenants from all major Australian covenanting programs (Table 1). Our initial motivation was to determine if data was available to answer the following three questions: (i) what proportion of conservation covenants within the major covenanting programs have been released; (ii) what proportion are known to have had their conditions breached; and (iii) what were the main reasons for the release or breach, and what factors could help predict these; and if so what are the main issues affecting the permanence of covenants?

Methods

Between October 2013 and January 2014, we asked individuals within the 13 major Australian covenanting organizations who were familiar with and had access to database records to provide the numbers of and reasons for covenant releases and breaches. We followed up responses with further questioning where needed. The programs involved cover all states and territories (with the exception of the Australian Capital Territory, where covenants are not present; Table 1).

Database records varied across organizations and programs – both in the detail (e.g., the type of impact caused the breach) and the style of recording (i.e., hard copy or electronic). Detailed information was not always available due to confidentiality, limited record-keeping, or the difficulty of retrieving data when resourcing restrictions precluded their ability to sift through hard copy records. Where only limited data was available, we asked program staff to instead provided estimates. The type of information provided by staff clearly fell into two categories: (1) "minimum bound estimates," where staff provided the known cases but indicated that the true number was likely greater but unknown; (2) "rough estimates," where staff were unsure of actual cases and could only provide a rough estimate. The description of the activities behind the covenant release and breach data were used to categorize these into common themes.

Table 1 Details of the covenanting programs included in this study

State	Covenant program	First covenant	Covenanting organization	Security
New South Wales	Conservation Agreement	1990	Office of Environment and Heritage	Multi-party
New South Wales	Trust Agreement	2005	Nature Conservation Trust of NSW	Multi-party
New South Wales	Registered Property Agreement	1997	Office of Environment and Heritage	Multi-party
New South Wales	Wildlife Refuge	1951	Office of Environment and Heritage	Single-party
Northern Territory	Conservation Covenant	2009	Parks and Wildlife Commission NT	Multi-party
Queensland	Nature Refuge and Coordinated Conservation Area	1994	Department of Environment and Heritage Protection	Multi-party
South Australia	Heritage Agreement	1994	Department of the Environment, Water and Natural Resources	Multi-party
Tasmania	Conservation Covenant	1999	Department of Primary Industries, Parks, Water and Environment	Multi-party
Victoria	Conservation Covenant	1986	Trust for Nature (Victoria)	Multi-party
Victoria	Section 69 Agreement	1987	Department of Environment and Primary Industries	Multi-party
Western Australia	Conservation Covenant	1971	The National Trust of Australia (WA)	Multi-party
Western Australia	Nature Conservation Covenant	1990	Department of Parks and Wildlife	Multi-party
Western Australia	Conservation Covenant	1980	Department of Agriculture and Food	Multi-party

Covenant releases and breaches

We considered covenants "released" if they had been signed over a particular piece of land in the past but had subsequently been removed from the land title (i.e., the covenant had been terminated in accordance with the relevant security provisions). Because obligations vary between programs, we considered a covenant "breached" if its obligations had not been met, but the covenant had remained in place. We did not count third-party damage (e.g., by neighbors) as a landholder breach, but recorded this information separately, as we consider this type of damage reasonably beyond the immediate control of the landholder and the administering body.

Results

Covenant releases

The single-party NSW Wildlife Refuge covenants had by far the highest number of releases, although this was based on the estimate provided by program staff (130 of 673). A total of eight out of 6,818 multiparty covenants (0.12%) had been released across Australia, with Victoria (4) and Western Australia (3) having the highest numbers of releases (Table 2).

For multi-party covenants, the reasons for release varied considerably, ranging from unauthorized timber removal to government acquisition or administrative error (Table 3). As examples, two early covenants were established on old farms, which were released after it became clear they had limited conservation value and were unsuitable for covenanting. Another covenant at Ironbark

Basin in Victoria was released when the land was transferred to the State Government for inclusion in a national park. Arguably, in this case "release" may not be the most appropriate term given the conservation values remained protected. Equivalent data for single-party Wildlife Refuges was unavailable, however indications from program staff suggests that these releases occurred predominantly at the request of the landholder.

Covenant breaches

Detailed breach data was not available from most programs, which precluded deeper quantitative analysis. Of the available data, 71 breaches were reported (Table 2), with most of these in Western Australia (42) and Tasmania (20). However, given the constraints on covenant monitoring by the programs (Fitzsimons & Carr 2014), these reported breaches should be interpreted as minimum bound estimates, with the true number likely to be greater.

Some 43 of the 71 breaches (60%) had insufficient information for classification (Table 4). Of those able to be categorized, as a percentage of all reported breaches, most arose from land clearing and/or development (13%), road construction (7%), forestry operations (7%), or unauthorized timber removal (7%). Some 25% of all breaches were attributed to a third party. In one third-party breach, forestry contractors working on a neighboring property cleared vegetation on a covenanted property where the boundary delineation was unclear; in another case, a third party had gained illegal entry to the property and collected firewood.

Table 2 Number of covenants, area covenanted, releases and breaches, by covenant type

State	Total number covenants in place	Area covenanted (ha)	Percentage of private land area[a] in the jurisdiction that is covenanted	Number released	Percentage released	Number breached	Percentage breached
Single party covenants							
New South Wales	673	1,889,791.52	2.65	130[b]	19.31	n/a	n/a
Multi-party covenants							
Western Australia	2,016	1,322,684.69	1.20	3	0.15	42[c]	2.08
South Australia	1,523	646,280	1.12	0	0	1[c]	0.07
Victoria	1,419	64,741	0.42	4	0.28	4[c]	0.28
New South Wales	672	170,595.35	0.24	0	0	4[c]	0.60
Tasmania	731	84,655	3.11	1	0.14	20	2.74
Queensland	455	3,439,875	2.20	0	0	0	0
Northern Territory	2	131,043.01	0.19	n/a	n/a	n/a	n/a
National total (multi-party only)	6,818	5,859,874.05	1.22	8	0.12	71	1.04
National total (single and multiparty covenants)	7,491	7,749,665.57	1.61	138	1.84	71	0.95

[a]includes indigenous land.
[b]Detailed records unavailable and the numbers represent staff member's rough estimate.
[c]Detailed records unavailable and the numbers here are cases specifically known to staff and represent minimum bounds.

Discussion

The importance of strong security provisions

Using Private Protected Areas (PPAs) to conserve biodiversity is a growing approach in conservation policy. By definition, PPAs require protection through legal or other effective means (Stolton *et al.* 2014), and by extension, their effectiveness as a permanent conservation mechanism relates directly to the ease in which that agreement can be released, amended or enforced.

Focusing on Australian covenants as a form of PPA, our case study found only a small number of multi-party covenants had been released, suggesting they are a conservation mechanism with high permanence. Moreover, our study also highlights a clear distinction in the proportion of releases between covenants with differing security provisions, with a relatively high proportion of single-party Wildlife Refuge releases (19%) compared with multi-party covenants (0.12%). Considering the extent of legal challenges that permanent agreements face (Rissman & Butsic 2011) and are likely to face in the future, this is a clear demonstration to policy-makers of the value of strong security provisions, whereby requiring authorization from multiple parties reduces the potential for release, and contributes towards ensuring these agreements meet their promise of in perpetuity protection (McLaughlin 2007). We thus emphasize the importance for policy makers to consider and prioritize multi-party provisions to secure their agreements. However, this extra security would have to be weighed up against the potential for these provisions to act as a deterrent to landholders entering the program (Kabii & Horwitz 2006).

Preparing for threats to agreements

Whilst strong security provisions may help prevent release, the early identification of threats to these agreements could help policy-makers prepare and adapt to emerging issues. Part of this requires understanding the reasons why covenants are being released. The data analyzed in our study showed no standout cause for multi-party covenant release and instead, each appears a product of individual circumstances. However, in the single party Wildlife Refuges program, the higher number of releases was attributed to landholders opting to withdraw. Further research is needed to understand why landholders are leaving the program, for example by investigating landholder commitment and satisfaction with the covenanting program (e.g., Selinske *et al.* 2015).

Beyond release, some breaches of obligations are a potential threat to the permanence of agreements, through damage to ecological values of the property which may

Table 3 Reported reasons for covenant releases

	Reason	No. reported cases
Multi-party releases	Site subsequently deemed unsuitable	2
	Acquired by state government for development	2
	Ceded to government as reserve	1
	Unauthorised timber removal	1
	Administrative error – unintended covenant	1
	Elderly landowner – unable to meet obligations	1
	Total multi-party releases	8
Single-party releases	Releases at landholder request	130[a]
	Total single-party releases	130

[a]Detailed records unavailable and the type and number of releases represent staff member's rough estimate.

Table 4 Summary of available information on covenant breaches and the responsible parties. Numbers represent minimum bound estimates

	Party responsible			
Reason	Landholder	Third party	Unknown	No. reported cases
Land clearing and/or development	6	3	–	9
Road construction	1	4	–	5
Forestry operations	–	5	–	5
Unauthorised timber removal (e.g., firewood)	–	5	–	5
Dumping rubbish	2	–	–	2
Management actions incomplete	1	–	–	1
Recreational vehicles	–	1	–	1
Unknown/insufficient information	–	–	43	43
Total reported breaches	10	18	43	71

in some extreme cases cause major loss in values, leading to covenant release. It is possible that the reasons behind breaches may be similar to releases, providing room for organizations to intervene early to prevent release. In our study, of those breaches with sufficient information, land clearing showed up as the biggest issue. Due to the limited available data, the extent of this issue is unclear, as are the reasons for clearing, but it highlights one of the key challenges for policy makers – how to minimize unwanted landholder behavior from a distance with minimal intervention. One approach could be for private land organizations to increase the level of enforcement and consider strengthening the compliance components within the legal agreement if needed (see Jay 2013). However, maintaining a strong and constructive relationship with landholders could help prevent the substantial costs associated with enforcement (Rissman & Butsic 2011) and as a preventative measure, an increased focus on landholder support may help clarify landholder understanding of their obligations (Stroman & Kreuter 2014) and help uncover the reasons behind this clearing.

In response to breaches, a number of organizations mentioned covenant amendment as a preferred method

of resolution to release, provided the property's ecological values remained protected. This fits with the findings of Rissman (2010), who noted that land trusts in the United States have an incentive to act moderately when obligations are not met. We did not look directly at amendments, and the data available from our study was insufficient to determine how many covenants have been amended, or even the nature of these changes (e.g., renegotiating boundaries or obligations). However, as amendments can relate to the permanence of covenant obligations and the effectiveness of these agreements for use in conservation policy, we highlight the need for programs to monitor and record the nature and extent of any amendments to permanent agreements and suggest this as an important area requiring further research.

Some organizations suggested that the turnover of conservation covenants to successor landholders may be developing into a policy issue, which has also been noted elsewhere (Collins 2000; Czech 2002; Rissman & Butsic 2011; Stroman & Kreuter 2014). These are landholders who, for example, have purchased or inherited the property from the original covenantor. Without being original parties to the covenant, their ownership of protected properties may result in higher rates of legal challenge

(Rissman & Butsic 2011) and/or breaching of conditions. It may be that successor covenantors prove an important predictor of covenant breach or requests for release, although understanding the reasons behind this requires further research. Policy-makers would be well placed to consider ways of engaging and supporting new owners, as well as elderly covenantors who may need additional support in order to meet their obligations (see also Fitzsimons & Carr 2014).

Although a significant policy challenge, dealing with current and future owners of protected properties is only one dimension of permanence. Our case study suggests that policy-makers also need to account for actors outside of the direct agreement. Most breaches in our study for which detailed information was available were attributed to damage from a third party (25% of all known breaches), also noted as an issue for easements in the United States (Rissman & Butsic 2011). This raises an important question for policy makers about who holds responsibility for monitoring, preventing and rectifying damage to covenanted properties resulting from trespass, particularly if the third party remains unidentified. Trespass is an issue for conservation areas in general, impacting both the public and private conservation estate.

As noted elsewhere, we also agree that the decoupling of above- and below-ground property rights is an important issue for conservation covenants (Adams & Moon 2013; Root-Bernstein et al. 2013). In Australia, covenants do not provide protection for underground resources, with mineral exploration and extraction rights remaining in government ownership. Although this study shows that mining activities have not yet resulted in covenant release, it is likely that in the near future coal extraction will be permitted on a Nature Refuge covenant in the Galilee Basin in Queensland (Lauder 2013). This is an important policy issue, not only because mineral extraction can result in the loss of ecological value, but also because of the potential loss of public investment (McLaughlin 2012) and faith in conservation that has played an important role in funding the development of the private conservation estate.

A need for improved monitoring and recording

It is likely that the growth in permanent conservation agreements will continue, particularly with their increasing use via new pathways such as biodiversity offsets, which are growing in prominence internationally and in all Australian jurisdictions (Bull et al. 2013). It is possible that this will also lead to an increase in the number of releases and breaches, making effective monitoring of these agreements essential for identifying issues, supporting

enforcement (Rissman & Butsic 2011), and evaluating their ecological contribution. Whilst our study showed few releases, detailed breach information was limited, with the number of breaches occurring largely unknown. This is surprising given the prominence of permanence as a key feature of the mechanism, but such fragmented and incomplete data is not unique to covenants, having also been noted before for easements in the United States (Wilson Morris & Rissman 2009).

The relevant policy questions therefore become where, how and what to monitor? Limited resourcing of covenanting organizations makes monitoring a particular challenge (Fitzsimons & Carr 2014), and organizations may be best to focus their efforts where and when the probability of breach is highest (Czech 2002). From this study, a starting point may be in areas with known concentrations of successor covenantors or hotspots for third-party trespass. Aerial photographs, remote-sensing and predictive modeling techniques offer opportunities to identify possible breaches remotely, which could be used where resourcing limitations impede the recommended annual site visits (LTA 2004). Where breaches are hard to detect remotely, indirect observations, self-reporting and direct questioning of landholders could be used (see Gavin et al. 2010), and more generally, specialized landholder questioning techniques could help obtain estimates of noncompliance (Nuno & St. John 2015; Thomas et al. 2015). When organizations collect breach data, we suggest other data should be recorded in addition to the location, actor (i.e., the landholder or a third party), and the type and extent of the damage. This should include both the landholder type (i.e., originator or successor) and where possible, the intention of the actor (i.e., accidental or intentional). Of course beyond identifying a breach, organizations must also ensure there are sufficient resources and capacity available for enforcement (Rissman & Butsic 2011).

Our study provides insights into the methodological challenges of multi-jurisdictional studies on conservation agreements. Obtaining sufficient and consistent breach data proved particularly difficult, due largely to organization resourcing constraints on its collection, differences in how breaches are monitored and recorded across organizations (i.e., centrally or regionally, electronically or in hard copy), and privacy concerns over sharing this type of information. There were also challenges in analyzing across different programs (e.g., what constitutes a breach under different legislation or landholder agreements). However, our study highlights an opportunity to share data, pool resources and collaborate across organizations to allow for more detailed quantitative and qualitative studies in the future. For this, support is needed from policy-makers for more consistency in covenant monitoring (e.g., LTA 2014), as well as a coordinated approach to recording and sharing breach and release data in ways that address confidentiality concerns. This data should be in digital form in centralized and secure databases, such as the National Conservation Easement Database in the USA (USEFC 2014), with data sharing provisions to allow for comparison across different agreement types, such as U.S. easements and Australian covenants. In Australia, the National Conservation Lands Database (DSWEPaC 2011) has the potential to be an equivalent portal, although its future viability is currently uncertain.

As the role of PPAs in protecting biodiversity grows, so does the need to ensure they remain an effective part of the conservation policy toolkit. The numbers of covenant releases and known breaches in our case study were low, suggesting that covenants may be an enduring mechanism for conservation, although we acknowledge the likely under-reporting and minimal data available for breaches. However, ongoing compliance monitoring of covenant breaches and releases will allow policy-makers to respond to issues as they arise, and will also enable future comparison of the permanence of PPAs to the public estate and other protected area categories. This data is key to understanding the permanence and long-term effectiveness of these agreements and crucial for improving the sustainability of conservation policy on private land.

Acknowledgments

This research was conducted with funding support from the Australian Research Council (ARC) Centre of Excellence for Environmental Decisions and ARC Discovery Project DP150103122. Ideas for this article arose at a Private Land Conservation Workshop, University of Melbourne, 11–14 June 2013. The authors wish to thank the three anonymous reviewers for their constructive and substantive comments on earlier versions of this manuscript, and the covenanting organizations for supporting this study (NSW OEH, Nature Conservation Trust of NSW, Parks and Wildlife Commission NT, Queensland DEHP, Tasmania DPIPWE, SA DEWNR, Trust for Nature (Victoria), Victoria DEPI, National Trust of Australia (WA), WA DPaW, WA DAF). Sarah Bekessy is supported by an ARC Future Fellowship (FT130101225).

References

Adams, V.M. & Moon, K. (2013). Security and equity of conservation covenants: contradictions of private protected area policies in Australia. *Land Use Pol.*, **30**, 114-119.

Adams, V.M., Pressey, R.L. & Stoeckl, N. (2014). Estimating landholders' probability of participating in a stewardship program, and the implications for spatial conservation priorities. *PLoS One*, **9**, e97941.

Australian Surveying and Land Information Group (AUSLIG). (1993). *Australia land tenure*. AUSLIG, Canberra.

Bull, J.W., Suttle, K.B., Gordon, A., Singh, N.J. & Milner-Gulland, E.J. (2013). Biodiversity offsets in theory and practice. *Oryx*, **47**, 369-380.

Centre for Environmental Management. (1999). *Mid-term review of the Natural Heritage Trust: National Reserve System Program*. Centre for Environmental Management, University of Ballarat.

Collins, D.G. (2000). Enforcement problems with successor grantors. Pages 157-165 in J.A. Gustanski & R.H. Squires, editors. *Protecting the land: conservation easements past, present and future*. Island Press, Washington, D.C.

Cox, R.L. & Underwood, E.C. (2011). The importance of conserving biodiversity outside of protected areas in Mediterranean ecosystems. *PLoS One*, **6**, e14508.

Czech, B. (2002). A transdisciplinary approach to conservation land acquisition. *Conserv. Biol.*, **16**, 1488-1497.

Department of Environment, Climate Change and Water NSW (DECCW) (2010). *Conservation Partnerships: a guide for landholders*. Department of Environment, Climate Change and Water NSW, Sydney.

Department of Sustainability, Environment, Water, Population and Communities (DSEWPaC) (2011). National Conservation Lands Database—data products. Available from http://nrmonline.nrm.gov.au/?utf8=%E2%9C%93 &search_field=all_fields&q=ncld/. Accessed 9 September 2015.

Department of the Environment (DotE) (2014). Collaborative Australian Protected Areas Database 2014. Available from http://www.environment.gov.au/land/nrs/science/capad/ 2014. Accessed 29 December 2015.

Dudley, N. (2008). Guidelines for applying protected area management categories. IUCN.

Ewing, K. (2008). Conservation covenants and community conservation groups: improving the protection of private land. *New Zeal. J. Environ. Law*, **12**, 315-337.

Figgis, P. (2004). *Conservation on private lands: the Australian experience*. IUCN, Gland, Switzerland, and Cambridge, UK.

Figgis, P., Humann, D. & Looker, M. (2005). Conservation on private land in Australia. *Parks*, **15**, 19-29.

Fishburn, I.S., Kareiva, P., Gaston, K.J. & Armsworth, P.R. (2009). The growth of easements as a conservation tool. *PLoS One*, **4**, e4996.

Fitzsimons, J. & Wescott, G. (2001). The role and contribution of private land in Victoria to biodiversity conservation and the protected area system. *Aust. J. Environ. Manage.*, **8**, 142-157.

Fitzsimons, J.A. (2006). Private Protected Areas? Assessing the suitability for incorporating conservation agreements over private land into the National Reserve System: a case study of Victoria. *Environ. Plan. Law J.*, **23**, 365-385.

Fitzsimons, J.A. (2015). Private protected areas in Australia: current status and future directions. *Nature Conserv.*, **10**, 1-23.

Fitzsimons, J.A & Carr, C.B. (2014). Conservation covenants on private land: issues with measuring and achieving biodiversity outcomes in Australia. *Environ. Manage.*, **54**, 606-616.

Gavin, M.C., Solomon, J.N. & Blank, S.G. (2010). Measuring and monitoring illegal use of natural resources. *Conserv. Biol.*, **24**, 89-100.

Gordon, A., Langford, W.T., White, M.D., Todd, J.A. & Bastin, L. (2011). Modelling trade offs between public and private conservation policies. *Biol. Conserv.*, **144**, 558-566.

Iftekhar, M.S., Tisdell, J.G. & Gilfedder, L. (2014). Private lands for biodiversity conservation: review of conservation covenanting programs in Tasmania, Australia. *Biol. Conserv.*, **169**, 176-184.

Jay, J.E. (2013). When perpetual is not forever: the challenge of changing conditions, amendment, and termination of perpetual conservation easements. *Harvard Environ. Law Rev.*, **36**, 1-78.

Jones, B.T.B., Stolton, S. & Dudley, N. (2005). Private protected areas in East and southern Africa: contributing to biodiversity conservation and rural development. *Parks*, **15**, 67-77.

Kabii, T. & Horwitz, P. (2006). A review of landholder motivations and determinants for participation in conservation covenanting programmes. *Environ. Conserv.*, **33**, 11.

Kiesecker, J.M., Comendant, T., Grandmason, T. *et al.* (2007). Conservation easements in context: a quantitative analysis of their use by The Nature Conservancy. *Front. Ecol. Environ.*, **5**, 125-130.

Korngold, G. (2007). Solving the contentious issues of private conservation easements: promoting flexibility for the future and engaging the public land use process. *Utah Law Rev.*, **4**, 1039-1084.

Land Trust Alliance (LTA). (2004). *Practice 11C: easement monitoring.* Land Trust Alliance.

Langholz, J.A. & Lassoie, J.P. (2001). Perils and promise of privately owned protected areas. *BioScience*, **51**, 1079-1085.

Lauder, S. (2013). Waratah Coal welcomes Galilee Basin mine approval despite environmental conditions. *ABC News.* Available from http://www.abc.net.au/news/2013-12-21/ waratah-coal-welcomes-approval-of-galilee-basin-mine/ 5170912/. Accessed 9 September 2015.

Mascia, M. & Pailler, S. (2011). Protected area downgrading, downsizing, and degazettement (PADDD) and its conservation implications. *Conserv. Lett.*, **4**, 9-20.

McLaughlin, N. (2007). Conservation easements: perpetuity and beyond. *Ecol. Law Quart.*, **34**, 673-712.

McLaughlin, N.A. (2012). Extinguishing and amending tax-deductible conservation easements: protecting the federal investment after Carpenter, Simmons and Kaufman. *Florida Tax Rev.*, **13**, 217-303.

Merenlender, A.M., Huntsinger, L., Guthey, G. & Fairfax, S.K. (2004). Land trusts and conservation easements: who is conserving what for whom? *Conserv. Biol.*, **18**, 65-76.

Nuno, A. & St. John, F.A.V. (2015). How to ask sensitive questions in conservation: a review of specialized questioning techniques. *Biol. Conserv.*, **189**, 5-15.

Owley, J. (2011). Changing property in a changing world: a call for the end of perpetual conservation easements. *Stanford Environ. Law J.*, **30**, 121-173.

Platt, S. & Ahern, L. (1995). Voluntary nature conservation on private land in Victoria—evaluating the Land for Wildlife programme. Pages 211-216 in Bennett, A.F., Backhouse, G. & Clark, T., editors. *People and nature conservation: perspectives of private land use and endangered species recovery.* The Royal Zoological Society of New South Wales, Mosman, NSW.

Pocewicz, A., Kiesecker, J.M., Jones, G.P., Copeland, H.E., Daline, J. & Mealor, B. A. (2011). Effectiveness of conservation easements for reducing development and maintaining biodiversity in sagebrush ecosystems. *Biol. Conserv.*, **144**, 567-574.

Rissman, A.R. (2010) Designing perpetual conservation agreements for land management. *Rangeland Ecol. Manage.*, **63**, 167-175.

Rissman, A.R. & Butsic, V. (2011). Land trust defense and enforcement of conserved areas. *Conserv. Lett.*, **4**, 31-37.

Rissman, A.R., Lozier, L., Comendant, T. *et al.* (2007). Conservation easements: biodiversity protection and private use. *Conserv. Biol.*, **21**, 709-718.

Rissman, A.R., Owley, J., Shaw, M.R. & Thompson, B.B. (2014). Adapting conservation easements to climate change. *Conserv. Lett.*, **8**, 68-76.

Root-Bernstein, M., Montecinos Carvajal, Y., Ladle, R., Jepson, P. & Jaksic, F. (2013). Conservation easements and mining: the case of Chile. *Earth's Futur.*, **1**, 33-38.

Selinske, M.J., Coetzee, J., Purnell, K. & Knight, A.T. (2015) Understanding the motivations, satisfaction, and retention of landowners in private land conservation programs. *Conserv. Lett.*, **8**, 282-289.

Stolton, S., Redford, K.H. & Dudley, N. (2014). *The futures of privately protected areas.* IUCN, Gland, Switzerland.

Stroman, D.A. & Kreuter, U.P. (2014). Perpetual conservation easements and landowners: evaluating easement knowledge, satisfaction and partner organization relationships. *J. Environ. Manage.*, **146**, 284-291.

Taylor, M.F.J., Sattler, P.S., Fitzsimons, J. *et al.* (2011). *Building nature's safety net 2011: the state of protected areas for Australia's ecosystems and wildlife.* WWF-Australia, Sydney.

Thomas, A.S., Gavin, M.C. & Milfont, T.L. (2015). Estimating non-compliance among recreational fishers: insights into factors affecting the usefulness of the randomized response and item count techniques. *Biol. Conserv.*, **189**, 24-32.

Todd, J.A. (1997). Victoria's conservation covenant program: how effective has it been at achieving private land conservation? Pages 173-175 in Hale, P. & Lamb, D., editors. Conservation Outside Nature Reserves Centre for Conservation Biology, The University of Queensland, Brisbane.

U.S. Endowment for Forestry and Communities. (2014). National Conservation Easement Database. Available from http://conservationeasement.us/. Accessed 9 September 2015.

von Hase, A., Rouget, M. & Cowling, R.M. (2010). Evaluating private land conservation in the Cape Lowlands, South Africa. *Conserv. Biol.*, **24**, 1182-1189.

Watson, J.E.M., Evans, M.C., Carwardine, J. *et al.* (2011). The capacity of Australia's protected-area system to represent threatened species. *Conserv. Biol.*, **25**, 324-332.

Wilson Morris, A. & Rissman, A.R. (2009). Public access to information on private land conservation: tracking conservation easements. *Wisconsin L. Rev.*, **6**, 1237-1282.

6

EU's Conservation Efforts Need More Strategic Investment to Meet Continental Commitments

Virgilio Hermoso[1,2], Miguel Clavero[3], Dani Villero[1], & Lluís Brotons[1,4,5]

[1] Centre Tecnològic Forestal de Catalunya (CEMFOR - CTFC), Crta. Sant Llorenç de Morunys, Km 2, 25280, Solsona, Lleida, Spain
[2] Australian Rivers Institute, Griffith University, Nathan, Qld 4111, Australia
[3] Estación Biológica de Doñana-CSIC, Américo Vespucio s.n., 41092, Sevilla, Spain
[4] CREAF, Cerdanyola del Vallés, 08193, Spain
[5] CSIC, Cerdanyola del Vallés, 08193, Spain

Keywords
Birds Directive; conservation investment; conservation policy; Habitats Directive; LIFE-Nature; REFIT.

Correspondence
Virgilio Hermoso, Centre Tecnològic Forestal de Catalunya (CEMFOR - CTFC), Crta. Sant Llorenç de Morunys, Km 2, 25280 Solsona, Lleida, Spain.
E-mail: virgilio.hermoso@gmail.com

Editor
Joern Fischer

Abstract

The European Union (EU) has made significant conservation efforts in the last two decades, guided by the Birds and Habitats Directives, currently under evaluation. Despite these efforts a large proportion of priority species are still in unfavorable condition and continue declining. For this reason, a thoughtful review of the implementation of conservation efforts in Europe is needed to identify potential causes behind this poor effectiveness. We compiled information on the distribution of all conservation funds under the LIFE-Nature, the main financial tool for conservation in Europe. We found that LIFE-Nature has not adequately covered continental conservation needs. The majority of funds have been directed toward nonthreatened species or regions of low conservation priority. Given the limited resources available, two key aspects are in urgent need for revision and improvement. First, the distribution of funds should be guided by continental and global conservation needs and planned at the EU scale. Second, new mechanisms are required to set conservation priorities in a dynamic fashion, rather than relying on fixed lists (i.e., the Directives' Annexes) that may rapidly become outdated. These improvements would require new mechanisms to set priorities and redistribution of conservation efforts, supported by adequate policy and a more effective top-down control on investment.

Fitness check to the European Union (EU)'s conservation policy

The EU is currently carrying out a fitness check of its conservation policy, as part of a broader Regulatory Fitness and Performance Program (REFIT; EC 2014a). The aim of the fitness check is to evaluate to which extent the EU policy is "fit for purpose" and adequate to face the challenges of a rapidly changing world. This process recalls attention on the Birds (79/409/EEC) and Habitats (92/43/EEC) Directives, which are the cornerstone of the EU's conservation policy and guide the implementation of conservation efforts in Europe. These Directives define clear conservation objectives and priorities aimed at achieving the EU 2020 Biodiversity target of halting and reversing the loss of biodiversity at the continental and global (*target 1* and *target 6*, respectively) scales (EC 2011a). The Directives also translate into the EU's policy conservation commitments from other international conventions that the EU has subscribed, such as the Convention on Biological Diversity. In order to guide conservation efforts, the Directives provide Annexes listing priority species and habitats that should be the focus of conservation management. Following these guidelines, the EU has made a significant conservation effort in the last two decades, which has involved the declaration of the Natura 2000 Network, the world's largest network of protected areas. In order to provide financial support to these conservation efforts, the EU founded in 1992 the LIFE program, which has become the main financial instrument of the EU conservation policies (EC 2011b). There were four successive LIFE programs in the period 1992–2013 fully completed to date. Each of these programs had specific objectives but with the common priority of demonstrating how to implement the Birds and Habitats Directives and reinforce the role of the Natura 2000 network at preserving the EU's biodiversity. Within all the different subprograms included in LIFE, LIFE-Nature is the most relevant from a conservation perspective because it is the most directly related to the implementation of on the ground conservation actions for priority species and habitats. It has also attracted a significant proportion of all LIFE funds, making it the most important subprogram within LIFE. In 2014, a new extension of LIFE was approved for the period 2014–2020, with an overall budget of €2.7 billion under the subprogram for Environment, which includes previous LIFE-Nature, and €0.9 billion under the subprogram for Climate Action.

However, despite these efforts, more than half of the species legally protected by the Directives were considered in unfavorable status in the last State of nature in the EU (European Environment Agency 2015). For this reason, assessments on whether efforts have covered continental conservation needs and promoted the achievement of targets are timely and urgently needed under the umbrella of the review process opened by REFIT. Here, we evaluate how conservation funds under the LIFE-Nature subprogram have been invested. With an accumulated experience of more than 20 years, this program represents a critical milestone for assessing whether conservation efforts made by the EU have targeted species and areas with higher continental conservation needs and have hence been relevant toward achieving the EU's conservation goals.

Assessing more than 20 years of LIFE-Nature investment

In order to get a complete picture of the implementation of the LIFE-Nature program, we reviewed information on the investment made, spatial distribution and species benefited under each of the 1,448 projects funded in the period 1992–2013 (Supporting Information Appendix 1). We then brought LIFE-Nature into the context of conservation needs in the EU by comparing the investment made across different IUCN conservation status categories. We use the IUCN assessments as estimates of conservation needs because they are the result of standardized evaluations of the conservation status of species and the threats affecting them at the European and global levels. National and regional-level IUCN assessments are

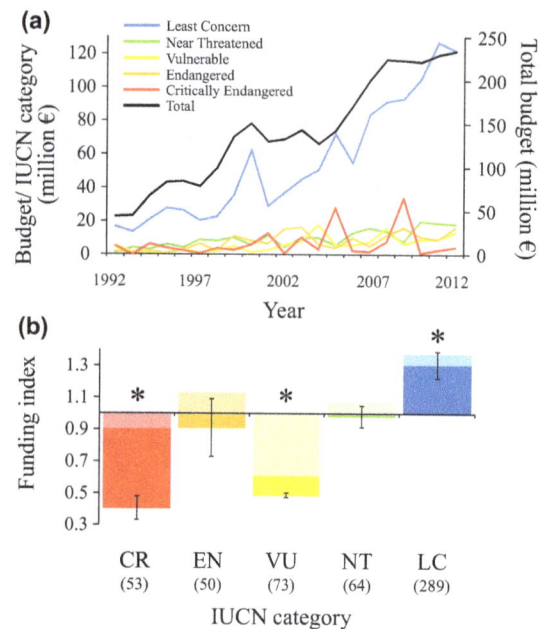

Figure 1 (a) Temporal distribution of LIFE-Nature funds in the period 1992–2013, split by IUCN classes. (b) Funding index (Average ± SE) for all threatened species listed under different IUCN classes. The funding index represents the ratio between the total budget received by each species and the budget that the species would have received if funds had been randomly distributed. Values close to 1 indicate that the species have received a similar budget than expected by random distribution of funds. Values above and below 1 indicate species that have received more or less budget than expected by random distribution of funds, respectively. To avoid the influence of species that have received large investment (see peaks in A due to a single species—Iberian Lynx, *Lynx pardinus*-), we deleted from each category the species with the highest budget in B (*Lynx pardinus*, *Margaritifera margaritifera*, *Ottis tarda*, *Gypaetus barbatus*, *and Ursus arctos* for the CR, EN, VU, NT, and LC categories, respectively). *indicates IUCN classes for which significant differences from random distribution of funds were found. The average values including all species are also shown for each class in faded colors. The number of species that have received LIFE-Nature funds in the studied period is also shown in parentheses.

also available, but continental assessments better convey the global extinction risk and then suit the EU's conservation commitments. Furthermore, these assessments are commonly used by the EU in the periodic revisions of the status of Europe's biodiversity and evaluation of conservation targets achievement (European Environment Agency 2015), which aligns our results with these other assessments.

The EU contributed with €1,625 million, of the total of €2,964 million invested into LIFE projects when including Member State contributions. These funds have been continuously increasing at an average annual rate of 7% from the beginning of the program (Figure 1a). About two-thirds of the LIFE-Nature investment has been spent on the implementation of on the ground conservation programs for European priority species and habitats. Some funds have also been invested in creating and consolidating Natura 2000 national networks, implementing environmental monitoring or improving waste management, among others. Regarding to projects focused on implementation of species conservation, a total of 666 species (203 species from the Birds Directive, 451 from the Habitats Directive, and 12 not listed in either of them) have benefited from LIFE-Nature. This represents 33% of all priority species listed in the Annexes of the Directives although only 10% of globally threatened species present in Europe according to IUCN (IUCN 2015).

Despite the significant financial contribution made by the EU, the distribution of LIFE-Nature funds has not covered continental conservation needs. The majority of funds have been directed toward species of low global/continental conservation concern (75% of all funds spent on Least Concern species; Figure 1a). This has resulted in overfunding of nonthreatened species even in relation to a random distribution of funds, which is far from an ideal funding schedule (Figure 1b). On the other hand, globally threatened species have been clearly underfunded. The 53 critically endangered (CR) species and the 73 vulnerable (VU) that benefited from LIFE-Nature funds have received, respectively, only 40% and 50% of the expected budget if funds had been distributed randomly (Figure 1b; excluding the Iberian lynx—*Lynx pardinus*—and great bustard—*Ottis tarda*—which received 30 and 6 times the average investment spent on CR and VU species, respectively, and then overinflated the funding index for their IUCN categories). As a result, only three of the 10 species with the largest LIFE-Nature investment (22% of the total budget) were globally threatened (IUCN 2015). Note that the estimates of under/over funding presented here are constrained to the pool of species that were cited as beneficiary of at least one LIFE-Nature project, most of them listed as priority species in the Directives' Annexes. Our assessment focuses then on analyzing whether funds have been distributed according to conservation needs within the pool of benefited species and not on assessing whether all continental conservation needs in the EU have been addressed.

The distribution of LIFE-Nature funds also shows poor spatial relation with continental conservation needs. The annual average investment by LIFE-Nature projects in EU regions was positively related to the proportion of their territory protected under Natura 2000, the number of species listed in the Directives' Annexes and the regional wealth, measured as their GDP. All these variables showed significant effects on the regional average

investment in a General Lineal Model (Appendix 1). The number of threatened vertebrates was the only variable in the model that did not show significant effects on that GLM model (Table S1). These modeling results reflect the spatial mismatch between conservation investment and the distribution of threatened vertebrate species, which we used as a surrogate for the spatial variation of continental conservation needs (Figures 2b and c). While some regions, especially in Northern and Central Europe, have received large proportions of LIFE-Nature funds despite having scarcely threatened biotas, several regions in Southern and Eastern Europe that hold high numbers of threatened species have been notably underfunded (Figure 2c). These spatial analyses were constrained to vertebrate species given the lack of full coverage by IUCN assessments and distribution for other taxonomic groups like invertebrates, plants, or fungi present in Europe. However, the spatial patterns of funds reported in this study are a good representation of the whole conservation effort done under LIFE-Nature, given that vertebrate species attracted 80% of all funds.

The LIFE-Nature experience and future policy in Europe

As recently acknowledged on a public consultation process carried out within REFIT, where more than half million European individuals and NGOs participated, the objectives established in the Birds and Habitats Directives are still sound from a conservation point of view (Fries-Tersch *et al.* 2015). The overall target of halting biodiversity loss is still recognized as critical because of the important services that European societies receive from it, contributing to the added value of conservation policies. However, we have shown that the financial mechanisms used to guide conservation efforts have failed to target the species and regions most in need of conservation action at the continental scale. LIFE-Nature proved an effective conservation tool in some cases when enough funds were available, as demonstrated by the recent reclassification of the Iberian lynx from CR to EN (Rodríguez & Calzada 2015). However, apart from some exceptional cases, the actual distribution of conservation funds that we report here could undermine the EU's capacity to achieve its 2020 biodiversity targets and compromises the EU's contribution to the conservation of biodiversity at the global level, and then compromise the effectiveness and efficiency of the EU's conservation efforts.

There are different nonexclusive reasons behind the poor coverage of continental conservation needs by LIFE-Nature funds. First, given that LIFE-Nature projects must focus on species listed in the Annexes of EU's Birds

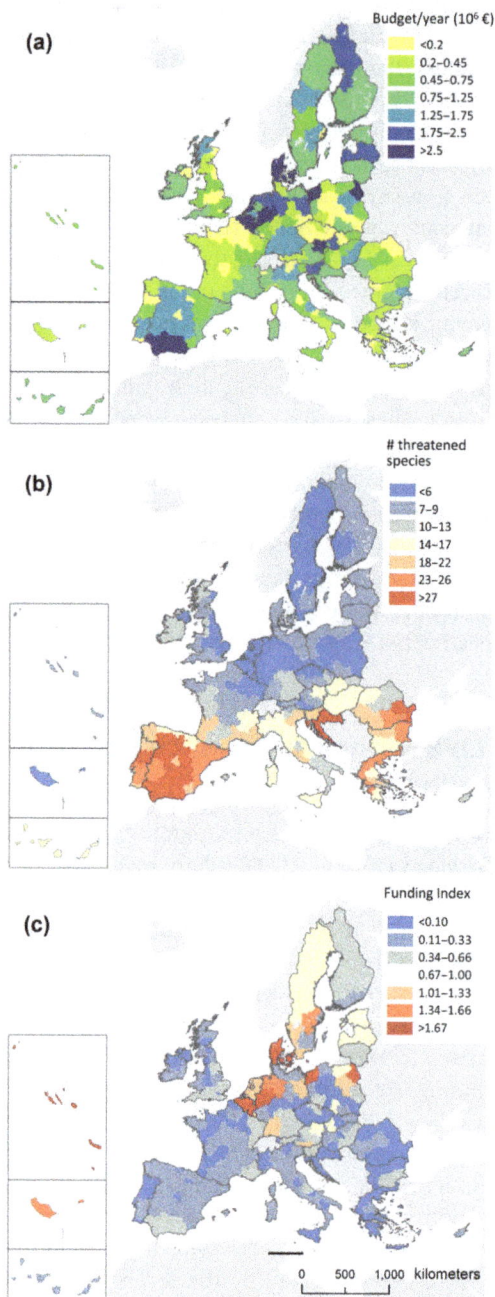

Figure 2 Distribution of LIFE-Nature funds and richness of threatened species (CR, EN, and VU) across NUTS regions in the EU (see Appendix 1 for more detail). (a) Average budget/year received by each region. (b) Number of threatened vertebrates in each region. Further details can be found in the Materials and Methods section. (c) Regional funding index, assessed as the ratio between the annual budget received by each region for conservation of vertebrate species and the annual budget that the region would have received if funds had been randomly distributed across all vertebrate species. Analyses were restricted to vertebrate species because these were the only for which spatial distribution data were available. Values above and below 1 indicate regions that have received more or less budget than expected by random distribution of funds, respectively.

and Habitat Directives, the poor funding coverage of threatened species might be due to the inconsistencies between the priorities set in the Directives' Annexes and global assessments of conservation status. A large proportion of the species listed in the Directives' Annexes are not globally threatened (72% of the 1,864 species listed), while many other threatened ones are not listed (e.g., Cardoso 2012; Hochkirch *et al.* 2012). This could have biased conservation funds toward nonthreatened species. However, this does not seem to be the only driver, given that even within the list of priority species funds were preferentially directed toward nonthreatened species (Figure 1). Second, the focus of conservation efforts might have been biased toward species that involve feasible actions from a logistic and economic point of view or toward species of highest social impact (e.g., investing in birds or mammals over fish or invertebrates). Finally, the pure bottom-up approach to project selection followed when distributing LIFE-Nature funds, might also constrain the capacity of different European regions from gaining access to them. In this sense, only those with the capacity to apply and co-fund projects would access LIFE funds. This bottom-up approach was recently highlighted as one of the main factors limiting the LIFE program's policy impact (EC 2011a) due to the restricted capacity of strategic investment subject to project applications received.

Where to from here?

Continental conservation objectives such as those aimed in the EU's Biodiversity Strategy need continental plans and commitments. Given the limited resources available for pursuing biodiversity conservation targets and the limited coverage of continental conservation needs that we report, some key aspects are in urgent need for revision and improvement. These should help increase the effectiveness, efficiency, and coherence of the application of the Directives, some of the key criteria under assessment in the REFIT process (EC 2014a). First, conservation efforts should focus on those species and areas most in need, as the most effective way forward toward halting biodiversity loss. This entails that continental conservation needs should prevail over socioeconomic or governance factors when deciding where to invest conservation funds. This investment should be guided by adequate planning at the continental scale as well. The Directives already set the internal mechanisms to ensure that conservation priorities are sound and based on the most up to date scientific knowledge and to reinforce the periodic revision of the biodiversity's status. For example, articles 10 and 18 in the Birds Directive and

Habitats Directives, respectively, establish that Member States and the Commission shall encourage the necessary research and scientific work to guide the achievement of the EU's conservation goals. Furthermore, they state that appropriate amendments as necessary for adapting Annexes to technical and scientific progress should be considered (articles 15 and 19 in the Birds and Habitats Directives, respectively). Some advances have been made in this direction under the current LIFE program, specifically under the Multiannual working program 2014–2017 (CE 2014b), where a window of opportunity has been opened to funding conservation projects for Endangered and Critically Endangered species not included in the Annexes. This would help overcome some of the gaps in the Annexes recently highlighted (e.g., Cardoso 2012; Hochkirch *et al.* 2012) by providing conservation opportunities to highly threatened species. However, this would not solve the problem alone if not accompanied by additional measures to ensure more strategic planning of investment.

Second, the distribution of conservation funds should be planned at the EU scale, overriding particular interests of national governments or limited capacity that may compromise the efficiency of conservation efforts (Donald *et al.* 2007; Kark *et al.* 2009). More strategic funding guidelines set by the EU would enhance the impact of the LIFE-Nature program and help overcome the current dependency of this program on project proposals highly biased by local/regional capacity. A whole-EU spatially explicit planning of conservation investment would require new mechanisms to triage conservation priorities. Some voices claim that more funds are urgently required to increase the relevance and effectiveness of conservation efforts in the EU (e.g., Rands *et al.* 2010; Hodge *et al.* 2015; Kati *et al.* 2015). We believe that larger funds would help address more species and habitats but that it will be more beneficial if it is accompanied by better guidance and central planning. Future planning could take advantage of the extensive development of systematic planning approaches based on cost-effective analyses that have become common practice to help decision making in conservation problems (Margules & Pressey 2000; Moilanen *et al.* 2009). These methods pose an objective and transparent approach for setting priorities, which could then be used to guide strategic investment under the LIFE-Nature program. For example, by applying these planning methods, the EU could identify priority species and regions where to focus LIFE-Nature investment that should be covered by project proposals (a more top-down control). These priorities should be periodically revised to account for achieved goals (and avoid recursive funding of some species/regions) and new ones expected to appear in this rapidly changing world. This would make LIFE-Nature investment more effective and able to respond to changing continental conservation needs, although it would imply a more top-down control on investment.

Moreover, both the new priorities and distribution of conservation efforts should address the traditional bias toward some groups of vertebrates (e.g., 80% of LIFE-Nature funds in the period 1992–2013) and more effectively target other underrepresented taxa, such as invertebrates and/or regions. In order to address this bias, it would be also needed to improve the poor knowledge on the conservation status of other taxonomic groups (e.g., invertebrates) to help better evaluate their conservation needs and act accordingly.

Further efforts are finally required to overcome the potential biases in funding derived from lack of local capacity to access and acquire LIFE-Nature funds and focus on vertebrates. In this regard, the "capacity building" projects recently created in the framework of the current LIFE program (EC 2011b) will provide financial support to enhancing local capacity to address conservation problems and getting access to LIFE funds (e.g., recruitment and training of personnel). All these measures should ideally drive a better distribution of conservation funds and more effective achievement of EU's conservation goals and enhance the added value of the Directives, one of the specific criteria under assessment in REFIT (EC 2014a). Finally, one of the aims of LIFE project is to serve as demonstration for future implementation of conservation actions in similar situations. This can only be done by securing a solid monitoring of implemented actions, which is not always the case (*Henle et al.* 2013). Without this critical piece of information, it will be difficult to learn from previous on successful and unsuccessful experiences.

Strengthening the EU for better conservation outcomes

Many of the challenges for the conservation of biodiversity highlighted here fit in a broader debate on the need for "more Europe versus less Europe." In an admittedly simplistic overview, the contrasting positions advocate either moving toward stronger European institutions and common policies or toward a more decentralized Europe where national governments would get back some of the governance capacity once entrusted to Brussels. We argue that from a conservation point of view, the "more Europe" option, i.e., common planning at the continental scale, is the only way forward for achieving continental conservation goals in an efficient way. Common goals and stronger collaboration across EU has been highlighted as the best strategy to face other environmental problems such as the ecological and socioeconomic impacts of invasive species (Hulme *et al.* 2009) or climate change (Lung *et al.* 2014). A common conservation strategy could also help tackle the poor effectiveness of the Directives at protecting biodiversity already listed as priority, such as long migratory birds (Sanderson *et al.* 2015), which might require of pan European (and even beyond) planning of conservation efforts. As our assessment shows, the European Directives are useful instruments for guiding the implementation of conservation efforts on the ground. As we demonstrate, conservation funds have been directed mainly toward areas with higher numbers of listed vertebrates and exclusively to Natura 2000 sites. However, this technically accurate implementation of the Directives has been constrained by the limited success at attending continental and global conservation priorities. We believe that most of the mechanisms that we propose here are already into place in the Directives (e.g., periodic review of priorities set in the Annexes) or derived policy, such as the LIFE Multiannual working program (e.g., consideration of threatened but not listed species within LIFE). There is also a growing debate on the suitability of a more top-down approach to distributing LIFE-Nature funds (EC 2011b) to reinforce the impact of this program. The REFIT process opens a window of opportunity to discuss the need for more strategic implementation of existing policy or introducing changes to enhance conservation practice in Europe that we should not miss. Leveraging this opportunity could allow the EU to embrace continental conservation needs and lead to more robust and efficient continental-scale conservation of biodiversity.

Acknowledgments

V.H. and M.C. were supported by two Ramón y Cajal contracts funded by the Spanish Ministry of Science and Innovation (RYC-2013-13979 and RYC-2010-06431, respectivey).

Supporting Information

Table S1. Results from GLM analyses.
Appendix 1. Material and methods.

References

Cardoso, P. (2012). Habitats Directive species lists: urgent need of revision. *Insect Conserv. Diver.*, **5**, 169-174.

Donald, P.F., Sanderson, F.J., Burfield, I.J., Bierman, S.M., Gregory, R.D. & Waliczky, Z. (2007). International conservation policy delivers benefits for birds in Europe. *Science*, **137**, 810-813.

EC. (2011a). Our life insurance, our natural capital: an EU biodiversity strategy to 2020. COM(2011) 244 final (visited September 30, 2015).

EC. (2011b). Regulation of the European Parliament and of the Council on the establishment of a Programme for the Environment and Climate Action (LIFE). 2011/0428 (COD). (visited January 25, 2016).

EC. (2014a). Communication from the commission to the European Parliament, the council, the European economic and social committee and the committee of the regions. Regulatory Fitness and Performance Programme (REFIT): State of Play and Outlook. COM(2014) 368 final (visited September 30, 2015).

EC. (2014b). Commission implementing decision of 19 March 2014 on the adoption of the LIFE multiannual work programme for 2014-17. (2014/203/EU) (visited January 25, 2016).

European Environment Agency. (2015). State of nature in the EU. Results from reporting under the nature directives 2007–2012. Available at: http://www.eea.europa.eu/publications/state-of-nature-in-the-eu (visited January 20, 2016).

Fries-Tersch, E., Sundseth, K. & Ballesteros, B. (2015). Report on the open public consultation of the "fitness check" one the Birds and Habitats Directives. Final report for the European Commission, DG Environment, Brussels, October 2015.

Henle, K., Bauch, B., Auliya, M. *et al.* (2013). Priorities for biodiversity monitoring in Europe: a review of supranational policies and a novel scheme for integrative prioritization. *Ecol. Indic.*, **33**, 5-18.

Hochkirch, A., Schmitt, T., Beninde, J. *et al.* (2012). Europe needs a new vision for a Natura 2020 network. *Conserv. Lett.*, **6**, 462-467.

Hodge, I., Hauck, J. & Bonn, A. (2015). The alignment of agricultural and nature conservation policies in the European Union. *Conserv. Biol.*, **29**, 996-1005.

Hulme, P.E., Pysek, P., Nentwig, W. & Vilà, M. (2009). Will threat of biological invasions unite the European Union? *Science*, **34**, 40-41.

IUCN. (2015). The IUCN Red List of Threatened Species. Version 2015-3 (http://www.iucnredlist.org). Downloaded on May 15, 2015

Kark, S., Levin, N., Grantham, H.S. & Possingham, H.P. (2009). Between-country collaboration and consideration of costs increase conservation planning efficiency in the Mediterranean Basin. *Proc. Natl. Acad. Sci.*, **106**, 15368-15373.

Kati, V., Hovardas, T., Dieterich, M., Ibisch, P.L., Mihok, B. & Selva, N. (2015). The challenge of implementing the European network of protected areas Natura 2000. *Conserv. Biol.*, **29**, 260-270.

Lung, L., Meller, A.J.A., van Teeffelen W. & Thuiller M. Cabeza (2014). Biodiversity funds and conservation needs in the EU under climate change. *Conserv. Lett.*, **7**, 390-400.

Margules, C.R. & Pressey, R.L. (2000). Systematic conservation planning. *Nature*, **405**, 243-253.

Moilanen, A., Wilson, K.A. & Possingham, H.P. (2009). *Spatial conservation prioritization: quantitative methods and computational tools*. Oxford University Press, Oxford, UK. 320 pp.

Rands, M.R.W., Adams, W.M., Bennun, L. *et al.* (2010). Biodiversity conservation: challenges beyond 2010. *Science*, **239**, 1298-1303.

Rodríguez, A. & Calzada, J. (2015). Lynx pardinus. The IUCN Red List of Threatened Species 2015: e.T12520A50655794. http://dx.doi.org/10.2305/IUCN.UK.2015-2.RLTS.T12520 A50655794.en.

Sanderson, F., Pople, R.G., Ieronymidou, C. *et al.* (2015). Assessing the performance of EU Nature legislation in protecting target bird species in an era of climate change. *Conserv. Lett.*, doi: 10.1111/conl.12196.

7

"Devil Tools & Tech": A Synergy of Conservation Research and Management Practice

Carolyn J. Hogg[1], Catherine E. Grueber[2,4], David Pemberton[3], Samantha Fox[3], Andrew V. Lee[3], Jamie A. Ivy[4], & Katherine Belov[2]

[1] Zoo and Aquarium Association Australasia, Mosman, NSW, Australia
[2] Faculty of Veterinary Science, University of Sydney, Sydney, NSW, Australia
[3] DPIPWE, Save the Tasmanian Devil Program, Hobart, Tasmania, Australia
[4] San Diego Zoo Global, San Diego, CA, USA

Keywords
Conservation genetics; genomics; implementation gap; research; Tasmanian devil.

Correspondence
Carolyn Hogg, Zoo and Aquarium Association, PO Box 20, Mosman, NSW, 2088, Australia.
E-mail: carolyn@zooaquarium.org.au

Editor
Mark W. Schwartz

Abstract

Biodiversity conservation continually presents new challenges, yet conservation resources are limited, and funding for applied conservation research projects more so. Recently, many have reported on the "research–implementation gap," whereby conservation research findings are infrequently translated into conservation actions. In this perspective, we describe our experiences working in a large multi-institutional, multi-disciplinary team as we attempt to bridge the research–implementation gap by developing conservation tools needed to address the conservation challenges faced by Tasmanian devils (*Sarcophilus harrisii*). We discuss our project's history, lessons learnt, outcomes, and future plans to provide insights that may help others develop multi-institutional projects, designed to target rapid and direct implementation of conservation research into management action. Key to our success is the needs-based prioritization of research measured against the management team's questions, recognition of the different needs of academia, industry and government, a collegiate approach, and willingness to embrace adaptive management. Challenges include developing a project which meets all strategic targets of different institutions, in addition to sourcing funds. Overall, our goal has been to establish an enduring research-management framework, to facilitate improved integration of scientific research into the management needs of Tasmanian devil conservation, and serve as a template for other species management projects.

Introduction: the conservation research–implementation gap

Preserving our planet's biodiversity has been described as a "wicked" problem characterized by a myriad of complexities and challenges (Game *et al.* 2014). To address this we need to have evidence-based approaches to conservation, where realistic solutions are provided by scientists who understand management (Braunisch *et al.* 2012). The disconnect between those that produce science and those that use this science to make management decisions is referred to as the research–implementation gap (Knight *et al.* 2008). Some reviewers have attributed this gap to a lack of implementation planning from researchers (Cook *et al.* 2013), while others highlight the challenges of translating technical and fundamental findings into direct management actions (Shafer *et al.* 2015). Cash *et al.* (2003) assessed those who have successfully bridged this gap and found there to be four key elements across successful programs: collaboration, communication, translation, and mediation. More recently it has been argued that conservation researchers and management practitioners need to work together better from the outset (e.g., Susskind *et al.* 2012; Hoban *et al.* 2013; Wood *et al.* 2015). The literature contains pleas from both parties for improved integration of conservation science and practice, highlighting the need for workers in both industries to come together. While Legge (2015) provides a practitioner's perspective, others provide recommendations to assist conservation scientists in translating their findings (e.g., Gordon *et al.* 2014; Wood *et al.* 2015; Keller *et al.* 2015), dissemination of successful case studies (Shafer *et al.* 2015), and recommendations on how to integrate scientists into management organizations (Cook *et al.* 2013).

In theory, management practitioners and conservation scientists working together enables researchers to benefit from developing research questions targeting real-world problems, and conservation management teams to benefit by obtaining the data they require to address those problems (Cook *et al.* 2013). In practice however, while recent work has discussed the benefits of collaboration and open communication between scientists and policy-makers, this is not always straightforward (Karl *et al.* 2007; Susskind *et al.* 2012). Management practitioners and conservation scientists often have different perspectives on problems and different measures of "success." However, because consultation between researchers and managers can expedite both the collection of data and its use, integrating communication from the early stages of a project's development allows the identification of synergies that can support the goals of all parties (Susskind *et al.* 2012). In this perspective, we draw on our own experiences working in a large multi-institutional, multi-disciplinary team to develop conservation tools needed to address the problems faced by the Tasmanian devil (*Sarcophilus harrisii*) to highlight how the four key elements described above can bridge the research–implementation gap (Cash *et al.* 2003). We do not claim to have all the solutions to the challenge of research implementation, but by providing a discussion of our project's history, lessons learnt along the way, outcomes, and our plans for the future, we hope that this project will provide insights that help others to develop positive collaborations for their projects.

A framework for fruitful collaboration: the "devil tools & tech project"

In Figure 1, we summarize some of the stages at which, in our experience, communication between management practitioners and research can lead to positive outcomes for both. In the following paragraphs, we describe how the development of our "tools & tech" project led to these observations.

Origins of the "tools & tech" project

The Save the Tasmanian Devil Program (STDP) is the official government response to the decline in Tasmanian devil populations due to the contagious and lethal devil facial tumor disease (DFTD), which was first observed in 1996 (Hawkins *et al.* 2006). The aim of the STDP is to ensure "an enduring and ecologically functional population of devils living wild in Tasmania" (STDP, 2014). There are a number of management actions surrounding this aim including development and maintenance of an insurance population, release of animals onto islands/fenced sites, annual monitoring of wild populations and recovery of wild devil populations (DPIPWE, 2010).

The insurance population (IP) commenced in 2006 (Hogg *et al.* 2015) and was designed to be a source population for releases onto islands and fenced sites. The STDP contracted the Zoo and Aquarium Association Australasia (ZAA) to manage the insurance population on their behalf. In order for the ZAA to successfully manage the IP, an understanding of genetic relationships between founders was needed, as many founding animals were sourced from the Tasmanian north-west and west coast, ahead of the disease front at the time (Hogg *et al.* 2015). The intent of the IP was to release devils back into Tasmania within 30 years, or 5 to 6 generations (CBSG, 2008), as it was believed that DFTD would cause the extinction of the devil in 25 years (McCallum *et al.* 2007). Species management modeling indicated that inbreeding would be a concern in the short-term if founder individuals were closely related (Rudnick & Lacy 2008). ZAA and the University of Sydney formed a collaboration to determine founder relatedness, as the university specialized in devil genetics and were undertaking parentage analyses for IP devils. At the same time, the first island release site was being established and parentage analyses on a broader scale would be needed to facilitate ongoing genetic management. Due to already low genetic diversity of the species, previously developed microsatellite markers (Jones *et al.* 2003); Cheng & Belov 2012) were unsuitable for founder analysis (Hogg *et al.* 2015), and so an assay for genotyping single nucleotide polymorphisms (SNPs) was proposed and developed (Wright *et al.* 2015). From these initial discussions the concept of the "devil tools & tech project" was conceived, with the underlying concept of bridging the gap between the lab bench and forest floor through a collaborative approach for applied management of the devil in the Tasmanian landscape. That is, an integrated approach of conservation research and management, with scientists and managers contributing equally.

On reflection, we believe the success of the "tools & tech" project lies in the fact that we started with one small, targeted project to determine whether the integrated approach would meet the needs of all parties. Once the process and "ground rules" were established we were able to build the overall project. For example, a quarterly meeting schedule was established for both face-to-face and on-line meetings where each party was tasked with

Traditional conservation research approach

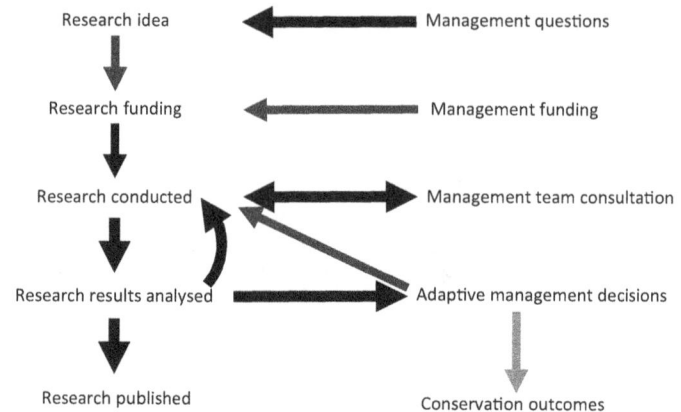

Figure 1 Differences between the tradtional conservation research approach and the Devil Tools & Tech approach, where there is full integration between research/practice/policy, with the final conclusion being a beneficial conservation outcome for the species. Darker heavier lines are indicative of a higher probability of success/occurrence, with thin gray lines indicative of lower probaility.

presenting a component of their work; and discussions for further work ensued. Further, we opted to work together on a specific project that was essential for management but which also required the development of new scientific methods and research. So, central to the success of the whole project is that both sides, "science" and "management," gain from the work when it is conducted in parallel, rather than flowing sequentially in one direction (Figure 1).

Our larger project grew via the incorporation of quality scientific data into management practice in real-time. As the release of devils onto Maria Island by the STDP was initiated, San Diego Zoo Global was seeking to fund a post-doctoral research position on a conservation-based Australian species and so "tools & tech" was born. We brought together people from each of the three disciplines: academia, government, and industry. It is important to consider the combination of three areas: research, practice, and policy (e.g., Hoban *et al.* 2013) and so in our opinion all three perspectives should be addressed and incorporated to make an integrated project truly successful.

Implementation

Excellent communication has been crucial to the success of our program. The lead personnel from each of the partner organizations communicate regularly to ensure that changes within research/practice/policy can be addressed and implemented. As an important outcome of this approach is applied management, strong communication ensures that the implementation of adaptive management processes does not compromise the integrity of the conservation research being undertaken. That is, when changes are to be made to management practices, we think about how these changes may impact any current long-term research projects and endeavor to minimize any impact. For example, when a cohort of devils involved in a long-term behavioral research project were needed for breeding, communication with researchers ensured that the project was completed prior to the transfer date.

Dialogue between the research and management teams was undertaken, and continues, at each stage of project development from the initial questions, sourcing of funds, conducting research, and analyzing preliminary data. By creating a feedback loop between the research and management teams, scientific discoveries and development of new tools and technologies can be directly implemented in the field in a shorter timeframe (Figure 1). As a result alternative avenues for applied conservation funding have been sourced as we are using new, innovative science to develop the conservation tools. Further, management teams can utilize the raw data prior to publication to ensure the rapid implementation of new discoveries.

The use of raw data requires a level of trust between the research and management teams, particularly when the data being utilized are produced by students, or other individuals, who are developing their skills. Working with conservation practitioners provides research students with industry exposure and an opportunity to receive training in the mechanisms of conservation implementation (e.g., meeting deadlines, permits, timely communication), providing them with additional transferrable skills that can help them in future careers in the conservation sector (Knight *et al.* 2008). A person with research experience, who is a member of the management team, is also part of the doctoral student advisory team. In addition, the STDP has developed a governance structure to share both samples and data acquired during daily management actions. These data and samples are managed through a data sharing agreement with the STDP under an oversight committee.

Outcomes

There are a number of direct management outcomes which have already been implemented as a result of the "tools & tech" project. These include identification of parentage of individuals held in group enclosures; determination of founder relationships to better inform breeding recommendations (Hogg *et al.* 2015); development of a new SNP-assay to assist with genomic analyses (Wright *et al.* 2015); and development of contraceptive tools to ensure even genetic representation across founding animals in fenced/island populations. A recent STDP initiative designed to augment and recover wild devil populations in the presence of DFTD (STDP, 2014) is also benefiting from the "tools & tech" approach. New research projects have been developed to assist in the implementation of this initiative, such as augmenting wild sites with devils selected for release based on their different genetic variants, and assessing the benefits of olfactory cues to dampen the potential dispersal of released individuals. A number of these management outcomes would not have been possible if it were not for the interaction with researchers and the support of

the academic sector, further highlighting the benefits of bridging the research–implementation gap.

Even though some conservation research outcomes have been already implemented into management (Figure 1), academic success indicators take longer to measure. Although the project is still in its early days, some academic success indicators include 20 research students and post-doctoral fellows, and publications which are just starting to emerge (e.g., Morris *et al.* 2015; Wright *et al.* 2015; Grueber *et al.* 2015). Other additional outcomes include raising the public profile of the "tools & tech" project and the broader plight of the Tasmanian devil, via the STDP newsletter, social media, print media, visitors/members to participating zoos and fauna parks and on-line presence (e.g., a documentary video about the project, produced by San Diego Zoo Global; http://bcove.me/t61imr6t [accessed 15 July 2015]).

Lessons learned

In order to better assist those who are embarking on the development of an integrated approach between management practitioners and conservation scientists we thought it useful to highlight some of the lessons we have learned. This is by no means an exhaustive list, but rather areas where we think our partnership is succeeding and how we were able to reach this goal.

From the start, we recognized that this project would be a significant investment for all parties, both intellectually and financially. This has underpinned the collegiate manner in which this project functions as we are all equal partners (Susskind *et al.* 2012). We did not commence the project with, and still do not have, a chief investigator/partner investigator mentality but rather each party from the outset raised what they needed from the project and what was important to the strategic goals of their institution. This way we were able to assess everything that was required of the partnership from the beginning and develop it in order to meet all partner needs. This is not a static process but an evolutionary one in recognition that institutional needs, particularly government and industry, change over time. Together, we have been successful as our organizations have had the capacity and resources to support the personnel involved in the project, as well as helping us to build capacity, e.g., San Diego Zoo Global post-doctoral fellowship.

We do have a species coordinator who, due to the management of the insurance meta-population and the purpose of the IP, works closely with all partners on different aspects of the project and so by default has become the project coordinator. This role ensures both regular communication between partners and timely communication of any changes and assists with mediation if required.

We developed an umbrella framework—"devil tools & tech"—with an overarching aim: to take the latest tools and technologies from the lab bench and make them directly applicable to the forest floor; to develop novel techniques and concepts for conservation work; and to trial their implementation in real world situations. The difference between our umbrella project and many other conservation research/management projects is that the questions were sourced from the management practitioners, allowing researchers to facilitate answers to specific management needs, which led to further management questions (Figure 1). In this way we are not answering one question, but multiple, sometimes overlapping questions, at the same time. Each research question we answer contributes to the larger management machine of the Save the Tasmanian Devil Program.

We promote a strong culture of collaboration and collegiality. This philosophy is strengthened by everyone's willingness to communicate and collaborate, across all the different sectors, and prioritize our time accordingly. We have regular on-line "face-to-face" meetings between lab and field teams, as we are located in different Australian states. Staff and students from each partner organization are encouraged to join these meetings, to work with and learn about the other partners and how their work fits into the bigger picture (see also Jenkins *et al.* 2012). A two-yearly meeting of all stakeholders, which includes staff/students from the lab, field, and zoos, is held to encourage better communication and give all partners an overview of how their efforts are working together to approach the goal of saving the devil.

The issue of devil management in the presence of DFTD lends itself well to the approach of pairing conservation research with management requirements so that the product, or sometimes by-product, of innovative scientific research can be directly utilized to inform management decisions. This is fundamental to our success—we do not want to do esoteric research, but to make sure there are direct management applications, while at the same time ensuring that the research is novel and innovative enough to be published and attract competitive funding. This is important as funding for applied conservation is often limited, and so an element of "thinking outside the square" is required to make this work. This sounds simple at face value, but was, and remains, perhaps one of the most challenging and time consuming aspects of the process to implement. For those managing other species, this integrated approach and sourcing of funds may be more complex, particularly if the management problem does not link well with research, does not attract public interest, have a sense of urgency; or the science already exists but just needs to be implemented.

We also initiated further collaborations with other members of our institutions to participate in the larger project. What commenced as a multi-institutional project, with a genomic focus, has now expanded into a multi-institutional, multi-disciplinary approach to provide management solutions for integrating Tasmanian devils back into the wild. To date, there have been no local extinctions of devils in the wild; therefore, waiting to reintroduce devils post extinction of both the devil and DFTD, as was originally suggested in 2006, is not possible (STDP *pers. comm.*). As all parties recognize the value of an adaptive management approach, we have been able to expand and evolve the "tools & tech" project to encompass changes in management focus. Another core characteristic of the "tools & tech" approach is that we did not rely on passive dissemination as is generally the case in the "conventional approach" (Figure 1). The transition from one small targeted project to a multi-faceted conservation research program directly servicing management actions and questions is the result of the approach taken here, rather than the conventional approach of disparate scientists working on different aspects of the conservation challenges that face the Tasmanian devil, irrespective of the tools needed for management.

Looking ahead

The issue of managing devils in the presence of DFTD lends itself well to working at the science–practice–policy interface; that is Tasmanian devils are an iconic species that the public and government are committed to saving. This has contributed significantly to the success of the "devil tools & tech" project and our ability to source funding. Further, our approach contains the key elements known to bridge the research implementation gap, communication, collaboration, translation of results and mediation (Cash *et al.* 2003). In light of our successes and lessons learnt, we are commencing discussions with other research and management teams in regard to establishing similar projects with other Australian native fauna. The development and implementation of the "devil tools & tech" project has been underway since 2011 and will continue in its present form until 2018. Even now, the current project partners are broadening the umbrella framework to expand beyond the initial scope and timeframes of the project, whilst at the same time embodying the philosophy of the project. This further work will continue into the future and endure beyond the original "tools & tech" lifespan. It is hoped that the framework we have built will persist for as long as government/academia/industry participate in the central aim of having an ecologically functional and persisting population of Tasmanian devils in the wild.

Acknowledgments

This manuscript benefitted from constructive feedback from Mark Schwartz and two anonymous reviewers. Our research into conservation management tools and technologies, including genetics, genomics, dampening dispersal and contraception of Australian threatened species, receives support from the Australian Research Council, San Diego Zoo Global, Save the Tasmanian Devil Program, Save the Tasmanian Devil Appeal, the Zoo and Aquarium Association, and the University of Sydney.

References

Braunisch, V., Home, R., Pellet, J. & Arlettaz, R. (2012). Conservation science relevant to action: a research agenda identified and prioritized by practitioners. *Biol. Conserv.*, **153**, 201-210.

Cash, D.W., Clark, W.C., Alcock, F., *et al.* (2003). Knowledge systems for sustainable development. *Proc Natl Acad Sci U S A*, **100**(14), 8086-8091.

Cook, C.N., Mascia, M.B., Schwartz, M.W., Possingham, H.P. & Fuller, R.A. (2013). Achieving conservation science that bridges the knowledge-action boundary. *Conserv. Biol.*, **27**(4), 669-678.

CBSG. (2008). *Tasmanian Devil PHVA final report.* IUCN/SSC Conservation Breeding Specialist Group, Apple Valley, MN.

Cheng, Y. & Belov, K. (2012). Isolation and characterisation of 11 MHC-linked microsatellite loci in the Tasmanian devil (Sarcophilus harrisii). *Conserv. Genet. Resour.*, **4**, 463-465.

Department of Primary Industries, Parks, Water and Environment. (2010). *Draft recovery plan for the Tasmanian devil (Sarcophilus harrisii).* DPIPWE, Hobart.

Game, E.T., Meijaard, E., Sheil, D. & McDonald-Madden, E. (2014). Conservation in a wicked complex world; challenges and solutions. *Conserv. Lett.*, **7**, 271-277.

Gordon, I.J., Evans, D.M., Garner, T.W.J., *et al.* (2014). Enhancing communication between conservation biologists and conservation practitioners: letter from the Conservation Front Line. *Anim. Conserv.*, **17**, 1-2.

Grueber, C.E., Peel, E., Gooley, R. & Belov, K. (2015). Genomic insights into a contagious cancer in Tasmanian devils. *Trends in Genetics*, **31**(9), 528-535.

Hawkins, C.E., Baars, C., Hesterman, H., *et al.* (2006). Emerging disease and population decline of an island endemic, the Tasmanian devil Sarcophilus harrisii. *Biol. Conserv.*, **131**(2), 307-324.

Hoban, S.M., Hauffe, H.C., Pérez-Espona, S., *et al.* (2013). Bringing genetic diversity to the forefront of conservation policy and management. *Conserv. Genet. Resour.*, **5**, 593-598.

Hogg, C.J., Ivy, J.A., Srb, C., *et al.* (2015). Influence of genetic provenance and birth origin on productivity of the Tasmanian devil insurance population. *Conserv. Genet.*, **16**(6), 1465-1473

Jenkins, L.D., Maxwell, S.M. & Fisher, E. (2012). Increasing Conservation impact and policy relevance of research through embedded experiences. *Conserv. Biol.*, **26**(4), 740-742.

Jones, M.E., Paetkau, D., Geffen, E. & Moritz, C. (2003). Microsatellites for the Tasmanian devil (Sarcophilus laniarius). *Mol. Ecol. Notes.*, **3**, 277-279.

Karl, H.A., Susskind, L.E. & Wallace, K.H. (2007). A dialogue not a diatribe: effective integration of science and policy through joint fact finding. *Environment*, **49**(1), 20-34.

Keller, D., Holderegger, R., van Strien, M. & Bolliger, J. (2015). How to make landscape genetics beneficial for conservation management? *Conserv. Genet.*, **16**, 503-512.

Knight, A.T., Cowling, R.M., Rouget, M., Balmford, A., Lombard, A.T. & Campbell, B.M. (2008). Knowing but not doing: selecting priority conservation areas and the research–implementation gap. *Conserv. Biol.*, **22**, 610-617.

Legge, S. (2015). A plea for inserting evidence-based management into conservation practice. *Anim. Conserv.*, **18**(2), 113-116.

McCallum, H., Tompkins, D.M., Jones, M., *et al.* (2007). Distribution and impacts of Tasmanian devil facial tumor disease. *Ecohealth*, **4**(3), 318-325.

Morris, K.M., Wright, B., Grueber, C.E., Hogg, C.J. & Belov, K. (2015). Lack of genetic diversity across diverse immune genes in an endangered mammal, the Tasmanian devil (Sarcophilus harrisii). *Mol. Ecol.*, **24**, 3860-3872.

Rudnick, J.A. & Lacy, R.C. (2008). The impact of assumptions about founder relationships on the effectiveness of captive breeding strategies. *Conserv. Genet.*, **9**, 1439-1450.

STDP. (2014). Save the Tasmanian Devil Program – Business Plan 2014–2019. Version 1.1. http://www.tassiedevil.com.au/tasdevil.nsf/files/82C18864F5819337CA2576CB0011569B/$file/STDP%20Business%20Plan%202014-19.pdf (downloaded 21/07/2015).

Shafer, A.B.A., Wolf, J.B.W., Alves, P.C., *et al.* (2015). Genomics and the challenging translation into conservation practice. *Trends Ecol. Evol.*, **30**, 78-87.

Susskind, L., Camacho, A.E. & Schenk, T. (2012). A critical assessment of collaborative adaptive management in practice. *J. Appl. Ecol.*, **49**(1), 47-51.

Wright, B., Morris, K., Grueber C.E., *et al.* (2015). Development of a SNP-based assay for genotyping the Tasmanian devil insurance population. *BMC Genomics*, **16**: 791-802.

Wood, K.A., Stillman, R.A. & Goss-Custard, J.D. (2015). Co-creation of individual-based models by practitioners and modellers to inform environmental decision-making. *J. Appl. Ecol*, **52**(4), 810-815.

8

Mismatch between Habitat Science and Habitat Directive: Lessons from the French (Counter) Example

Martin Jeanmougin[1], Camille Dehais[2], & Yves Meinard[3]

[1] Centre d'Ecologie et des Sciences de la Conservation (CESCO - UMR7204), Sorbonne Universités-MNHN-CNRS-UPMC, Muséum national d'Histoire naturelle, CP135, 43 rue Buffon, 75005, Paris, France
[2] GERECO, 30 avenue Leclerc, F-38217, Vienne, France
[3] Université Paris-Dauphine, PSL Research University, CNRS, UMR [7243], LAMSADE, 75016, FRANCE

Keywords

Conservation policies; decision analysis; habitats conservation; knowledge gaps; legitimacy; phytosociology; policy evaluation.

Correspondence

Yves Meinard, Université Paris-Dauphine, PSL Research University, CNRS, UMR [7243], LAMSADE, Place Lattre de Tassigny, F-75775 Paris Cedex 16, France.
E-mail: yves.meinard@lamsade.dauphine.fr

Editor

Douglas MacMillan

Abstract

The European Habitat Directive encompasses a conservation policy devoted to conserve habitats rather than single species. This ambition has strong ecological justifications, and inspires other initiatives such as the IUCN red list of ecosystems. Evaluating this policy is therefore pivotal to identify and reproduce best practices. However, the habitat aspect of this policy has so far not been systematically assessed. To make up for this lacuna, we take advantage of decision-aiding methodologies to introduce a new normative framework. According to this framework, a conservation policy is positively evaluated if it contributes to conservation, is science-based, operational, and legitimate. Based on an exploration of the published literature and unpublished reports and databases, we identify knowledge gaps plaguing the European habitat conservation policy. We argue that, due to these knowledge gaps, the contribution of this policy to the conservation of habitats is unproven, it is not science-based, not operational and not legitimate. Our study draws heavily on the French implementation. Analyzing this example, we highlight knowledge gaps that carry lessons for European conservation policies as a whole, but also for conservation initiatives focused on habitats in a broader geographical and political context. We then identify concrete means to strengthen habitats conservation policies.

Introduction

Natura 2000 (N2000) is the world's largest network of conservation sites (Evans 2012), covering more than 18% of the European Union (EU) land's area. The Birds and Habitat Directives (BD and HD; Supporting Information [SI], SI-Table 1-B1-B2) require that Member States codify European protections in National laws and actively implement them within this network. The corresponding conservation actions focus on "Species of Community Interest" (SCI) and "Habitats of Community Interest" (HCI) listed in annexes of the Directives (HCI are presented in "EUR28": SI-Table 1-A1; SI-Table 4).

The habitat aspect of this policy embodies the largely justified ambition to overcome species-focused approaches by targeting ecosystems (Keith *et al.* 2013). Beyond the acknowledged importance of assessing public policies (Ferraro & Pattanayak 2006, SI-Table 1-D1), its evaluation therefore has a particular significance for conservation. However, although several studies address the impact of N2000 on various taxa (birds: Pellissier et al. 2013; Sanderson et al. 2015; terrestrial vertebrates: Maiorano et al. 2015; bats: Lisón et al. 2013)), its habitat aspect has not been systematically assessed (SI-Table 1-A2).

We address this lacuna through an analysis mostly focused on France, used as an example to draw lessons for Europe as a whole. Indeed, according to the subsidiarity principle (SI-Table 1-B3), N2000 is orchestrated at EU-scale, but Member States implement it as they see fit. This diversity of implementations provides opportunities to learn from local experiences, which is our approach here.

Table 1 Indeterminacy of translation between the functional ecology literature and HCI practice

The ontology of functional ecology		The ontology of HCI practitioners
Reference	Categories used to describe habitats	Nonexhaustive list of HCI and non-HCI categories (including syntaxons) to which the habitat could belong
Andueza *et al.* 2010	Grassland rich in grasses, Grassland rich in Forbs	6430 Hydrophilous tall herb fringe communities of plains and of the montane to alpine levels
Ansquer *et al.* 2009	Grazed grasslands	6510 Lowland hay meadows (*Alopecurus pratensis, Sanguisorba officinalis*) *Agrostietea stoloniferae* Th. Müll. & Görs 1969
de Vries *et al.* 2012	Unimproved grassland, semi-improved and improved grasslands	*Plantaginetalia majoris* Tüxen ex von Rochow 1951
Gardarin *et al.* 2014	Grazed and mown permanent grasslands	
Garnier *et al.* 2004; Cortez *et al.* 2007	A successional sere following vineyard abandonment in the Mediterranean region of France	*Agropyretalia intermedii-repentis* Oberd., Th.Müll. & Görs in Th. Müll. & Görs 1969 *Artemisietea vulgaris* Lohmeyer, Preising & Tüxen ex von Rochow 1951 *Crataego monogynae-Prunetea spinosaee* Tüxen 1962 6220 * Pseudo-steppe with grasses and annuals of the *Thero-Brachypodietea*
Vile *et al.* 2006 ; Garnier *et al.* 2007 ; Fortunel *et al.* 2009	Agroecosystems	6430 Hydrophilous tall herb fringe communities of plains and of the montane to alpine levels 6510 Lowland hay meadows (*Alopecurus pratensis, Sanguisorba officinalis*) *Agrostietea stoloniferae* Th. Müll. & Görs 1969 *Plantaginetalia majoris* Tüxen ex von Rochow 1951 *Agropyretalia intermedii-repentis* Oberd., Th.Müll. & Görs in Th. Müll. & Görs 1969 *Artemisietea vulgaris* Lohmeyer, Preising & Tüxen ex von Rochow 1951 *Crataego monogynae-Prunetea spinosae* Tüxen 1962 *Polygono arenastri-Poetea annuae* Rivas Mart. 1975 corr. Rivas Mart., Báscones, T.E.Diáz, Fern.Gonz. & Loidi 1991 *Sisymbrietea officinalis* Gutte & Hilbig 1975
Quétier *et al.* 2007, Lavorel & Gargulis 2012, Lavorel *et al.* 2010	Arable rotation	*Stellarietea mediae* Tüxen, Lohmeyer & Preising ex von Rochow 1951
	Fertilized hay meadow	CORINE Biotopes 81
	Unfertilized hay meadow without *Festuca paniculata*	*Agrostietea stoloniferae* Th. Müll. & Görs 1969 6430 Hydrophilous tall herb fringe communities of plains and of the montane to alpine levels 6170 Alpine and subalpine calcareous grasslands *Festucetum paniculatae austro-occidentale centuretosum* Lacoste 1970
	Unfertilized hay meadow with *Festuca paniculata*	6170 Alpine and subalpine calcareous grasslands *Festucetum paniculatae austro-occidentale centuretosum* Lacoste 1970
	Grazed pasture without Festuca paniculata	6170 Alpine and subalpine calcareous grasslands *Plantaginetalia majoris* Tüxen ex von Rochow 1951
	Grazed pasture with *Festuca paniculata*	6170 Alpine and subalpine calcareous grasslands *Plantaginetalia majoris* Tüxen ex von Rochow 1951 *Festucetum paniculatae austro-occidentale centuretosum* Lacoste 1970
	Never mown >2000m grasslands	6170 Alpine and subalpine calcareous grasslands 6110 * Rupicolous calcareous or basophilic grasslands of the *Alysso-Sedion albi*
	Steep grazed slopes	6170 Alpine and subalpine calcareous grasslands *Festucetum paniculatae austro-occidentale centuretosum* Lacoste 1970 *Plantaginetalia majoris* Tüxen ex von Rochow 1951

Continued

Table 1 Continued

The ontology of functional ecology		The ontology of HCI practitioners
Reference	Categories used to describe habitats	Nonexhaustive list of HCI and non-HCI categories (including syntaxons) to which the habitat could belong
Storkey et al. 2013	Cereals and *Brassica* cultivated plots, Annually established seed mix	*Stellarietea mediae* Tüxen, Lohmeyer & Preising ex von Rochow 1951
	Floristically enhanced grass	*Agrostietea stoloniferae* Th. Müll. & Görs 1969
	Natural regeneration of the naturally occurring arable flora	CORINE Biotopes 81
		Stellarietea mediae Tüxen, Lohmeyer & Preising ex von Rochow 1951
		Sisymbrietea officinalis Gutte & Hilbig 1975
		Artemisietea vulgaris Lohmeyer, Preising, & Tüxen ex von Rochow 1951

We used the references enlisted by Garnier *et al.* (2015; see Table 6.1) to identify a representative list of publications in functional ecology dealing with the influence of plant traits on ecosystem processes (column 1). We reported the categories used in these articles to describe habitats (column 2) and identified in each case a series of HCI and non-HCI categories (including syntaxons) to which each category could correspond (column 3). This table illustrates that, on the basis of the information given in published articles, it is impossible to translate unequivocally the language used by functional ecologists to describe habitats into the one used by HCI practitioners.

To develop our evaluation, we first introduce a new normative framework inspired by recent advances in decision-aiding methodologies. Although the above-mentioned assessments exclusively focus on the impact of N2000 on conservation targets, this framework encompasses several dimensions of evaluation. Then, exploring the literature and unpublished reports and databases, we identify knowledge gaps (KG) plaguing HD's habitat policy. Gaps between research and practice have been studied on numerous conservation-related topics (Knight *et al.* 2008; Matzek *et al.* 2014), but the habitat case is poorly documented. We unveil important problems in this area, pertaining not only to how ecological advances can translate into practice, but also to capacity building and institutional design. We then use our framework to investigate the implications of these KG for the policy. Finally, we articulate concrete recommendations, and highlight the global significance of our study.

Policy analytics as evaluation framework

Numerous normative frameworks have been developed to rationalize evaluations and associated decision-aiding (De Marchi *et al.* 2016). Among them, "policy analytics" states that decision-aiding should be "value-adding" (i.e. help policies reaching their objectives), science-based, operational, and legitimating (Tsoukias *et al.* 2013). We propose to use and adapt this framework because these criteria are of particular relevance to the evaluation of HD's habitat policy, for the following reasons:

— HD aims to "maintain or restore natural habitats" (SI-Table 1-B1). Assessing its contribution to habitat

conservation is therefore crucial to evaluate it (criterion C1).
— It is largely admitted that scientific evidence and knowledge should contribute to assess and, in turn, to improve the effectiveness of conservation actions (Knight *et al.* 2008; Dicks *et al.* 2014; Senior *et al.* 2015). The evaluation should therefore investigate whether the policy is science-based (C2).
— Because impractical policies are pointless, the evaluation should assess whether the policy is operational (C3).
— Because N2000 sites witness highly diverse human activities (Tsiafouli *et al.* 2013), it is pivotal that N2000 initiatives be considered legitimate by local stakeholders and the general public (C4).

These criteria should not be seen as absolute requirements, because optimizing one criterion can have detrimental consequences on others. For example, a policy exceedingly reliant on cutting-edge science could become nonoperational because too demanding in technology and highly specific skills. Similarly, in some situations, non-scientific knowledge can contribute more efficiently than science to conservation (Mazzocchi 2006).

We accordingly only assume that, for each criterion C1-4, all other things held equal, the better the criterion is satisfied, the more positively the policy should be evaluated (Figure 1a). We now highlight knowledge gaps relating to HD's habitat policy, and investigate their consequences for C1-4.

(a)

A conservation policy…

Ex: Bird Directive has a positive impact on populations of target species (Sanderson et al. 2015)
Counterex: Various forest conservation programs failed to stop deforestation in Sub-Saharan Africa (Contestabile, 2012)

…contributes to conservation
C1

…is based on cutting-edge science
C2

Ex: Norms concerning the concentration of pollutants in the air (e.g. Clean Air Act in the US) are based on up-to-date scientific measurements
Counterex: Data on which lists of Bird Species of Community Interest are outdated (Cardoso, 2012)

…is positively evaluated

Ex: The French wetland protection policy uses a simplified pedological classification enabling experts to map wetlands on the basis of very simple pedological samplings
Counterex: *Pelophylax lessonae* and *P. kl esculentus* are impossible to sort out in the field but have different protection statuses (S1-Table1-B6)

C3

…is operational

C4

…is legitimate

Ex: To ensure legitimacy of management on N2000 sites, the French law (S1-Table1-B5) mandates setting up steering committee with representatives of local users
Counterex: Early attempts at implementing N2000 in France, which have been abandoned because, by bypassing local stakeholders, it came under heavy fire (Alphandéry & Fortier, 2001)

(b)

HD's habitat conservation policy…

Lack of data on the state of habitats of community interest **(KG1)**

…contributes to conservation
C1

…is based on cutting-edge science
C2

Gap between research and practice **(KG2)**

…can **not** be positively evaluated

Gap between teaching and practice **(KG3)**

C3

…is operational

C4

…is legitimate

Deficiencies in the epistemic structure **(KG4)**

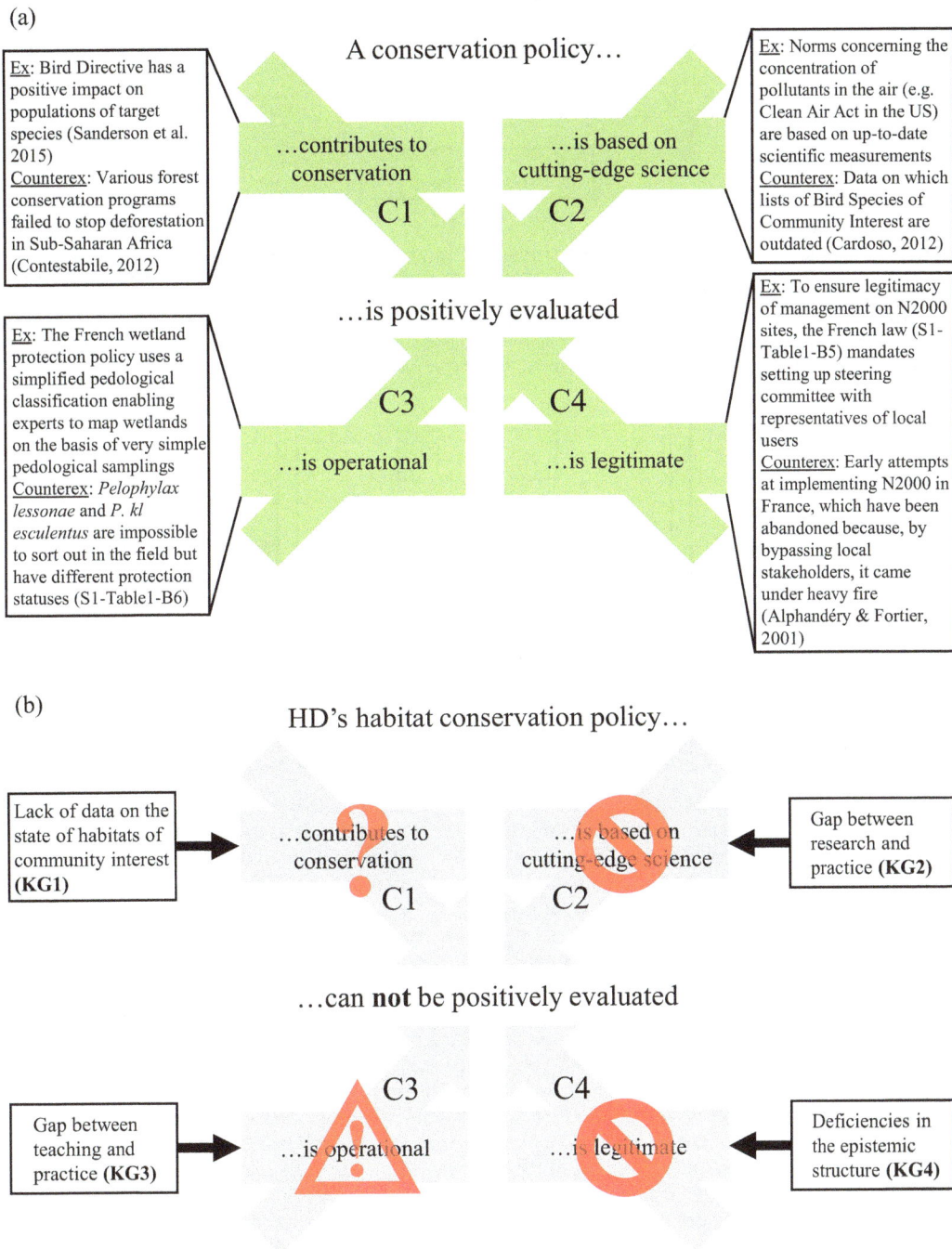

Figure 1 (a) Our evaluation framework, based on "policy analytics," explained and exemplified. C1-4 stand for the four criteria of evaluation of the policy under scrutiny. The four criteria should be understood as "all other things held equal" conditions: for any of these criteria, all other things held equal, the better the criterion is satisfied, the more positively the policy is evaluated. Notice that, in a general setting, maximizing the satisfaction of one criterion can impair the satisfaction of one or several of the others. Elaborating whether and how the various criteria can be aggregated in such cases falls beyond the scope of the present article. (b) Our framework applied to habitat conservation policy as part of the Habitat Directive. KG1-4 stand for the four knowledge gaps (see section Knowledge gaps). Because of KG1, C1 is indeterminate. Because of KG2, C2 is unsatisfied. Both notions apply to EU as a whole. Because of KG3, C3 is unsatisfied in France, where phytosociology is marginal in biology curricula. The French example should set off alarm bells for EU as a whole. KG4 testifies for a fragility of the European system as a whole, implying that C4 is not satisfied. Based on C1-4, one cannot positively evaluate HD's habitat conservation policy.

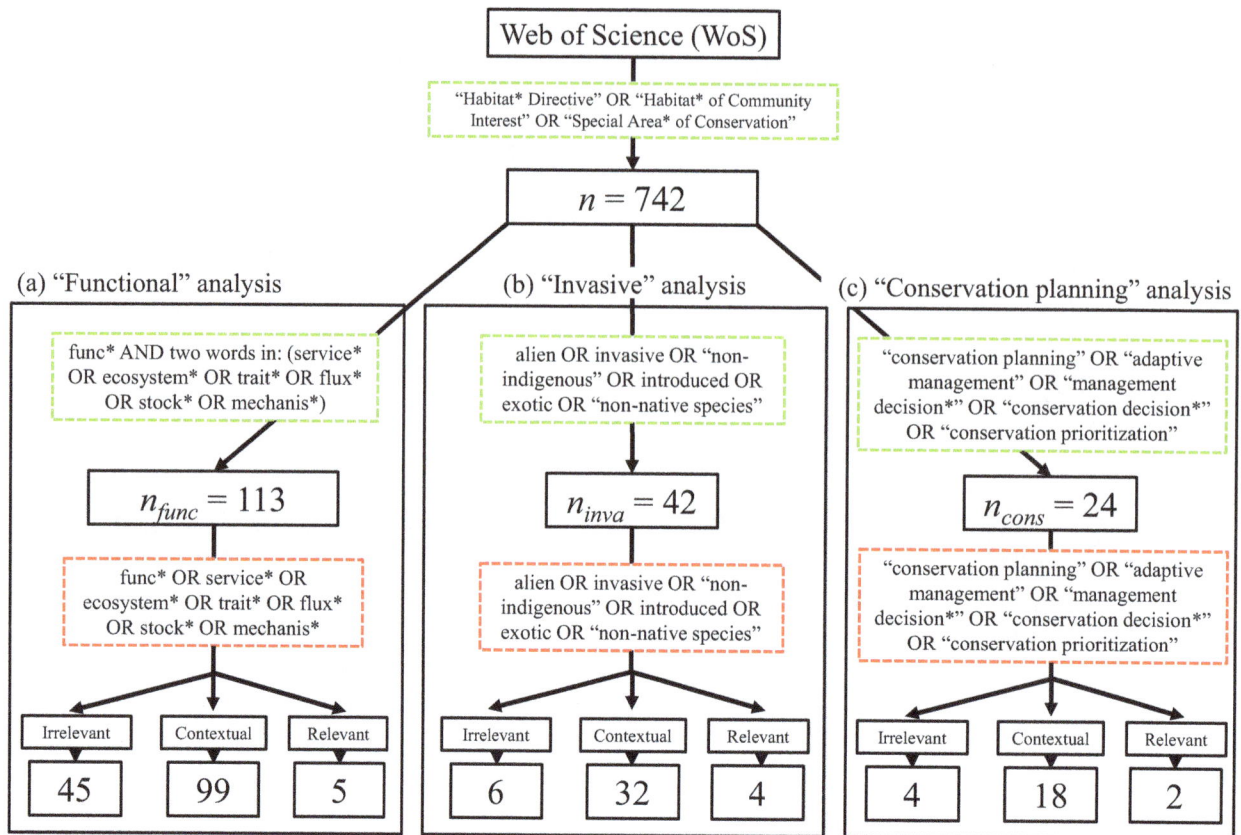

Figure 2 Flow chart explaining the bibliographical analyses performed on Web of Science Core Collection (Thomson Reuters, NY) and their results. Dashed boxes represent search on bibliographical corpus. Green dashed boxes mean that search is done on title, abstract and keywords of each publication. Red dashed boxes mean that search is done on entire text. We first identified a corpus of articles (n) dealing with habitats conservation in the N2000 sense, by selecting all the articles containing occurrences of "Habitat(s) Directive" or "Habitat(s) of Community Interest" or "Special Area(s) of conservation" in their abstracts, titles or keywords in the Web of Science Core Collection (timespan: 1970–2016; extensive list of articles can be found in SI-Table 5). On this first corpus, we then performed three parallel analyses to detect the available scientific knowledge on HCI pertaining to three prominent branches of ecology and conservation: functional ecology (a), invasive ecology (b), and conservation planning (c). The first step of each of the three analyses selects a branch-specific corpus by selecting articles containing occurrences of keywords of the corresponding branch in their title, abstract or keywords (corpuses n_{func}, n_{inva}, n_{cons}, respectively). In a second step, we detect occurrences of keywords of the concerned discipline in the whole text of each article in corpuses n_{func}, n_{inva}, n_{cons}. In a third step, we categorize these occurrences. Occurrences of branch-specific keywords are termed "irrelevant" when the detected words are actually homonyms of the searched words, used in another sense (e.g. searching for occurrences of "function" to detect occurrences of keywords of functional ecology, we select an article where the term "function" is used to refer to a mathematical function). Occurrences of branch-specific keywords are termed "contextual" when they are used to contextualize the study (e.g. in the introduction or the conclusion of the article) or when they appear in articles that do not mention any HCI in their main text. Occurrences are termed "relevant" when they appear in the formulation of scientific results dealing with both HCI and notions pertaining to the concerned branch. The sum of irrelevant, contextual and relevant occurrences can be greater than the size of the corresponding corpus because some words can appear multiple times in the same publication.

Notice that, in analysis (a), among the n_{func} articles, none is published in what we considered to be influential journals specialized in functional ecology (*Ecosystems*, *Functional Ecology*, *Biogeochemistry*), influential journals in ecology with sections devoted to functional ecology (*New Phytologist*, *Ecology*, *Journal of Ecology*, *Oikos*, *Oekologia*, *Ecology Letters*, *Global Change Biology*, *Basic and Applied Ecology*) or important review journals (*Trends in Ecology and Evolution*, *Annual Reviews of Ecology, Evolution and Systematics*, *Biological Reviews*). As addendum to analysis (b), all invasive species names from DAISIE, TAXREF v9 and EPPO Lists of Invasive Alien Plants (see SI-Table 3) were also searched in the general corpus ($n = 742$) on title, abstract and keywords. Linnaean binomial Latin names were used for each species because complete names (with authors and date) are not always given in articles. All synonyms were also checked using the French Taxonomic Reference Source TAXREFv9 (SI-Table 1-C5). This supplementary analysis does not reveal any other relevant articles than those previously detected by the principal analysis with n_{inva}.

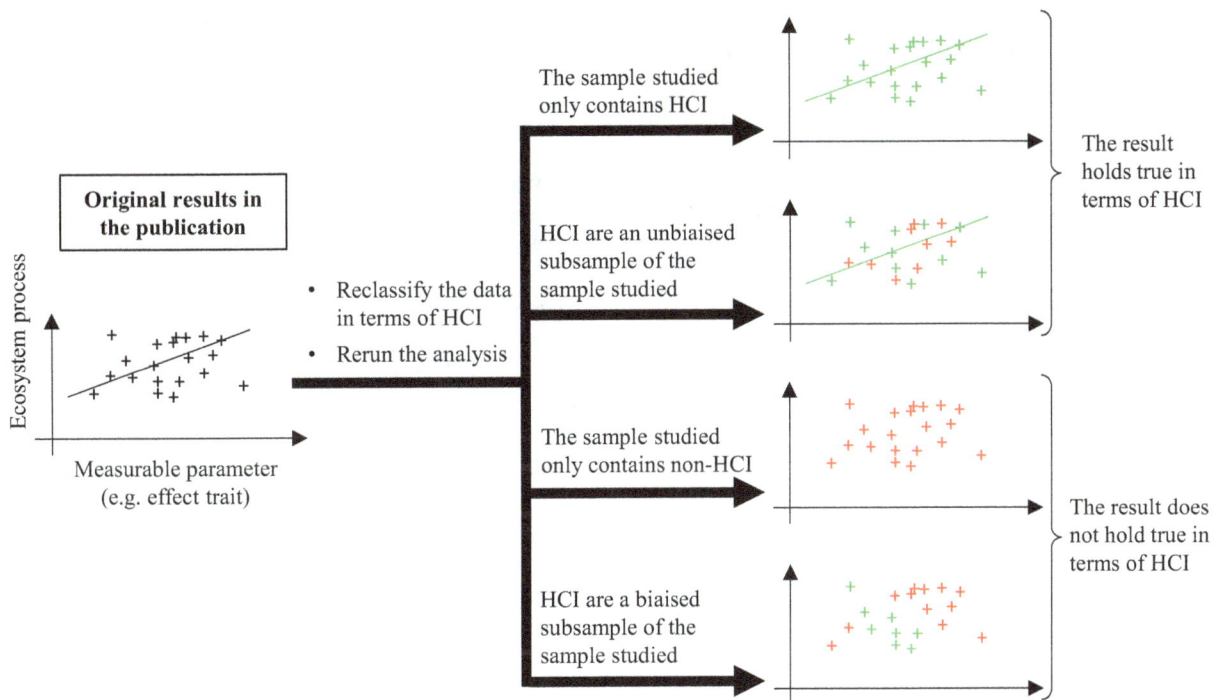

Figure 3 Schematic explanation of the analyses needed to translate typical results published in scientific journals into results usable in the field by N2000 practitioners, using results in functional ecology as a focal example. The left panel schematizes the results of a fictive scientific article, demonstrating a linear relation between a given ecosystem parameter and a given ecosystem process, for a series of habitats represented by black crosses and described in the original article using one or several of the categories exemplified in Table 1, column 2. Green and red crosses in the right panel respectively represent habitats that would be categorized as HCI and non-HCI using HD practitioners' categories, as exemplified in Table 1, column 3.

Evaluating HD's habitat policy

Knowledge gaps

Lack of data on HCI conservation status (KG1)

HD requires that Member States periodically perform conservation status assessments of HCI along three ecological parameters: range, coverage, and structure and function (article 17 evaluation: A17E). In each country, each HCI is evaluated in each biogeographical region (SI-Table 1-D2) where it occurs. The latest A17E (SI-Table 1-A3) spans over 2007–2012.

The methods and results of A17E are gathered by the European Topic Center on Biological Diversity (ETC-BD: SI-Table 1-D3) in a European-wide database (SI-Table 1-C1). France produced an additional, more detailed, database (SI-Table 1-C2). According to these databases, in France, evaluations based on a "complete survey or a statistically robust estimate" (in the official terminology: SI-Table 1-A4) are marginal (6.3%). 52.5% of evaluations are admitted to be based on data that are "partial" and 38% are "based on expert opinion." The lack of data is especially patent for structure and function: 85% of

evaluations are expert opinions (SI-Table 1-C2). This can be compared with evaluations of population size and population trends of Bird SCI in France in the BD reporting (SI-Table 1-C3), among which 44% are complete surveys and 7% are expert opinions. At European level, complete or statistically robust surveys are more important but in minority (16.6%), experts opinion are less dominating but still important (22.7%), and significantly more so (χ^2-test: $P < 0.001$) for structure and function (43.4%).

For each biogeographical zone/country/HCI combination, A17E provides an aggregate conservation status for all sites inside and outside N2000. A handful of countries (France, UK, Belgium, Germany, and Austria: SI-Table 1-A4) also produce evaluations at site level. However, these initiatives lack coordination (Maciejewski et al. 2016). The site-level French database (SI-Table 1-C4) contains information encoded in categories different from the ones used in A17E, with admitted ambiguities of translations between the two (SI-Table 1-A5). Besides, standardized evaluation methods are available only for a minority of HCI (in France: 52 HCI out of 132).

To sum-up, the data on the conservation status of HCI are of two sorts, too heterogeneous to be aggregated:

- A17E, which lumps together all sites at biogeographic/country level and is weakly based on complete or statistically robust estimates.
- Evaluations at site level, which are scarce and non-standardized.

Gap between research and practice (KG2)

Following the CORINE (SI-Table 1-A6) and Palearctic habitat classifications (Devilliers & Devilliers-Terschuren 1993), the denomination and descriptions of HCI (detailed in regional manuals such as the French "Habitat Books": SI-Table 1-A7) are largely based on phytosociological categories ("syntaxons"; Evans 2010).

To establish whether conservation actions based on HCI or phytosociological categories can take advantage of ecological advances, we explored the peer-reviewed literature in three disciplines that we take, for complementary reasons, to be the ones that should be the most useful for HD practitioners:

(a) *Functional ecology* because European texts emphasize the importance of preserving ecological functioning, and the A17E structure and function evaluation patently lacks data (3.1.1). We identified 113 articles in the literature about ecological functioning on N2000 habitats (Figure 2a). However, key terms of functional ecology are used there mainly to describe the context; only five articles refer both to HCI and functional ecology in their results. This illustrates that functional ecology and N2000 do not have a shared ontology: functional ecologists do not use HCI categories to elaborate and design their projects and never articulate results in this language. To be able to use these results, practitioners would have to reanalyze raw data to reclassify them in HCI terms, and redo analyses to see whether published results persist when translated (Figure 3).

(b) *Invasive ecology* because other European policies target invasive species (Beninde et al. 2015) and A17E takes them as an important threat to habitats (SI-Table 1-A4). Definitions of habitats in practitioner's manuals (SI-Table 1-A7) are however based on lists of "index species" often including invasive species. Depending on the invasive database used, 8–28% of the HCI present in France have, among their "index species," at least one species considered invasive in France (SI-Table 3). Robust knowledge on the impact of invasive species on HCI is therefore greatly needed, especially for HCI with index invasive species. However, to date, only four articles display results on the impact of invasive species on HCI (Figure 2b).

(c) *Conservation planning* because Popescu et al. (2014) identified it as the dominant discipline in the N2000 literature. We identified only 24 articles pertaining to this discipline on habitats (suggesting that the literature identified by Popescu et al. (2014) mainly dealt with the species aspects of N2000), and only two referred both to HCI and conservation planning in their results (Figure 2c).

The three corpuses, therefore, illustrate the same problem: despite their potential usefulness for HD practitioners, published results are not usable because they are not expressed in the categories that practitioners use, and translating them would be a scientific task on its own. As a consequence, it is not surprising that, among the 3117 evaluations performed as part of A17E, only 312 (10%) refer to articles published in journals listed in the *Journal of Citation Reports* (SI-Table 2).

Gap between teaching and practice (KG3)

Associated with the fact that academics rarely use phytosociological categories, phytosociology tends to disappear from ecology curricula. In France, all biology universities have marginalized phytosociology teaching (Bouzillé 2007); only four 1-week-long continuing formations are dispensed in phytosociology per year (SI-Table 1-D4). Phytosociology teaching remains more developed for example in Italy, Spain, or Eastern countries (e.g. SI-Table 1-D5). But the durability of this situation is uncertain if phytosociological categories vanish from the scientific literature.

Deficiencies in the structure of the epistemic community (KG4)

One of the strengths of N2000 is that numerous international experts were involved in its construction through seminars and experts groups (Evans 2012; Popescu et al. 2014). However, the French A17E database reveals problems in the structure of these panels. Although A17E involved more than 800 contributors, writing was largely confined to expertise institutions (81%), whereas research and teaching institutions played a marginal role (<5%). Among the 319 evaluations by expert institutions, all but one were performed by "Conservatoires Botaniques," institutions representing botanists and phytosociologists. The relevance of these actors for A17E is indisputable but others, including researchers and managers, also have insights to contribute, and the panel is biased against them (Figure 4). Social and economic scientists or experts could also contribute by improving the identification of threats impacting habitats and policies contradicting N2000 actions. Going beyond

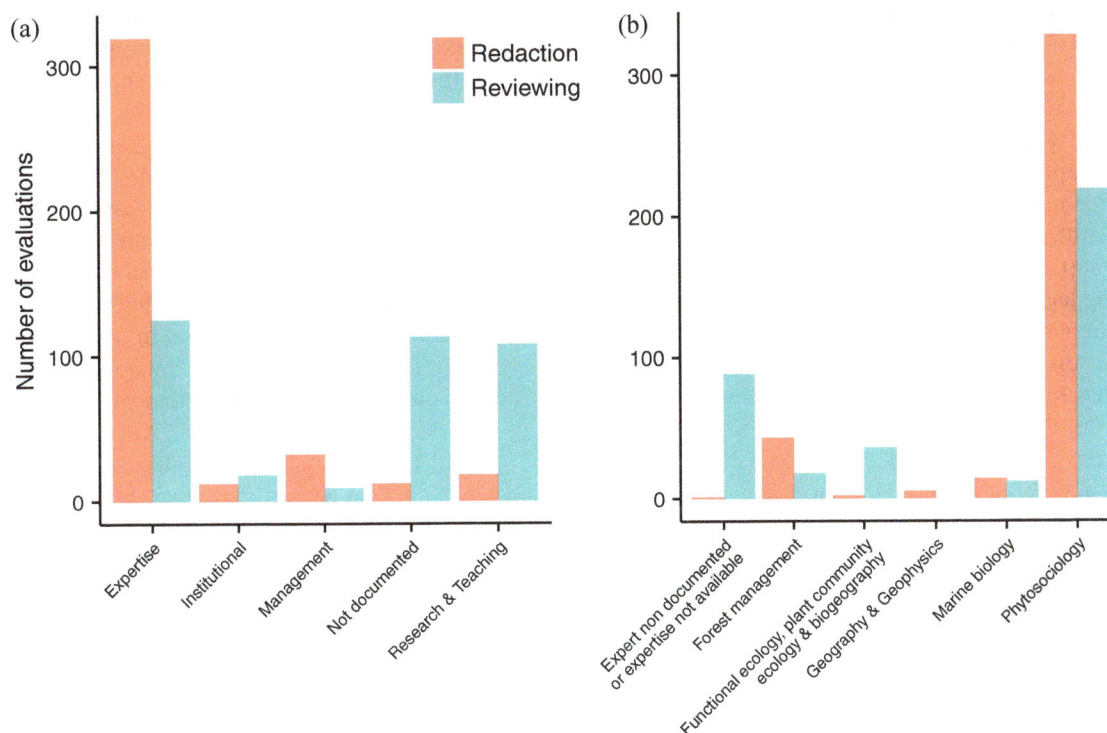

Figure 4 Institutional and disciplinary balance in the composition of the panel involved in the latest A17E in France. These information are from the French A17E database (SI-Table 1-C2). The database provides names and institutional affiliations for most of the experts involved in the evaluation. Part (a) of the graph reproduces the data from this database. To construct part (b), we used the disciplinary affiliation presented by the various experts on the websites of their institutions. "Conservatoires Botaniques" are the only exception, because most of the experts from these institutions do not have a personal page presenting them in the website of the "Conservatoire Botaniques" to which they belong. Since "Conservatoires Botaniques" are institutions devoted to botany and phytosociology, we classified experts from Conservatoires Botaniques in the disciplinary category "phytosociology."

European guidelines, the French database, therefore, unveils problematic biases invisible at European level.

Moreover, as far as reports indicate (SI-Table 1-A8), it has not been verified that the various A17E evaluators agreed in their understanding of "conservation status". The European guidelines (SI-Table 1-A4) do not require any such verification. Similarly, the construction of site scale methods (SI-Table 1-A9-A19) admits that "conservation status" is unproblematic for the panel of experts involved. Recent discussions (Epstein *et al.* (2016) on species, Boitani *et al.* (2015) on ecosystems, Maciejewski *et al.* (2016) on habitats), however, prove that the notion of conservation status is still open to largely divergent interpretations—despite purported clarifications in an internal European note (SI-Table 1-A20).

How knowledge gaps undermine HD's habitat policy (Figure 1)

Assessing HD's contribution to habitats conservation requires sorting out its proper effect from confounding factors such as historical trends or unrelated policies (Ferraro & Pattanayak 2006). This requires data on conservation status in sites inside and outside the network. But data on conservation status at site level are scarce and A17E lumps together statuses inside and outside the network (KG1). The contribution of N2000 to conserving habitats is therefore unproven due to a lack of data (C1 indeterminate).

KG2 shows that N2000 practice cannot take advantage of an important part of published scientific results. Although knowledge relevant to conservation is not confined to the academic world (Burgman *et al.* 2011), and A17E and HD practices are based on an extensive unpublished literature, they accordingly appear insulated from a large part of cutting-edge science (C2 unsatisfied).

In terms of operationality (C3), countries like France are additionally plagued by a lack in capacity building in phytosociology (KG3), and one can consequently hardly expect to find enough phytosociological competences in the field to produce robust management practices. In countries where phytosociology teaching remains important, the situation is less problematic, but due to the links between research and teaching, its durability is fragile.

KG also have implications for legitimacy (C4). Indeed, the procedural organization of collective scientific expertise, and in particular a balance between various knowledge-holders (in terms of institutions, discipline or background), is increasingly considered pivotal to achieve the pluralist knowledge liable to inform policies while bestowing legitimacy on them (Montana & Borie 2016). KG4 is therefore detrimental to the legitimacy of HD.

To sum up, C1 is indeterminate (due to KG1) and C2 unsatisfied (due to KG2) for Europe as a whole. Concerning C3, due to KG3 in France it is unsatisfied, and in countries where phytosociology is more developed the situation remains problematic. Concerning C4, KG4 testifies for a fragility of the European system as a whole. Based on C1-4, one therefore cannot positively evaluate this policy.

Improving European habitat conservation policies

Updating scientific basis

To ensure that HD's habitat policy becomes more science-based, a two-steps European-wide initiative is needed:

(1) First, the identifiability of HCI without referring to syntaxons should be assessed, and new characterizations clarified in up-to-date guidelines. The point is not to discard phytosociology, because valorizing phytosociological knowledge about species assemblages and repartition (Ewald 2003; Biondi 2011) is undoubtedly a cogent strategy for HD. A pervasive usage of phytosociological vocabulary, however, insulates HD from ecological advances. For some HCI, such as the easily identifiable "Dunes with *Hippophae rhamnoides*" (2160), a clarification of EUR28 will suffice. For others, deeper reshufflings are necessary. In France, an ongoing work coordinated by the "Muséum National d'Histoire Naturelle" considers that 92 HCI (out of 130 analyzed) need clarifications (V. Gaudillat, personal communication), because their definitions raise unsettled questions.

(2) Second, uncertainties concerning the rarity status of HCI should be assessed. The outdatedness of the knowledge justifying lists of HCI is currently tackled in a piecemeal, informal way. For example, *Heleochloion schoenoidis* communities in Corsica are part of HCI 3170*, but they are now known to be common, and are informally excluded in cartographies of HCI (SI-Table 1-A21). A systematic review

and resolution of similar problems are needed. The workload will vary from one country to another. In the UK, the national vegetation classification (SI-Table 1-A22) involved a country-wide sampling, providing reliable data on relative rarity. By contrast, in France, the vegetation prodrome (SI-Table 1-A23) does not display completeness assessments, and only 10% of the territory is covered by vegetation maps (SI-Table 1-A24), which makes current rarity assessments pointless. In the Mediterranean area, arguably the most biodiverse in France (Blondel 2010), there is no vegetation catalog available, and small-scale studies suggest that phytosociological knowledge is vastly lacunar. A complete bibliographical synthesis in the Préalpes d'Azur (SI-Table 1-A25) hence shows that, for 16 of the 24 HCI inventoried, no post-1980 source was available, for three, no sources at all and only one was concerned by a post-2000 source. Up-to-date phytosociological knowledge appears practically nonexistent in this region.

In this process, scientific journals will have a prominent responsibility. Publishing and valorizing catalogues of plant communities could help increase scientific quality control and ensure a wider diffusion. This would also incite researchers to develop collaborations with local knowledge-holders, which would strengthen the link between researchers and practitioners.

Ecology journals could also require that authors integrate in their manuscripts a discussion of the translatability of their results in terms of HCI. If this translation is possible, articulating it will be helpful for practitioners. If it is impossible, making it clear prevents misinterpretations.

Building capacity

Strengthening scientific bases is, however, only one aspect of the improvements needed. By adopting HD, Member States committed to equip themselves with the capacity to implement it. In France, reference institutions for habitats are "Conservatoires Botaniques." Their botanical competence is indisputable but they have neither the obligation nor the financial and human means to perform teaching (SI-Table 1-B4) and have more limited access to scientific bibliographical databases than research institutions. This condemns France to competence scarcity. To perform the two tasks above, the EU and Member States should learn from this counterexample. They should provide financial and organizational support to retrieve the competence to learn, critically assess and renew phytosociological knowledge.

Reforming A17E

Our analysis also highlights A17E reforms that should be moved to the top of the ETC-BD agenda:

(1) The structuration of epistemic communities involved should be rationalized, e.g. through the work of a dedicated committee involving ecologists, social scientists, philosophers and practitioners;

(2) Evaluations at site level should become mandatory, inside and outside the network;

(3) Requirements should be specified in terms of quantitative data content and publication for sources used.

Our recommendations are all costly initiatives. But numerous N2000 funding instruments (SI-Table 1-A26) could contribute by integrating them in their guidelines. In particular, they represent opportunities for Life-Nature, the main EU conservation financial instrument, to fix its inability to address conservation priorities (Hermoso *et al.* 2016).

Conclusions

The weaknesses of HD's habitat policy that we highlight should not overshadow the undeniable strengths of N2000: its coherence with international environmental agreements (Beresford *et al.* 2016), its contribution to conserving species (Sanderson *et al.* 2015), or the breadth of the network. Our point is not to vilify the policy, but to identify how to improve it. In this respect, our analysis highlights knowledge gaps that have a more general bearing than N2000. In particular, problems encountered if categories used by practitioners and researchers differ carry lessons for conservation initiatives focused on habitats whatever their geographical and political context. The same goes for associated problems of capacity building and biased epistemic communities. Accordingly, our concrete recommendations can be transposed in other contexts, such as e.g. the emergent IUCN red list ecosystem initiative (SI-Table 1-A27).

Acknowledgments

We thank P. Castets, S. Coq, S. Gerber, R. Julliard, J.-J. Lazare, M. Lelièvre, M. Martin, G. Meunier, J. Rouchier, E. Porcher, A. Tsoukias, and anonymous reviewers.

Supporting Information

SI-Table 1. Unpublished sources cited in the article.
SI-Table 2. List of evaluations in A17E referring to peer-reviewed sources in scientific journals.

SI-Table 3. List of habitats of community interest present in France, for the identification of which the presence of alien invasive species plays a key role because they are considered "index species" in the Habitat Books (SI-Table 1-A7).

SI-Table 4. Complete list of HCI, with codes and official denomination; and list of the HCI present in France, in the different biogeographic region.

SI-Table 5. Extensive list of articles analyzed in Figure 2.

References

Alphandéry, P. & Fortier, A. (2001). Can a territorial policy be based on science alone? The system for creating the Natura 2000 network in France. *Sociol. Ruralis*, **41**, 311-328.

Beninde, J., Fischer, M.L., Hochkirch, A. & Zink, A. (2015). Ambitious advances of the European Union in the legislation of invasive alien species. *Conserv. Lett.*, **8**, 199-205.

Beresford, A.E., Buchanan, G.M., Sanderson, F.J., Jefferson, R. & Donald, P.F. (2016). The contributions of the EU nature directives to the CBD and other multilateral environmental agreements. *Conserv. Lett.* doi:10.1111/conl.12259.

Biondi, E. (2011). Phytosociology today: methodological and conceptual evolution. *Plant Biosyst. - Int. J. Deal. Asp. Plant Biol.*, **145**, 19-29.

Blondel, J. (2010). *The Mediterranean region: biological diversity through time and space*. Oxford University Press, Oxford.

Boitani, L., Mace, G.M. & Rondinini, C. (2015). Challenging the scientific foundations for an IUCN red list of ecosystems. *Conserv. Lett.*, **8**, 125-131.

Bouzillé, J.-B. (2007). *Gestion des habitats naturels et biodiversité*. Lavoisier, Paris, France.

Burgman, M., Carr, A., Godden, L., Gregory, R., McBride, M., Flander, L. & Maguire, L. (2011). Redefining expertise and improving ecological judgment. *Conserv. Lett.*, **4**, 81-87.

Cardoso, P. (2012). Habitats directive species lists: urgent need of revision. *Insect Conserv. Divers.*, **5**, 169-174.

Contestabile, M. (2012). Forest conservation: failed protection regimes. *Nat. Clim. Chang.*, **2**, 839-839.

De Marchi, G., Lucertini, G. & Tsoukiàs, A. (2016). From evidence-based policy making to policy analytics. *Ann. Oper. Res.*, **236**, 15-38.

Devilliers, P. & Devilliers-Terschuren, J. (1993). *A classification of palaearctic habitats*. Council of Europe, Strasbourg.

Dicks, L.V., Hodge, I., Randall, N.P., *et al.* (2014). A Transparent process for "evidence-informed" policy making. *Conserv. Lett.*, **7**, 119-125.

Epstein, Y., López-Bao, J.V. & Chapron, G. (2016). A Legal-ecological understanding of favorable conservation status for species in Europe. *Conserv. Lett.*, **9**, 81-88.

Evans, D. (2010). Interpreting the habitats of Annex I: past, present and future. *Acta Bot. Gallica*, **157**, 677-686.

Evans, D. (2012). Building the European Union's Natura 2000 network. *Nat. Conserv.*, **1**, 11-26.

Ewald, J. (2003). A critique for phytosociology. *J. Veg. Sci.*, **14**, 291-296.

Ferraro, P.J. & Pattanayak, S.K. (2006). Money for nothing? A call for empirical evaluation of biodiversity conservation investments. *PLOS Biol*, **4**, e105.

Hermoso, V., Clavero, M., Villero, D. & Brotons, L. (2016). EU's conservation efforts need more strategic investment to meet continental conservation needs. *Conserv. Lett.*

Keith, D.A., Rodríguez, J.P., Rodríguez-Clark, K.M., *et al.* (2013). Scientific foundations for an IUCN red list of ecosystems. *PLoS ONE*, **8**, e62111.

Knight, A.T., Cowling, R.M., Rouget, M., Balmford, A., Lombard, A.T. & Campbell, B.M. (2008). Knowing but not doing: selecting priority conservation areas and the research—implementation gap. *Conser. Biol.*, **22**, 610-617.

Lisón, F., Palazón, J.A. & Calvo, J.F. (2013). Effectiveness of the Natura 2000 Network for the conservation of cave-dwelling bats in a Mediterranean region: cave-dwelling bats and the Natura 2000 Network. *Anim. Conserv.*, **16**, 528-537.

Maciejewski, L., Lepareur, F., Viry, D., Bensettiti, F., Puissauve, R. & Touroult, J. (2016). État de conservation des habitats: propositions de définitions et de concepts pour l'évaluation à l'échelle d'un site Natura 2000. *Rev. D'Ecologie Terre Vie*, **71**, 3-20.

Maiorano, L., Amori, G., Montemaggiori, A., *et al.* (2015). On how much biodiversity is covered in Europe by national protected areas and by the Natura 2000 network: insights from terrestrial vertebrates: Biodiversity Conservation in Europe. *Conserv. Biol.*, **29**, 986-995.

Matzek, V., Covino, J., Funk, J.L. & Saunders, M. (2014). Closing the Knowing–Doing gap in invasive plant management: accessibility and interdisciplinarity of scientific research. *Conserv. Lett.*, **7**, 208-215.

Mazzocchi, F. (2006). Western science and traditional knowledge: despite their variations, different forms of knowledge can learn from each other. *EMBO Rep.*, **7**, 463-466.

Montana, J. & Borie, M. (2016). IPBES and biodiversity expertise: regional, gender, and disciplinary balance in the composition of the interim and 2015 multidisciplinary expert panel: IPBES and biodiversity expertise. *Conserv. Lett.*, **9**, 138-142.

Pellissier, V., Touroult, J., Julliard, R., Siblet, J.P. & Jiguet, F. (2013). Assessing the Natura 2000 network with a common breeding birds survey. *Anim. Conserv.*, **16**, 566–574.

Popescu, V.D., Rozylowicz, L., Niculae, I.M., Cucu, A.L. & Hartel, T. (2014). Species, habitats, society: an evaluation of research supporting EU's Natura 2000 Network. *PLoS ONE*, **9**, e113648.

Sanderson, F.J., Pople, R.G., Ieronymidou, C., *et al.* (2015). Assessing the performance of EU nature legislation in protecting target bird species in an era of climate change: impacts of EU nature legislation. *Conserv. Lett.*, **9**, 172-180.

Senior, M.J.M., Brown, E., Villalpando, P. & Hill, J.K. (2015). Increasing the scientific evidence base in the "high conservation value" (HCV) approach for biodiversity conservation in managed tropical landscapes. *Conserv. Lett.*, **8**, 361-367.

Tsiafouli, M.A., Apostolopoulou, E., Mazaris, A.D., Kallimanis, A.S., Drakou, E.G. & Pantis, J.D. (2013). Human activities in Natura 2000 sites: a highly diversified conservation network. *Environ. Manage.*, **51**, 1025-1033.

Tsoukias, A., Montibeller, G., Lucertini, G. & Belton, V. (2013). Policy analytics: an agenda for research and practice. *EURO J. Decis. Process.*, **1**, 115-134.

9

Perverse Market Outcomes from Biodiversity Conservation Interventions

Felix K. S. Lim[1], L. Roman Carrasco[2], Jolian McHardy[3], & David P. Edwards[1]

[1] Department of Animal and Plant Sciences, University of Sheffield, Sheffield S10 2TN, UK
[2] Department of Biological Sciences, National University of Singapore, Singapore
[3] Department of Economics, University of Sheffield, Sheffield S1 4DT, UK

Keywords

Agricultural intensification; biodiversity loss; deforestation; land-sparing land-sharing; leakage effect; market feedbacks; perverse outcomes; PES schemes; protected areas.

Correspondence

Felix K. S. Lim, Department of Animal and Plant Sciences, University of Sheffield, Sheffield, S10 2TN, UK.
E-mail: f.lim@sheffield.ac.uk

Editor

Douglas MacMillan

Abstract

Conservation interventions are being implemented at various spatial scales to reduce the impacts of rising global population and affluence on biodiversity and ecosystems. While the direct impacts of these conservation efforts are considered, the unintended consequences brought about by market feedback effects are often overlooked. Perverse market outcomes could result in reduced or even reversed net impacts of conservation efforts. We develop an economic framework to describe how the intended impacts of conservation interventions could be compromised due to unanticipated reactions to regulations in the market: policies aimed at restricting supply could potentially result in leakage effects through external or unregulated markets. Using this framework, we review how various intervention methods could result in negative feedback impacts on biodiversity, including legal restrictions like protected areas, market-based approaches, and agricultural intensification. Finally, we discuss how conservation management and planning can be designed to ensure the risks of perverse market outcomes are detected, if not overcome, and we address some knowledge gaps that affect our understanding of how market feedbacks vary across spatial and temporal scales, especially with teleconnectedness and increased international trade.

Introduction

With increasing global population and affluence, the global demand for timber, food, and other natural resources is rising, with crop demands projected to increase by 100–110% from 2005 levels by 2050 (Tilman *et al.* 2011). Meeting this demand will drive further forest degradation from logging and deforestation for agriculture and timber plantations, especially in the tropics (Hansen *et al.* 2013). Habitat loss in the tropics is the biggest driver of biodiversity and ecosystem function losses (DeFries *et al.* 2004). There is thus an urgent need to better manage tropical land-use change to reduce the loss of biodiversity and ecosystem functions, while addressing the issue of rising timber and food demands, for instance, via changing diets or reducing food waste (Erb *et al.* 2016).

Management strategies are commonly implemented to reduce conversion of natural habitats to other land uses, and therefore to stem the loss of biodiversity. These include legal restrictions on land-use by establishing protected areas (PAs) (Oliveira *et al.* 2007), market-based conservation efforts such as certification and payment for ecosystem services (PES) schemes (Chobotová 2013), and improvements in technology and agricultural intensification (Tilman *et al.* 2011).

Often, however, there are indirect and unintended consequences of conservation measures. Trade-offs are inevitable with changes in land use (DeFries *et al.* 2004), and this includes when implementing conservation efforts. Many unintended consequences are often overlooked when assessing the effectiveness of biodiversity conservation actions (Larrosa *et al.* 2016), in part because indirect environmental and ecological impacts of land-use changes and conservation actions typically occur over longer time-scales and larger distances than directly measured outcomes. Unintended consequences can have positive or negative effects on the overall (net) outcomes of interventions. Positive feedback effects include protection or restrictions on wildlife harvests diminishing demand (Pain *et al.* 2006) and unintended crowding-in effects from market-based conservation policies (Wunder 2013). By contrast, negative feedbacks could include demand driving leakage of deforestation into unprotected areas in response to establishing PAs (Ewers & Rodrigues 2008) and an increase in demand brought about by improving cost-efficiency of agriculture (Rudel *et al.* 2009).

Given that negative feedbacks compromise conservation efforts, we focus on their emergence via the influence of market forces. Knowledge of the socio-ecological system responses is crucial for decision makers to minimize negative unintended feedbacks. Typologies that classify feedbacks between deletion (removal of pre-existing feedbacks), addition, and flows (changes in magnitude of pre-existing feedbacks) have recently been developed (Larrosa *et al.* 2016). Under this typology, market feedbacks could be considered flow feedbacks and PAs establishment addition feedbacks.

Market feedbacks in response to the initial conservation intervention can undermine conservation efforts (Armsworth *et al.* 2006), and although important, they are often not considered in policies (Jantke & Schneider 2011; Miller *et al.* 2012; St John *et al.* 2013). Understanding how agents in a market respond to policy changes is very important in determining whether a certain scheme will have the intended consequences, or instead be counterproductive (Galaz *et al.* 2015). Changes in policy typically affect the incentives of agents, resulting in changes in their behaviors. In many cases, the reaction of agents to the new incentives resulting from policies in biodiversity conservation will have a large influence over whether the policy is successful, or if perverse incentives will lead to damaging unintended consequences instead. Arguably, unintended consequences of environmental and conservation policies through these channels have not attracted sufficient scrutiny to date (Milner-Gulland 2012). It is therefore essential to understand how a policy will alter the pattern of incentives of agents in the market. Indeed, successful policies will be designed to ensure the incentives of agents in the market are compatible with the intended aims of the policy.

In this review, we focus on the potential perverse market outcomes of conservation efforts. We first develop a theoretical model to explain how market regulations respond to conservation efforts. Reframing conservation in an economic context allows us to understand how

market feedbacks could lead to perverse outcomes. We then apply the framework to different conservation interventions and discuss how conservation policies and management practices might be adapted to minimize the risk of negative consequences.

Economic underpinnings of unintended consequences via market feedbacks

Conservation places restrictions on resource use

Conservation interventions can be viewed as external impacts on the resource market (e.g., timber), which can result in a shift in the equilibrium or lead to disequilibrium. In many cases, conservation actions revolve around restricting access to a resource. Placing logging bans and quotas on timber harvests, for example, restricts trade of the commodity (i.e., timber). Establishing PAs also restricts land availability and access to resources. Imposing such restrictions limits the quantity of supply (q_1, Figure 1A), and the market is no longer in equilibrium. The market responds with a rise in the price to p_1, above the initial equilibrium level, at which suppliers would like to offer more than q_1 to the market (q_1, Figure 1A). This represents the basic economic model of a quota (Goolsbee *et al.* 2016).

However, there is some question about the effectiveness of such policies in practice. For instance, applications of this theory to include leakages (Murray *et al.* 2004; Jonsson *et al.* 2012) involve the expansion of supply back toward the quantity at the free-market equilibrium. Related arguments of how illegal trade can expand consumption in the presence of import quotas have also been established in the international trade literature (e.g., Falvey 1978). In other words, some of the restriction in quantity due to the quota may be undone by illegal trade. Perverse "illegal" (black) market incentives reduce the impact of the conservation policy.

The imposition of a quota also produces an artificially high price in the formal market, meaning that inefficient (high-cost) suppliers can co-exist with efficient ones. This presents the question—which suppliers supply the formal market and which supply the black market? In addressing this question, we highlight a novel further possible perverse impact of a quota policy. Compared to a free market where efficiency considerations determine which firms supply the formal market, the allocation of supply rights under a quota policy is now determined by the regulatory authority. With high prices within the formal market, there is less incentive to be efficient, and inefficient suppliers could end up supplying the market. If the regulatory authority does not observe efficiency and

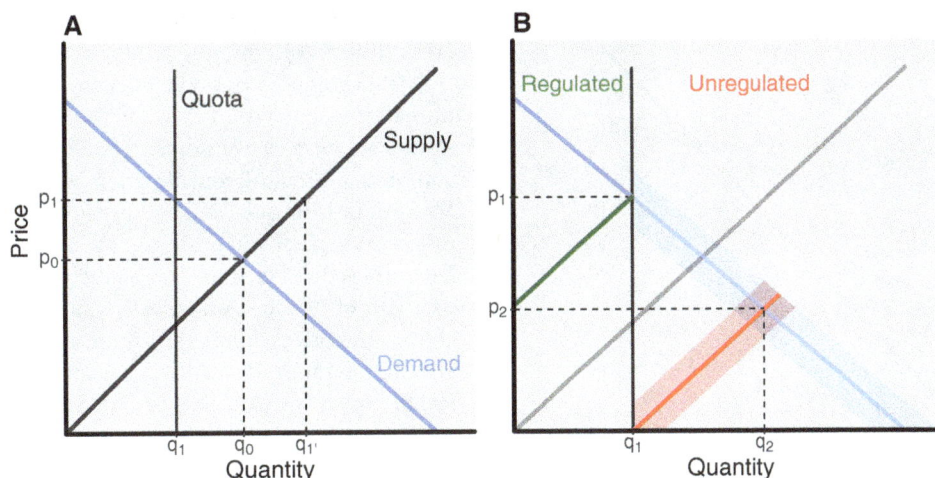

Figure 1 Conceptual framework describing possible effects of an output quota (A) on the creation of an illegal market (B). (A) The market is initially in equilibrium at (q_0, p_0). Setting a quota restricts the quantity of resource traded to q_1 raising the price to p_1. Firms can respond to this disequilibrium by creating an illegal market. Assuming, for simplicity, there are no additional costs to illegal supply relative to regulated supply, firms will expand supply through the illegal market moving up the supply curve beyond q_1 with total supply rising toward the pre-quota level, q_0. (B) If, for instance, the inefficient firms (green) supply the formal market, supplying q_1, and the efficient firms (red) supply the illegal market, supplying q_2-q_1, the creation of the illegal market can result in an overall increase in the quantity traded $(q_2 > q_0)$ despite the quota. We might also expect shifts in demand and supply (shaded) within the unregulated market due to externalities.

allocate supply rights to the most efficient firms, or, if the authority is able to exploit the power of allocating rights to pursue their own agenda (e.g., corruptly supplying rights to "friendly", possibly high-cost, firms), then one perverse result of the imposition of a quota might be a decline in market efficiency. In such a case, the inefficient firms supply the formal market and the efficient firms supply the unregulated market—there is a reorganization of supply (Figure 1B) akin to the rationing rules on the demands side used, for instance, in Davidson & Deneckere (1986). The total output across the formal and black markets in this case could be greater than under the initial equilibrium before the quota was introduced (q_2, Figure 1B). The quota might be ineffective in terms of reducing trade, and may even result in an increase in trade.

The degree to which the quantity traded exceeds the quota will depend on which firms supply each market, as well as the costs and benefits to firms and consumers from trading in the illegal market. For instance, supply in the illegal market will shift downward, reducing the leakage effect, if the costs associated with supplying the legal market are lower than supplying the illegal market. However, where the legal market has high costs associated with meeting regulatory standards, by avoiding these costs, firms trading in the illegal sector might offset extra costs associated with the illegal market (e.g., concealment costs).

Improving land-use practices and efficiency to reduce land-use

Conservation measures could also involve improving efficiency and technology through agricultural intensification and new crop varieties, to reduce the pressure to convert more land. While the direct impact of such measures could be an increase in production with a reduced need for land, there may also be unintended consequences. Although there is uncertainty as to whether intensification would improve the cost-efficiency of production, if improved, it could lead to a decrease in the costs of resources: suppliers are willing to supply more at any given price. This is associated with a rightward shift in the supply curve and consequently, in the equilibrium (q_0, p_0 – q_2, p_2, Figure 2), resulting in an overall increase in resources being traded (e.g., Villoria *et al.* 2014). The size of this shift in equilibrium does depend on the price elasticity demand of the product—with a more pronounced effect in the case of an elastic demand (where demand varies strongly with prices). Conservation measures aimed at regulating the supply of, for instance, agricultural production might instead put additional pressure on remaining available resources.

These perverse outcomes could be exacerbated in markets where the global demand is supplied though multiple substitutes, such as different types of vegetable oil crops (e.g., oil palm, rapeseed). Assuming markets for both commodities are the same (perfect substitutes),

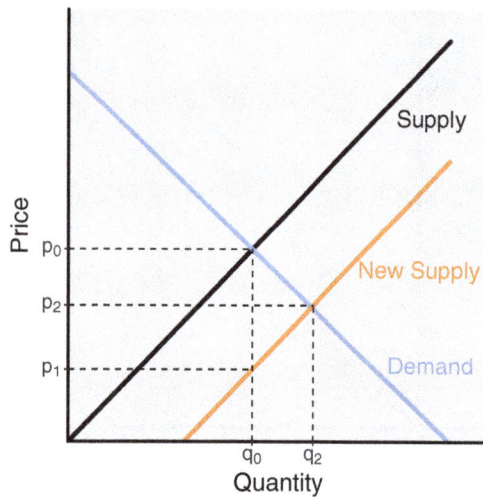

Figure 2 Conceptual framework describing how prices and quantities of resources traded vary when crops are improved. Agricultural intensification can shift supply rightward (black to orange) resulting in an overall shift in equilibrium from (q_0, p_0) to (q_2, p_2).

initial equilibria for both commodities should also be identical ($q_0 p_0$, Figure 3A). Improving yields in one crop lowers prices for any given quantity proportionally, represented by a shift in supply curve from $Supply_A$ to $Supply_{AInt}$, and causes a shift in equilibrium from $q_0 p_0$ to $q_A p_A$ (Figure 3B). Additionally, we can expect decreased demand for the less efficient substitute, denoted by the downward shift in $Demand_B$, and a decrease in quantity of crop B (from q_0 to q_B): consumers are likely to favor the cheaper crop beyond price p_A. We can therefore expect higher quantities of crop A traded, and a surplus of crop B not traded. Overall, however, there could still be a net increase in agricultural land use (deforestation for crop A – forest recovery on abandoned land from crop B), and a net loss of old-growth forests across a larger region (Carrasco *et al.* 2014).

Examples of conservation measures and market feedbacks

Legal restrictions on land use

Legal protection and restrictions on land use are widely implemented globally and include establishing PAs and regulating logging and other resource harvest quotas. Such conservation measures rely on regulation by an authority, usually governmental, to ensure the impacts on ecosystems and biodiversity are minimal, or at least compensated. However, legally enforced (i.e., sufficiently funded) conservation policies are often only effective within their designated areas, typically at local spatial scales, and could lead to a displacement of destructive

activity and land use into unprotected and unregulated areas (Ewers & Rodrigues 2008).

Establishing PAs could be effective in directly reducing human impacts within targeted forests, but in many instances might be driven more by markets than by conservation or ecological considerations (Rayner *et al.* 2014). Many PAs lack additionality because they are situated in locations passively protected by their distance to markets, unproductive soils, steep gradients, etc. The establishment of PAs in economically valuable areas could increase land prices across remaining areas (Polasky 2006), and shift deforestation and land-use changes into unprotected forests instead. This could create incentives for an unregulated market with consequences as outlined in the framework (Figure 1B).

In the tropics, such leakage effects result in high rates of clearing and degradation of forested areas surrounding PAs. For instance, while deforestation rates in the Peruvian Amazon were as low as 2% within PAs, they were up to 18 times higher outside PAs (Oliveira *et al.* 2007). Similarly, protection of mature forests in Costa Rica reduced the rate of mature forest loss by 50%, but resulted in cropland expansion redirected into unprotected natural habitats, including wetlands, native reforestation, and young secondary forests, due to the lack of legal protection of these areas (Fagan *et al.* 2013). Furthermore, import of timber and agricultural products into the country increased, displacing land-use change internationally (Jadin *et al.* 2016). Another potential perverse outcome is an acceleration of land-grabs before regulations are put in place. This situation was observed in Tanzania where accelerated land conversion occurred in anticipation of PA expansion (Baird *et al.* 2009).

Legal restrictions against resource extraction, much like PAs, can also have displacement effects into unregulated areas, rather than decreased harvests as intended: a similar restriction to a quota is placed on resource quantity, which could result in an informal market arising with a re-organization of supply and expansion of total output (Figure 1B). Reduction in deforestation rates across multiple countries was, for instance, associated with displacement via international trade (Meyfroidt *et al.* 2010).

Basic economic principles can be used to show how endogenous market feedbacks (i.e., changes within the market, Figure 1) could undermine conservation efforts and benefits, and change conservation priorities (Murray *et al.* 2004; Armsworth *et al.* 2006). More recently, studies have integrated sub-models of resource extraction and biodiversity impacts, fluctuations in household utility and market prices, and spatially explicit distributions of biodiversity and resources to highlight the impacts on land-use change (Bode *et al.* 2015; Renwick *et al.* 2015).

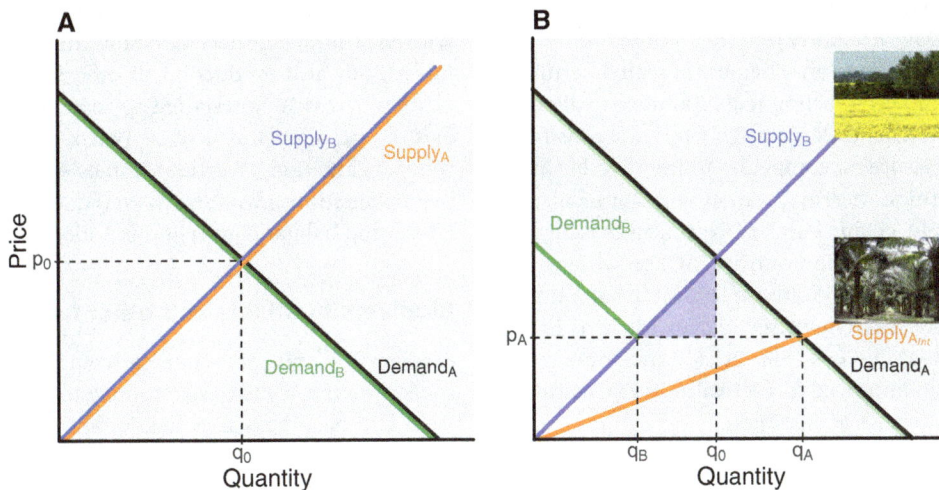

Figure 3 Conceptual framework describing the possible effects of agricultural intensification on demand and land-use of substitute crops (B). (A) Assuming, for simplicity, that demand and supply conditions of each commodity are identical, both markets will have the same initial equilibrium (q_0,p_0). (B) Agricultural intensification of one crop (A, orange) can increase amount traded, from q_0 to q_A. Additionally, demand for the substitute crop B could decrease (green), as consumers are likely to favor the cheaper substitute above price p_A. Amount of crop B traded decreases to q_B, resulting in either a surplus (shaded region) not traded. This surplus will ultimately lead to either innovation to use the surplus, or land abandonment from crop B. Improving oil palm yields (pictured lower inset) in the tropics, for example, could lead to a decreased demand in rapeseed oil (upper inset), allowing secondary forest regrowth in temperate regions, but the increased demand for palm oil could increase tropical deforestation.

Market-mediated conservation measures

Market-based approaches to conservation policies are increasingly seen as efficient, effective means of managing resources, while promoting conservation (Chobotová 2013). The impacts of biodiversity are controlled through use of markets, and practices that promote conservation are incentivized over practices with negative environmental outcomes. These could be used as complements to legally mandated conservation measures (Lambin *et al.* 2014). Nevertheless, as highlighted in our framework, these approaches still allow a quota to be set: a governing authority defines a formal, regulated market and determines who supplies within this market, potentially re-organizing supply with possible perverse consequences (Figure 1B). Furthermore, market-based measures revolve around incentivizing suppliers of the formal market, and do not necessarily penalize suppliers of the informal market. Much like legally mandated measures, market-based interventions could favor an unregulated market (with uncertified resources at lower prices, q_2p_2, Figure 1B) alongside the regulated market (with certified resources at higher prices, q_1p_1, Figure 1B). These perverse outcomes can occur at both local and transnational scales, because policies are typically narrowly focused and do not account for their wider consequences.

PES schemes, such as the United Nations' Reducing Emissions from Deforestation and forest Degradation (REDD+), although not widely implemented, are in-creasingly popular (Wunder 2013). They provide a means of internalizing the externalities from loss of ecosystem services and enhancing conservation efforts by compensating suppliers who help improve or protect ecosystem services via habitat protection or restoration.

PES schemes could promote more sustainable practices within the market, and allow authorities to decide who supplies the regulated market. However, this neither directly reduces the overall demand for a resource, nor does it penalize suppliers to the informal market. The incentive of supplying the informal market could therefore remain high (Figure 1B), and the market might favor suppliers of the informal market—we could ultimately witness a displacement of land-degrading activity into areas not regulated by the PES scheme. This leakage effect could be exacerbated as prices within the regulated market increase (p_1, Figure 1B).

Perverse incentives could also occur within PES schemes if they are not implemented and managed well (Wunder 2013). When suppliers are only rewarded by favorable practices within designated areas (e.g., for additional management practices like afforestation), we could observe a leakage of effect, where destructive activity displaced into areas not enrolled in PES schemes but belonging to the same owner are neglected (Atmadja & Verchot 2012). Managing PES schemes also becomes increasingly difficult in situations where a single approach is implemented to achieve multiple objectives: PES schemes are frequently also viewed as poverty alleviation and development tools (Daw *et al.* 2011).

Sustainability certification schemes (hereafter certification schemes) and eco-labeling (e.g., Rainforest Alliance, Roundtable on Sustainable Palm Oil) rely on consumer activism and pressure on companies to improve business practices and ethics, thereby promoting sustainability in the global supply chain. Forestry certification schemes like the Forest Stewardship Council (FSC) are among the most developed schemes (Auld *et al.* 2008): the amount of FSC-certified forests has increased to over 186 Mha in the span of about two decades (FSC 2016), and some studies have reported improved forest health in FSC-certified forests (e.g., Kalonga *et al.* 2015).

Certification schemes, however, have the potential to create similar effects in terms of a re-organization of supply and associated consequences for expansion of output as those arising from a quota. For instance, if the scheme gives certified suppliers exclusive access to the consumers with a high willingness to pay (q_1, Figure 1), but is not tied to firm efficiency, then less efficient firms could end up among the suppliers in the certified market, displacing efficient firms into the uncertified market. Indeed, certification usually results in a rise in price of certified commodities (from p_0 to p_1, Figure 1B), thereby restricting access to wealthier consumers. Prices of certified timber within Malaysia, for instance, were up to 56% higher than uncertified timber (Kollert & Lagan 2007). Therefore, only relatively wealthy consumers can afford certified products, while less wealthy purchasers continue buying unregulated and uncertified products (q_2p_2, Figure 1B). Additionally, if prices of certified-sustainable goods are too high, the market demand could be lower than the supply and we could observe lower uptake than expected (Edwards & Laurance 2012). FSC schemes, for example, have increased in popularity over the last decade, but were concentrated in newly developed countries across the tropics, and usually do not include developing nations with larger native forests (Auld *et al.* 2008).

Importantly, because certification schemes do not penalize the informal market (i.e., no additional costs for supplying the unregulated market), we can expect the unregulated market to thrive. While certification schemes like FSC directly reduce poor logging practices within certified forests (formal market; from q_0 to q_1, Figure 1B), they could also result in leakage of (illegal) logging into unmanaged forests (from q_0 to q_2, Figure 1B), making the overall management of resources and deforestation more difficult. Indeed, illegal timber products account for 50–90% of forestry products across the tropics (Nellemann 2012). Similar leakage effects could also emerge from other certification schemes. RSPO certification might be effective in promoting sustainable agricultural practices within certified oil palm plantations, but we could also witness a leakage effect not only affecting unprotected forests, but also production of other crops. RSPO certification across Indonesia led to increased conversion of existing rice cropland (Koh & Wilcove 2008) and jungle rubber plantation (Warren-Thomas *et al.* 2015) into oil palm plantations, resulting in an indirect displacement of efforts and habitat conversion in Indochina.

Biodiversity offsets and other trading schemes

Biodiversity trading schemes could be classified as market-driven measures to reduce biodiversity loss, but have also been passed as legislations in some countries. These are typically enforced on companies and developers to allow for economic growth and development, while indirectly reducing human pressures on biodiversity and the environment (Froger *et al.* 2015). Such schemes have been legislated in a number of countries (e.g., Australia) or regions (e.g., California), and widely embraced and adopted by private land developers and companies, including mining and oil companies (Edwards *et al.* 2014), as a means of measuring and reducing their impact on biodiversity loss.

Biodiversity offsets and trading schemes, in essence, place restrictions on some areas (reserves) while allowing others to be converted for use. Fundamentally, these methods mimic legally mandated conservation efforts (e.g., PAs), where the amount of land use is restricted (Figure 1A). Such offsets can only be effective at a very local level (i.e., within reserves themselves), and reserves need to have higher conservation values than areas being converted to achieve a no-net loss outcome. Enforcing restrictions on land-use, as with other conservation measures, does not affect the demand for land, timber or nontimber forest resources, and could result in a displacement of efforts outside the managed (regulated) area. Where reserves are of high economic value, land purchases in biodiversity offsetting programs could also alter supply and demand of resources, resulting in increased land rent and therefore biodiversity loss in unprotected areas (Armsworth *et al.* 2006).

Land sparing and high-yielding crop varieties

The land-sparing versus land-sharing framework, which considers the trade-offs between agricultural or timber demands and the desire to protect biodiversity, has been widely applied in the debate of how best to meet growing resource demands (Phalan *et al.* 2011). Notwithstanding the limitations of both strategies (Fischer *et al.* 2011; Erb *et al.* 2016), a large number of data-based assessments suggest that the land-sparing approach of high-yield farming with habitat conserved elsewhere, if managed correctly with strong governance and effective protection of remaining forests, might be more effective in promoting biodiversity conservation while meeting demand (Phalan *et al.* 2011; Erb *et al.* 2016). Agricultural intensification is a necessary condition for land-sparing, but not a sufficient condition for reducing the need to convert more forest to farmland (Erb *et al.* 2016).

Agricultural intensification, however, does not reduce the incentives associated with expansion; agricultural area has been observed on occasion to increase with intensification (Ewers *et al.* 2009; Rudel *et al.* 2009). Projections of land-use changes have also suggested the possibility of further loss of forests with improved crop yields (Kaimowitz & Angelsen 2008; Phelps *et al.* 2013; Villoria *et al.* 2014). This is especially so in passive land-sparing scenarios, where remaining forests are not managed or protected effectively, and hence easily targeted for agricultural expansion (Phalan *et al.* 2016). A rebound effect known as Jevon's paradox could arise, where the increased efficiency and reduced costs of crop production instead lead to increased demands. The magnitude of this effect could vary, depending on elasticity demand of the product (Hertel 2011; Villoria *et al.* 2014). For agricultural intensification to be effective in reducing land-use change, an active land-sparing framework is necessary, with heavy reliance on the role of PAs and effective governance (Phalan *et al.* 2016), which many countries might lack (Fischer *et al.* 2011).

Increasing agricultural productivity could make crops cheaper and more profitable to produce over time, and increase its uses and demand as a cheaper substitute for other less cost-efficient crops (Villoria *et al.* 2014), even at transnational scales (Figure 3). If this results in favoring more cost-efficient tropical crops (e.g., oil palm), this could then exacerbate agricultural expansion across the tropics (Carrasco *et al.* 2014). There could be an increase in land abandonment and reforestation within low-profit areas, i.e., a decrease in land use from q_0 to q_B (Figure 3), but this is coupled with increased expansion and deforestation (q_0 to q_A, Figure 3) within areas of higher market value. The benefits of increased forest regeneration in marginal areas for agriculture across the Neotropics (e.g., highlands), for instance, would be outweighed by the negative impacts on biodiversity from increased deforestation in the lowland tropical forest (Aide *et al.* 2013).

Managing the effects of market outcomes

Assessing perverse outcomes in studies

Conservation interventions need to work toward incorporating steps to monitor and minimize perverse outcomes (Larrosa *et al.* 2016), but little has been done to overcome these outcomes. Some studies have, however, looked into incorporating and evaluating unintended feedbacks into their analyses of PAs and incorporated spatial information, theoretical models, and biodiversity maps to project spatially explicit predictions of areas more vulnerable to leakage (Bode *et al.* 2015; Renwick *et al.* 2015). Others have identified and measured leakage of conservation policies such as REDD+, using econometric or general equilibrium models (e.g., Murray *et al.* 2004; Gan and McCarl 2007). These models center on identifying the market feedbacks incurred from the conservation action, and understanding how they translate into indirect impacts on resources, i.e., through the unregulated market.

A number of factors also need to be considered when assessing, predicting, and managing these perverse outcomes. Since costs associated with land-use change vary across space due to multiple factors (social, political, and environmental), we would also expect the magnitude of market outcomes and impacts on biodiversity to vary between regions (Armsworth *et al.* 2006; Chaplin-Kramer *et al.* 2015) and across different spatial scales. While some studies acknowledge this, few have incorporated spatial information in their models (e.g., Bode *et al.* 2015). Not accounting for spatial variation results in often-erroneous assumptions of homogeneity across landscapes.

As with spatial scales, actions tend to be implemented across short timescales, but the effects of land conversion and land use are long term: time lags in responses and impacts on habitat and biodiversity (e.g., extinction debts and forest regeneration) might not be captured in static analyses (Ghazoul *et al.* 2015). Displacement costs of policies and actions could at times be intergenerational (Roca 2003), and while the immediate effects of some measures might seem positive, by not making assessments over longer temporal scales we do not consider other socio-economic factors and market feedbacks that might be detrimental (Hill *et al.* 2015). The benefits of PES schemes and other long-term measures are also often based on the assumption that other conditions in the market are constant, but land-use regimes could be implemented alongside other regulations and socio-economic changes and shocks (Müller *et al.* 2014), which will impact on the effectiveness of the regime. More emphasis needs to be placed on dynamic effects in planning long-term measures.

Another aspect of market-based outcomes often not addressed in studies is the interaction between distant parts of the world (teleconnectedness; Carrasco *et al.* 2014). Given the importance of global markets and transnational trade, overlooking the effects of teleconnectedness could lead to a considerable underestimation of the indirect impacts on land-use change (Renwick *et al.*

2015): since legislations, policies, and other conservation measures are usually localized, studies tend to focus only on local and national effects. The consequences of these conservation policies and actions are, however, usually spread across much larger spatial scales and between countries and continents (Liu *et al.* 2013). Reducing land-use in one area, without a reduction in resource demand, could lead to agricultural expansion and land conversion in another, and countries with large gains in forest cover might observe increases in imports of wood and agricultural products (Gan & McCarl 2007; Meyfroidt *et al.* 2013). For instance, regulations to increase forest regeneration within Vietnam (Meyfroidt & Lambin 2009) and China (Viña *et al.* 2016) led to an increased import of timber products. Projections of land-use changes should also account for the possible influence of alternative and complementary markets. Oil palm, for instance, can be a cheaper substitute to other oil-producing crops, including soy and rapeseed, and changes in prices and quantities of one crop could affect demands of each substitute crop (Figure 3), and ultimately increase land-use across the tropics (Carrasco *et al.* 2014).

Reducing the risk of perverse incentives within formal markets

Our framework also helps highlight how policies might be designed to help mitigate these unwanted effects. Conservation policies should recognize where conditions and incentives exist for officials to be corrupt regarding the selection of suppliers and make this a focal point for anti-corruption investigation. Policies also need to implement mechanisms to increase transparency and address information asymmetries by employing competitive tendering mechanisms allowing the efficient firms to reveal themselves (Smith & Walpole 2005). Third-party auditing, for instance, may be a potentially effective means of increasing transparency and minimizing probability of leakage and of perverse behavior within the formal market (Cook *et al.* 2016). Measures like these would limit the potential for corruption to dictate the exploitation of land and resources.

Reducing the risk of informal markets emerging

Our framework also points at the emergence of an informal market as an important source of perverse outcomes from conservation efforts (Figure 1B). Conservation policies therefore need to be more inclusive of the entire market to identify and manage leakage effects; the quota-policy framework only pays attention to the formal market. Conservation policies that, for instance, incorporate and account for trade and import of

agricultural and forestry products represent a step toward being more inclusive and could potentially minimize the likelihood and scale of leakage. Effective spatial planning and targeting specific areas to intensify agriculture, while ensuring designated areas are kept protected for conservation, is another way to minimize leakage effects (Phalan *et al.* 2016).

Managing conservation efforts should also focus on minimizing the risks of informal markets emerging. This involves a thorough understanding and projection of price and market condition changes in response to the initial conservation measure, as well as a working knowledge of the various actors in the formal and informal markets. Monitoring changes in prices of resources and understanding how they relate to emerging illegal markets is one way to better pre-empt and manage unintended feedbacks. Efforts toward better detection and punishment of illegal market operations will increase the costs of these trades and this will reduce the viability of suppliers in this market reducing the extent of the leakage (a downward shift in the supply curve in the unregulated sector in Figure 1B).

It is also important that we identify and monitor the key actors most likely to supply the informal market, and potential leakage sinks. This should allow for the more efficient detection of unintended and deleterious changes in land use. Measures to detect leakage should also focus on flows of unregulated or illegal products: trade flows may be used as a means of identifying transnational leakage and displacement of deforestation practices in response to conservation efforts (Meyfroidt & Lambin 2011). Achieving this can be challenging: illegal timber, for instance, is often laundered through legal plantations and mills (Nellemann 2012). Using satellite imagery could be another way of monitoring areas more likely to be cleared, and minimizing displacement and leakage effects. Empirical studies suggest, for instance, that buffer zones and forested areas surrounding PAs are more prone to being cleared (Pfeifer *et al.* 2012), and focusing monitoring efforts in such areas could lower the chances of forest loss. Spatially explicit econometric analyses might also be effective in identifying key areas likely to undergo land conversion. Importantly, monitoring and management should not be restricted within national boundaries, but should also include transnational leakage.

Conclusion

Most conservation measures tend to focus only on the primary and direct outcomes on nature and biodiversity, while indirect consequences are overlooked. This could lead to an overestimation of the true effects of the intervention, and the promotion of conservation actions that yield minimal to no overall conservation benefit: rather than reducing biodiversity loss, they could instead be counterproductive. Applying economic principles, we highlight the possible perverse consequences that are often not accounted for. This allows us to acknowledge these counterproductive impacts and, ideally, to seek ways to mitigate these effects through further regulations or extending the spatial extent of action, working toward optimal management strategies. Appreciating, if not understanding, the vulnerability and sensitivity of biodiversity conservation efforts to market feedbacks is a first step toward designing and managing conservation interventions more effectively.

Acknowledgments

F.K.S.L. acknowledges support from the Grantham Centre for Sustainable Futures.

References

Aide, T.M., Clark, M.L., Grau, H.R., *et al.* (2013). Deforestation and reforestation of Latin America and the Caribbean (2001-2010). *Biotropica*, **45**, 262-271.

Armsworth, P.R., Daily, G.C., Kareiva, P. & Sanchirico, J.N. (2006). Land market feedbacks can undermine biodiversity conservation. *Proc. Natl. Acad. Sci. U. S. A.*, **103**, 5403-5408.

Atmadja, S. & Verchot, L. (2012). A review of the state of research, policies and strategies in addressing leakage from reducing emissions from deforestation and forest degradation (REDD+). *Mitig. Adapt. Strateg. Glob. Chang.* **17**, 311-336

Auld, G., Gulbrandsen, L.H. & Mcdermott, C.L. (2008). Certification schemes and the impacts on forests and forestry. *Annu. Rev. Environ. Resour.*, **33**, 187-211.

Baird, T.D., Leslie, P.W. & McCabe, J.T. (2009). The effect of wildlife conservation on local perceptions of risk and behavioral response. *Hum. Ecol.*, **37**, 463-474.

Bode, M., Tulloch, A.I.T., Mills, M., Venter, O. & Ando, A.W. (2015). A conservation planning approach to mitigate the impacts of leakage from protected area networks. *Conserv. Biol.*, **29**, 765-774.

Carrasco, L.R., Larrosa, C., Milner-Gulland, E.J. & Edwards, D.P. (2014). A double-edged sword for tropical forests. *Science*, **346**, 38-40.

Chaplin-Kramer, R., Sharp, R.P., Mandle, L., *et al.* (2015). Spatial patterns of agricultural expansion determine impacts on biodiversity and carbon storage. *Proc. Natl. Acad. Sci. U. S. A.*, **112**, 7402-7407.

Chobotová, V. (2013). The role of market-based instruments for biodiversity conservation in Central and Eastern Europe. *Ecol. Econ.*, **95**, 41-50.

Cook, W., van Bommel, S. & Turnhout, E. (2016). Inside environmental auditing: effectiveness, objectivity, and transparency. *Curr. Opin. Environ. Sustain.* 18, 33-39.

Davidson, C. & Deneckere, R. (1986). Long-run competition in capacity, short-run competition in price, and the Cournot long-run competition in capacity, short-run competition in price, and the Cournot model. *RAND J. Econ.*, **17**, 404-415.

Daw, T., Brown, K., Rosendo, S. & Pomeroy, R. (2011). Applying the ecosystem services concept to poverty alleviation: the need to disaggregate human well-being. *Environ. Conserv.*, **38**, 370-379.

DeFries, R.S., Foley, J.A. & Asner, G.P. (2004). Land-use choices: balancing human needs and ecosystem function. *Front. Ecol. Environ.*, **2**, 249-257.

Edwards, D.P. & Laurance, S.G. (2012). Green labelling, sustainability and the expansion of tropical agriculture: critical issues for certification schemes. *Biol. Conserv.*, **151**, 60-64.

Edwards, D.P., Sloan, S., Weng, L., Dirks, P., Sayer, J. & Laurance, W.F. (2014). Mining and the African environment. *Conserv. Lett.*, **7**, 302-311.

Erb, K.-H., Fetzel, T., Haberl, H., *et al.* (2016). Beyond inputs and outputs: opening the black-box of land-use intensity. Pp. 93-124 in *Soc. Ecol.* Volume 5 of the series *Human-environment interactions*. Springer International Publishing.

Ewers, R.M. & Rodrigues, A.S.L. (2008). Estimates of reserve effectiveness are confounded by leakage. *Trends Ecol. Evol.*, **23**, 113-116.

Ewers, R.M., Scharlemann, J.P.W., Balmford, A. & Green, R.E. (2009). Do increases in agricultural yield spare land for nature? *Glob. Chang. Biol.*, **15**, 1716-1726.

Fagan, M.E., DeFries, R.S., Sesnie, S.E., *et al.* (2013). Land cover dynamics following a deforestation ban in northern Costa Rica. *Environ. Res. Lett.*, **8**, 34017.

Falvey, R.E. (1978). A note on preferential and illegal trade under quantitative restrictions. *Q. J. Econ.*, **92**, 175–178.

Fischer, J., Batáry, P., Bawa, K.S., *et al.* (2011). Conservation: limits of land sparing. *Science*, **334**, 593.

Froger, G., Ménard, S. & Méral, P. (2015). Towards a comparative and critical analysis of biodiversity banks. *Ecosyst. Serv.*, **15**, 152-161.

FSC. (2016). FSC: facts & figures, April 6, 2016 [WWW Document]. URL https://ic.fsc.org/en/facts-figures

Galaz, V., Gars, J., Moberg, F., Nykvist, B. & Repinski, C. (2015). Why ecologists should care about financial markets. *Trends Ecol. Evol.*, **30**, 571-580.

Gan, J. & McCarl, B.A. (2007). Measuring transnational leakage of forest conservation. *Ecol. Econ.*, **64**, 423-432.

Ghazoul, J., Burivalova, Z., Garcia-Ulloa, J. & King, L.A. (2015). Conceptualizing forest degradation. *Trends Ecol. Evol.*, **30**, 622-632.

Goolsbee, A., Levitt, S. & Syverson, C. (2016). *Microeconomics*. Macmillan, New York.

Hansen, M.C., Potapov, P. V, Moore, R., *et al.* (2013). High-resolution global maps of 21st-century forest cover change. *Science*, **342**, 850-853.

Hertel, T.W. (2011). The global supply and demand for agricultural land in 2050: a perfect storm in the making? *Am. J. Agric. Econ.*, **93**, 259-275.

Hill, R., Miller, C., Newell, B., Dunlop, M., & Gordon, I. J. (2015). Why biodiversity declines as protected areas increase: the effect of the power of governance regimes on sustainable landscapes. *Sustainability Science*, **10**(2), 357-369. http://link.springer.com/article/10.1007%2Fs11625-015-0288-6

Jadin, I., Meyfroidt, P. & Lambin, E.F. (2016). International trade, and land use intensification and spatial reorganization explain Costa Rica's forest transition. *Environ. Res. Lett.*, **11**, 35005.

Jantke, K. & Schneider, U.A. (2011). Integrating land market feedbacks into conservation planning — a mathematical programming approach. *Environ. Model. Assess.*, **16**, 227-238.

Jonsson, R., Mbongo, W., Felton, A. & Boman, M. (2012). Leakage implications for European timber markets from reducing deforestation in developing countries. *Forests*, **3**, 736-744.

Kaimowitz, D. & Angelsen, A. (2008). Will livestock intensification help save Latin America's tropical forests? *J. Sustain. For.*, **27**, 6-24.

Kalonga, S. K., Midtgaard, F., & Eid, T. (2015). Does forest certification enhance forest structure? Empirical evidence from certified community-based forest management in Kilwa District, Tanzania. *International Forestry Review*, **17**(2), 182-194. http://www.bioone.org/doi/abs/10.1505/146554815815500570.

Koh, L.P. & Wilcove, D.S. (2008). Is oil palm agriculture really destroying tropical biodiversity? *Conserv. Lett.*, **1**, 60-64.

Kollert, W. & Lagan, P. (2007). Do certified tropical logs fetch a market premium? A comparative price analysis from Sabah, Malaysia. *For. Policy Econ.*, **9**, 862-868.

Lambin, E.F., Meyfroidt, P., Rueda, X., *et al.* (2014). Effectiveness and synergies of policy instruments for land use governance in tropical regions. *Glob. Environ. Chang.*, **28**, 129-140.

Larrosa, C., Carrasco, L.R. & Milner-Gulland, E.J. (2016). Unintended feedbacks: challenges and opportunities for improving conservation effectiveness. *Conserv. Lett.*, **9**, 316-326.

Liu, J., Hull, V., Batistella, M., *et al.* (2013). Framing sustainability in a telecoupled world. *Ecol. Soc.*, **18**, 1-17.

Meyfroidt, P. & Lambin, E.F. (2009). Forest transition in Vietnam and displacement of deforestation abroad. *Proc. Natl. Acad. Sci. U. S. A.*, **106**, 16139-16144.

Meyfroidt, P., & Lambin, E. F. (2011). Global forest transition: prospects for an end to deforestation.

Meyfroidt, P., Lambin, E.F., Erb, K.-H. & Hertel, T.W. (2013). Globalization of land use: distant drivers of land change and geographic displacement of land use. *Curr. Opin. Environ. Sustain.*, **5**, 438-444.

Meyfroidt, P., Rudel, T.K. & Lambin, E.F. (2010). Forest transitions, trade, and the global displacement of land use. *Proc. Natl. Acad. Sci. U. S. A.*, **107**, 20917-20922.

Miller, B.W., Caplow, S.C. & Leslie, P.W. (2012). Feedbacks between conservation and Social-Ecological Systems. *Conserv. Biol.*, **26**, 218-227.

Milner-Gulland, E.J. (2012). Interactions between human behaviour and ecological systems. *Philos. Trans. R. Soc. Lond. B. Biol. Sci.*, **367**, 270-278.

Müller, D., Sun, Z., Vongvisouk, T., Pflugmacher, D., Xu, J. & Mertz, O. (2014). Regime shifts limit the predictability of land-system change. *Glob. Environ. Chang.*, **28**, 75-83.

Murray, B.C., McCarl, B. & Lee, H.-C. (2004). Estimating leakage from forest carbon sequestration programs. *Land Econ.*, **80**, 109.

Nellemann, C., INTERPOL Environmental Crime Programme (eds). (2012). *Green carbon, black trade: illegal logging, tax fraud and laundering in the world's tropical forests*. A Rapid Response Assessment. United Nations Environment Programme GRID-Arendal.

Oliveira, P.J.C., Asner, G.P., Knapp, D.E., *et al.* (2007). Land-use allocation protects the Peruvian Amazon. *Science*, **317**, 1233-1236.

Pain, D.J., Martins, T.L.F., Boussekey, M., *et al.* (2006). Impact of protection on nest take and nesting success of parrots in Africa, Asia and Australasia. *Anim. Conserv.*, **9**, 322-330.

Pfeifer, M., Burgess, N.D., Swetnam, R.D., Platts, P.J., Willcock, S. & Marchant, R. (2012). Protected areas: mixed success in conserving East Africa's evergreen forests. *PLoS One*, **7**, e39337.

Phalan, B., Green, R.E., Dicks, L.V., *et al.* (2016). How can higher-yield farming help to spare nature? *Science*, **351**, 450-451.

Phalan, B., Onial, M., Balmford, A. & Green, R.E. (2011). Reconciling food production and biodiversity conservation: land sharing and land sparing compared. *Science*, **333**, 1289-1291.

Phelps, J., Carrasco, L.R., Webb, E.L., Koh, L.P. & Pascual, U. (2013). Agricultural intensification escalates future conservation costs. *Proc. Natl. Acad. Sci. U. S. A.*, **110**, 7601-7606.

Polasky, S. (2006). You can't always get what you want: conservation planning with feedback effects. *Proc. Natl. Acad. Sci. U. S. A.*, **103**, 5245–5246.

Rayner, L., Lindenmayer, D.B., Wood, J.T., Gibbons, P. & Manning, A.D. (2014). Are protected areas maintaining bird diversity? *Ecography (Cop.)*., **37**, 43-53.

Renwick, A.R., Bode, M. & Venter, O. (2015). Reserves in context: planning for leakage from protected areas. *PLoS One*, **10**, e0129441.

Roca, J. (2003). Do individual preferences explain the Environmental Kuznets curve? *Ecol. Econ.*, **45**, 3-10

Rudel, T.K., Schneider, L., Uriarte, M., *et al.* (2009). Agricultural intensification and changes in cultivated areas, 1970-2005. *Proc. Natl. Acad. Sci. U. S. A.*, **106**, 20675-20680.

Smith, R.J. & Walpole, M.J. (2005). Should conservationists pay more attention to corruption? *Oryx*, **39**, 251-256.

St John, F.A.V., Keane, A.M. & Milner-Gulland, E.J. (2013). Effective conservation depends upon understanding human behaviour. Pp. 344-361 in *Key Top. Conserv. Biol. 2.* John Wiley & Sons, Oxford.

Tilman, D., Balzer, C., Hill, J. & Befort, B.L. (2011). Global food demand and the sustainable intensification of agriculture. *Proc. Natl. Acad. Sci. U. S. A.*, **108**, 20260-20264.

Villoria, N.B., Byerlee, D. & Stevenson, J. (2014). The effects of agricultural technological progress on deforestation: what do we really know? *Appl. Econ. Perspect. Policy*, **36**, 211-237.

Viña, A., McConnell, W.J., Yang, H., Xu, Z. & Liu, J. (2016). Effects of conservation policy on China's forest recovery. *Sci. Adv.*, **2**, 1-7.

Warren-Thomas, E., Dolman, P.M. & Edwards, D.P. (2015). Increasing demand for natural rubber necessitates a robust sustainability initiative to mitigate impacts on tropical biodiversity. *Conserv. Lett.*, **8**, 230-241.

Wunder, S. (2013). When payments for environmental services will work for conservation. *Conserv. Lett.*, **6**, 230-23

10

Analysis of Trade-Offs Between Biodiversity, Carbon Farming and Agricultural Development in Northern Australia Reveals the Benefits of Strategic Planning

Alejandra Morán-Ordóñez[1,2], Amy L Whitehead[1], Gary W Luck[3], Garry D Cook[4], Ramona Maggini[5], James A Fitzsimons[6,7], & Brendan A Wintle[1]

[1] Quantitative & Applied Ecology Group, School of Biosciences, The University of Melbourne, Parkville, VIC 3010, Australia
[2] Centre Tecnològic Forestal de Catalunya, Ctra. Antiga St. Llorenç km 2, 25280 Solsona, Spain
[3] Institute for Land, Water and Society, Charles Sturt University, Albury, NSW 2640, Australia
[4] CSIRO Ecosystem Sciences, Private Mail Bag 44, Winnellie, NT 0822, Australia
[5] ARC Centre of Excellence for Environmental Decisions, NERP Environmental Decisions Hub, Centre for Biodiversity & Conservation Science, University of Queensland, Brisbane, Qld 4072, Australia
[6] The Nature Conservancy, Suite 2-01, 60 Leicester Street, Carlton, VIC 3053, Australia
[7] School of Life and Environmental Sciences, Deakin University, 221 Burwood Highway, Burwood, VIC 3125, Australia

Keywords
Biodiversity conservation; carbon storage; conservation planning; economic development; intensive agriculture; MaxEnt; land-use change; spatial conservation prioritization; species distributions; zonation.

Correspondence: Alejandra Morán-Ordóñez, CTFC Centre Tecnològic Forestal de Catalunya, Ctra. Antiga St. Llorenç km 2, 25280 Solsona, Spain.
E-mail: alejandra.moran@ctfc.es

Editor
Lewison Rebecca

Abstract

Australia's northern savannas are one of the few remaining large and mostly intact natural areas on Earth. However, their biodiversity and ecosystem values could be threatened if proposed agricultural development proceeds. Through land-use change scenarios, we explored trade-offs and synergies among biodiversity conservation, carbon farming and agriculture production in northern Australia. We found that if all suitable soils were converted to agriculture, habitat at unique recorded locations of three species would disappear and 40 species and vegetation communities could lose more than 50% of their current distributions. Yet, strategically considering agriculture and biodiversity outcomes leads to zoning options that could yield >56,000 km^2 of agricultural development with a significantly lower impact on biodiversity values and carbon farming. Our analysis provides a template for policy-makers and planners to identify areas of conflict between competing land-uses, places to protect in advance of impacts, and planning options that balance agricultural and conservation needs.

Introduction

Land-use change is a major driver of habitat degradation and species extinction worldwide (Sala et al. 2000; Foley et al. 2005; Fischer & Lindenmayer 2007). In Australia, nearly 50% of natural vegetation has been cleared or severely modified since 1788, leading to the extinction of numerous species and critically endangering many others (Bradshaw 2012; Lindenmayer & Possingham 2013).

Land clearing is spatially uneven across Australia. For example, only one third of natural vegetation remains in south-eastern Australia, while the tropical savannas of northern Australia currently occupy 99% of their original extent (Woinarski et al. 2011; Bradshaw 2012). While extensive pastoral activity has modified the composition and structure of much of the northern savanna forests and woodlands (e.g., Woinarski et al. 2011), most of the landscape remains at least structurally intact (Woinarski et al. 2006, 2007, 2011; Andersen et al. 2012). However, recent rapid declines in fauna populations across much of northern Australia due to a range of factors (including feral predators and other invasive species, overgrazing, altered fire regimes) indicate the desperate and immediate need for strategic land management approaches in the region (Woinarski *et al.* 2011, 2015).

In September 2014, the Australian Government identified options to promote the economic development of northern Australia over a 20-year period (Joint Select Committee on Northern Australia 2014). Among the proposed initiatives is the staged development of irrigated agriculture schemes "to help double Australia's agricultural output" (Joint Select Committee on Northern Australia 2014, p. 6). Announcement of a $5B Northern Australian Infrastructure Facility indicates that the government is taking northern economic development very seriously (Commonwealth of Australia 2015). To put this proposal in a geographic context, the northern savannas occupy an area approximately the size of France and Germany combined, with about 20% being deemed highly suitable for agricultural intensification based on soil properties (Wilson *et al.* 2009, 2013). This highlights the magnitude of the economic opportunity, but also the scale of the threat to northern Australia's biodiversity.

There has also been substantial uptake of carbon-emission reduction initiatives linked to payments for improved fire management (Cook *et al.* 2012; Russell-Smith *et al.* 2013; Walton & Fitzsimons 2015), known in Australia as "carbon farming." In tropical savannas, carbon emission reductions are achieved by using low-intensity, early-season burns to minimize the amount of fuel burnt in large-scale, high-intensity late-season wildfires (Cook & Meyer 2009; Bradshaw *et al.* 2013). Cooler, early-season burns are generally considered to be commensurate with biodiversity conservation objectives. Carbon farming in northern Australia could promote biodiversity conservation and socio-economic development of local indigenous communities (Woinarski *et al.* 2011; Cook *et al.* 2012; Fitzsimons *et al.* 2012; Russell-Smith *et al.* 2013).

The prospect of a major shift in land-use from relatively low-impact rangeland grazing to relatively high-impact irrigated intensive agriculture presents opportunities and challenges for regulators, industry, carbon-farming investors, conservation organizations and other stakeholders. Regulators need to balance the financial and food security benefits of expanded agriculture against the potential negative impacts on other industries such as fishing, prawning, several types of tourism, biodiversity, carbon-farming options and areas of indigenous cultural significance. Governments, conservation organizations, and broader society need to identify the most irreplaceable areas that may be lost to agricultural development and its offsite impacts, determine how to protect whole of landscape and hydrological functioning (which is so important for the savannas Woinarski *et al.* 2007), and use appropriate conservation and partnership mecha-

nisms to conserve landscapes with significant ecological values. Similarly, carbon-farming investors could secure commitments from current lease-holders to minimize carbon emissions which could have some auxiliary benefits for biodiversity. Balancing these competing objectives and identifying satisfactory solutions requires a strategic approach to land-use planning and management, which could be achieved under a legislated strategic planning process (e.g., Strategic Assessments under the Federal *Environment Protection and Biodiversity Conservation Act 1999* -EPBC Act).

Here, we analyze trade-offs between biodiversity, carbon, and agricultural intensification in northern Australia using maps of agricultural intensification potential, carbon-farming potential, and geographic distributions for 611 species and 43 vegetation communities. Through systematic spatial prioritization, we identify the strategies available to conservation practitioners, regulators or carbon-farming investors to maximize their respective objectives. Our analysis highlights the importance of considering the threat of land-use change in a spatially explicit way, and the relatively high biodiversity benefits that can be achieved at relatively low economic opportunity cost by a regulator who systematically balances biodiversity and economic development options. Our results are vital for planners and policy-makers considering developing northern Australia, but also hold important lessons for other relatively undeveloped regions that are slated for future land-use change.

Methods

Study area

The study area comprises the tropical savanna of northern Australia, extending between latitudes 10–20°S and covering approximately 960,000 km^2. *Eucalyptus* open forests and woodlands with a grassy understorey dominate the landscape, with *Acacia* and *Melaleuca* woodlands, and hummock and tussock grasslands occurring in some areas. The region contains four nationally threatened ecological communities and 199 threatened species listed under the EPBC Act. Cattle grazing, mining and nature-based tourism are the main industry sectors (Woinarski *et al.* 2007), while protected areas currently cover about 18% of the region.

Mapping biodiversity, carbon, and agricultural opportunity values

To characterize the biodiversity values of the northern savannas, we mapped the distributions of 611 species (tetrapods and plants) and 43 vegetation communities.

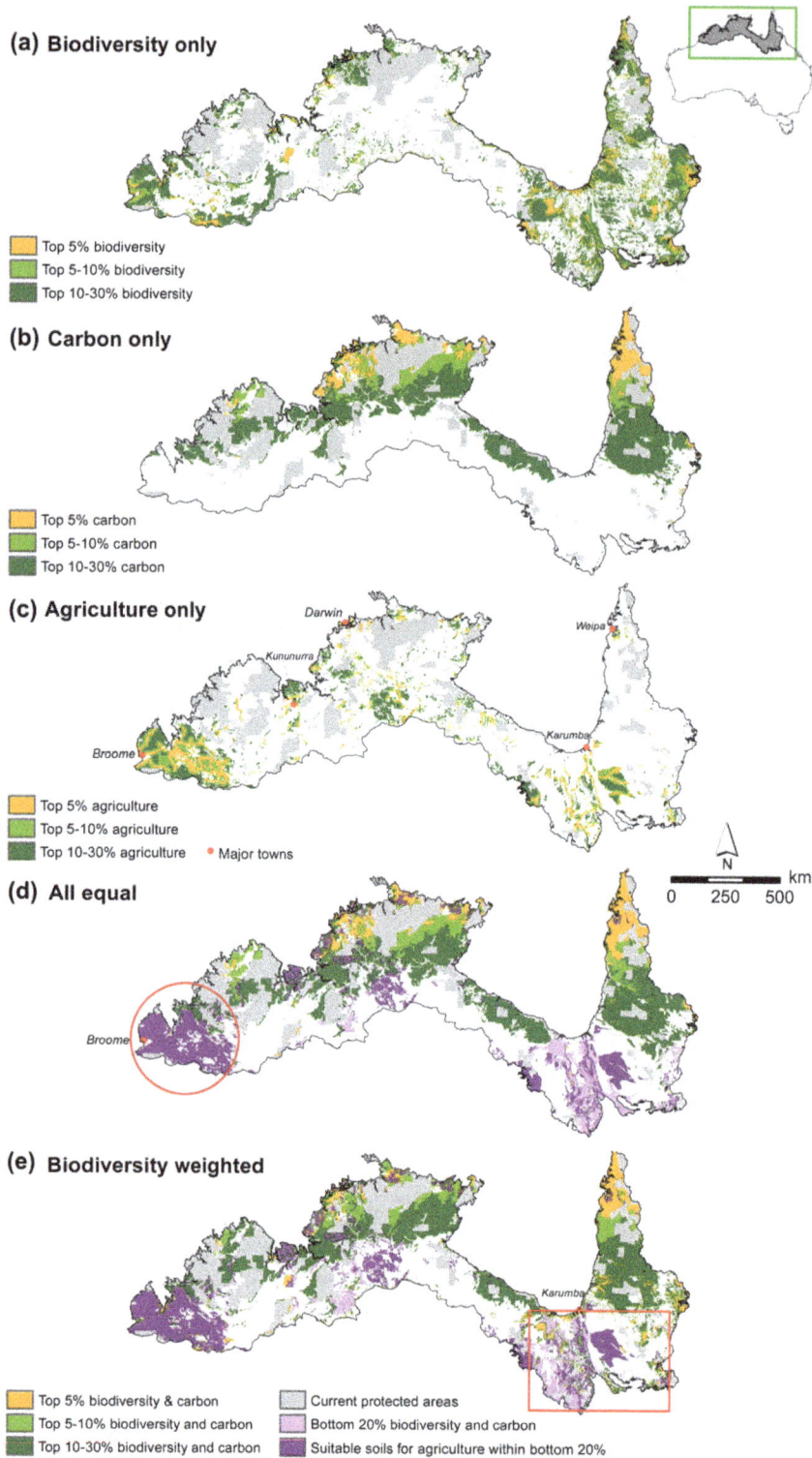

(a) Biodiversity only

Top 5% biodiversity
Top 5-10% biodiversity
Top 10-30% biodiversity

(b) Carbon only

Top 5% carbon
Top 5-10% carbon
Top 10-30% carbon

(c) Agriculture only

Top 5% agriculture
Top 5-10% agriculture
Top 10-30% agriculture   • Major towns

(d) All equal

(e) Biodiversity weighted

Top 5% biodiversity & carbon   Current protected areas
Top 5-10% biodiversity and carbon   Bottom 20% biodiversity and carbon
Top 10-30% biodiversity and carbon   Suitable soils for agriculture within bottom 20%

Figure 1 Priority maps for northern Australia based on five different scenarios: (a) a *biodiversity-only* scenario that ranks sites based on their value for 654 biodiversity features, without taking into account carbon storage or agricultural potential; (b) a *carbon-only* scenario that ranks sites based on their value for carbon storage, without accounting for biodiversity or agriculture; (c) an *agriculture-only* scenario that ranks sites based on agricultural potential (i.e., those with the most suitable soils for the development of irrigated agriculture as a function of their accessibility), without accounting for carbon storage potential or biodiversity values; (d) a scenario where all biodiversity features together are weighted the same as agriculture and carbon storage (*all-equal*); and (e) a scenario where all biodiversity features together are weighted 10-fold than agriculture and carbon storage (*biodiversity-weighted*). All scenarios take into consideration the biodiversity values within existing protected areas. See Methods and Supporting Information for further details.

Our aim was to include as many species as possible in order to characterise potential impacts of proposed land-use change on biodiversity as a whole. We developed species distribution models (SDM) for 356 species and also included published distribution maps for an additional 27 threatened species (Maggini *et al.* 2013). We generated presence-absence maps from point occurrence data for another 228 species for which there were insufficient point records to build SDMs. Carbon-security potential was mapped using a spatial layer of above-ground forest carbon stock in tonnes per hectare (Cook *et al.* 2015). Places of high carbon-security potential are primarily places in which there is an opportunity to reduce the loss of carbon stored in the landscape through sympathetic land management practices. We did not account for the carbon-security potential of "future" irrigated agriculture since first, this would be negligible compared with that lost from native woodlands and soils when native vegetation is cleared (Guo & Gifford 2002, Cook *et al.* 2010, Luo *et al.* 2010) and second, little is known about which particular crops are going to be promoted and where exactly they will be located (making impossible their spatially explicit assessment). Agricultural value (also referred to as irrigated agriculture from hereon) was mapped using an agricultural opportunity layer that integrates areas of high suitability of soils for irrigated annuals, irrigated perennials and irrigated improved pastures (Wilson *et al.* 2009, 2013). This layer identifies the areas at potential risk of land cover conversion from natural vegetation (savanna forest/woodland) to agriculture (i.e., to any of these three types of irrigated agricultural practices). The map of agricultural opportunity was further refined to identify areas close to existing roads that may be prioritized for development. See Appendix A1 for full details of data sources and handling, species modelling and mapping methods.

Spatial prioritization analysis

We used Zonation v4.0 (Appendix A2; Moilanen *et al.* 2005, 2012) to conduct a spatial prioritization across northern Australia of the three land-uses: biodiversity conservation, carbon storage and/or irrigated agriculture. We explored five alternative scenarios: *biodiversity-only*, *carbon-only*, *agriculture-only*, *all-equal* and *biodiversity-weighted*. The *biodiversity-only* scenario ranked each 1-km^2 grid cell in the landscape in terms of its biodiversity values on the basis of the distribution maps for all 654 biodiversity features. Zonation uses the relative rarity of species and vegetation communities to identify the most complementary set of cells to conserve at every level of landscape loss from 0% to 100%. The last cells to be lost are considered the most irreplaceable cells (here those

with highest biodiversity value). In this scenario all biodiversity features were assigned equal weight, independent of their threat status, under the assumption that any species or vegetation communities could become threatened with the major land-use changes being proposed for the study region. The *carbon-only and agriculture-only* scenarios ranked the landscape using the carbon-storage potential and agricultural opportunity spatial layers to identify the zones of highest vegetative carbon stocks and agricultural potential, respectively (regardless of their biodiversity values).

In each of the above three scenarios we used replacement cost analysis (Cabeza & Moilanen 2006), a feature in Zonation where the removal order of cells can be artificially altered, to account for the existing protected area network in northern Australia. This allowed us to identify high priority locations (best 5%, 10%, and 30% of the landscape) for biodiversity conservation outside the existing reserves (i.e., priority sites for *expanding* the level of protection for biodiversity in the landscape to conserve the features outlined above; Appendix A2). To identify possible areas of synergy or conflict between the three land-uses, we measured and mapped the zones of overlap between high priority locations for biodiversity outside protected areas and the high priority areas for carbon storage and agricultural potential, respectively. We also ran a *biodiversity-only* analysis unconstrained by the current distribution of protected areas, to measure the concordance between irreplaceability and existing reserves.

The *all-equal* and *biodiversity-weighted* Zonation scenarios combined maps of biodiversity, carbon-storage and agricultural potential to explore options for balancing or trading-off between competing land-uses. In these analyses, the value of each 1-km^2 cell for agriculture is introduced to Zonation as a cost, which, when all else is equal, favours the conservation of cells with lower agricultural value. In the *all-equal* scenario, biodiversity features were all weighted equally (1/654 each) and carbon storage and agricultural suitability were weighted 1.0 and −1.0, respectively. This implies that that the net value of all biodiversity features is equal to the net value of carbon storage, which is equal to the net value of agriculture in the region. It also assumes that biodiversity conservation and carbon storage are compatible land-uses, while intensive irrigated agriculture is incompatible with both. We acknowledge this is a simplified view of the system since our assumptions are not universally true (e.g., carbon farming may not necessarily be compatible with biodiversity conservation and agriculture can provide habitat and resources for biodiversity; Thomas *et al.* 2013; Luck *et al.* 2015); however, with this analysis we sought to find a solution that maximizes both biodiversity and carbon objectives simultaneously (Venter *et al.* 2009; Thomas *et al.* 2013), while avoiding areas that are potentially good for agriculture (Moilanen *et al.* 2011). The *biodiversity-weighted* scenario was built under assumptions similar to the *all-equal* scenario except it prioritizes biodiversity values more than carbon outcomes and agriculture suitability (i.e., it weighted biodiversity features ten times more with a weight of 10/654 for each feature than carbon storage and agricultural suitability). For all scenarios it was assumed that, outside of the current protected area system, any land-use could be possible, regardless of underlying tenure.

Results

High-priority areas for biodiversity conservation outside current protected areas are mostly concentrated in the north, east and south-western regions of northern Australia (Figure 1a). Carbon storage increased from south to north, corresponding with the rainfall gradient from semi-arid to subtropical and tropical regions in the study area (Figure 1b). The best areas for developing irrigated agriculture based on soil conditions are scattered across the landscape, although there are three highly ranked areas that stand out for their large and continuous extent in the southwest, center, and southeastern parts of the study area (Figure 1c). The map outputs from the *all-equal* and *biodiversity-weighted* analyses (Figures 1d–e) show how prioritizing biodiversity and carbon-storage tends to push agriculture further south in the region, leaving relatively irreplaceable sites for biodiversity conservation less impacted in the north, where carbon stocks are highest. The similarity between Figures 1(b) and (d)–(e) arises because many of the important sites for biodiversity in the north of the region are correlated with high carbon-security opportunity in those areas, and the high biodiversity areas in the south of the region are somewhat exchangeable in terms of species composition with areas further north (red circle Figure 1d). Interestingly, the ranking map output of the *biodiversity-weighted* scenario (Figure 1e) resembled more the *all-equal scenario* (with most biodiversity and carbon valuable sites in the north) than the *biodiversity-only* scenario (Figure 1a). The key difference between the *biodiversity-weighted* scenario and the *all-equal* scenario is that the former identifies some irreplaceable sites for biodiversity in the south-eastern part of the study region (red rectangle Figure 1e).

Opportunities for conservation gains

Comparing the prioritization analysis for biodiversity that ignores the current distribution of conservation reserves with the analysis that constrains the solution to include existing conservation reserves, shows that the current

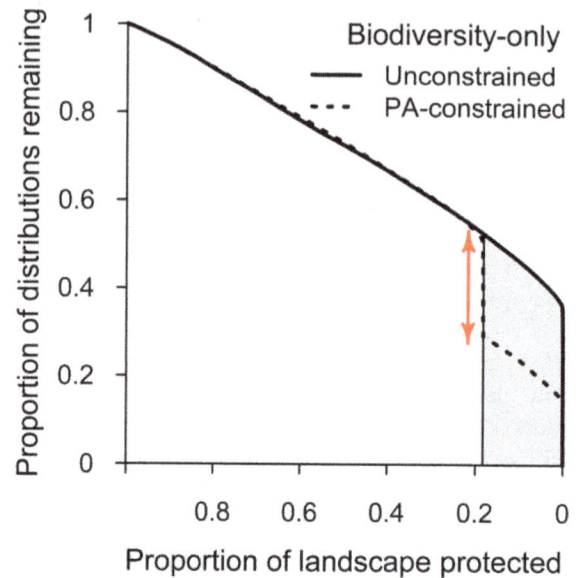

Figure 2 Representation of the distribution ranges of the 654 biodiversity features within current network of protected areas in northern Australia. The *Y*-axis represents the extent of suitable geographic range available for the biodiversity features, ranging from 0 – no suitable conditions available for the features – to 1 all suitable range available to the features. The *X*-axis represents the proportion of total landscape protected. The solid and dotted black lines represent the average performance of the 654 biodiversity features (read from the *Y*-axis) under the unconstrained and constrained solutions of the *biodiversity-only* scenario, respectively. The constrained solution artificially alters the order of cell removal in Zonation, forcing the existing protected area network into the top fraction of the landscape. The unconstrained solution identifies the areas that are most important to capture biodiversity values, irrespective of their current protection status. The grey shaded area delimits the extent of the current network of protected areas in northern Australia (read from *X*-axis). The difference between the solid and dotted lines read from the *Y*-axis (red arrow), indicates the opportunity for conservation under the unconstrained solution compared to what it is currently protected by the reserve system (i.e., average gain in distribution ranges of the 654 biodiversity features).

protected area network captures on average (across all species and vegetation communitites), 29% of the distributions of all 654 biodiversity features (Figure 2). If arranged to optimize representation, the same area of land (around 18.3% of the study area) could have represented up to 50% of the distributions of the same biodiversity features (based on an unconstrained prioritization of biodiversity values). Our analysis highlights a significant opportunity to dramatically increase representativeness with a minor expansion of the reserve system or other forms of protection by being more strategic about where new conservation areas are placed. For example, by expanding the protected area network to capture an additional 5% of northern Australia, we could

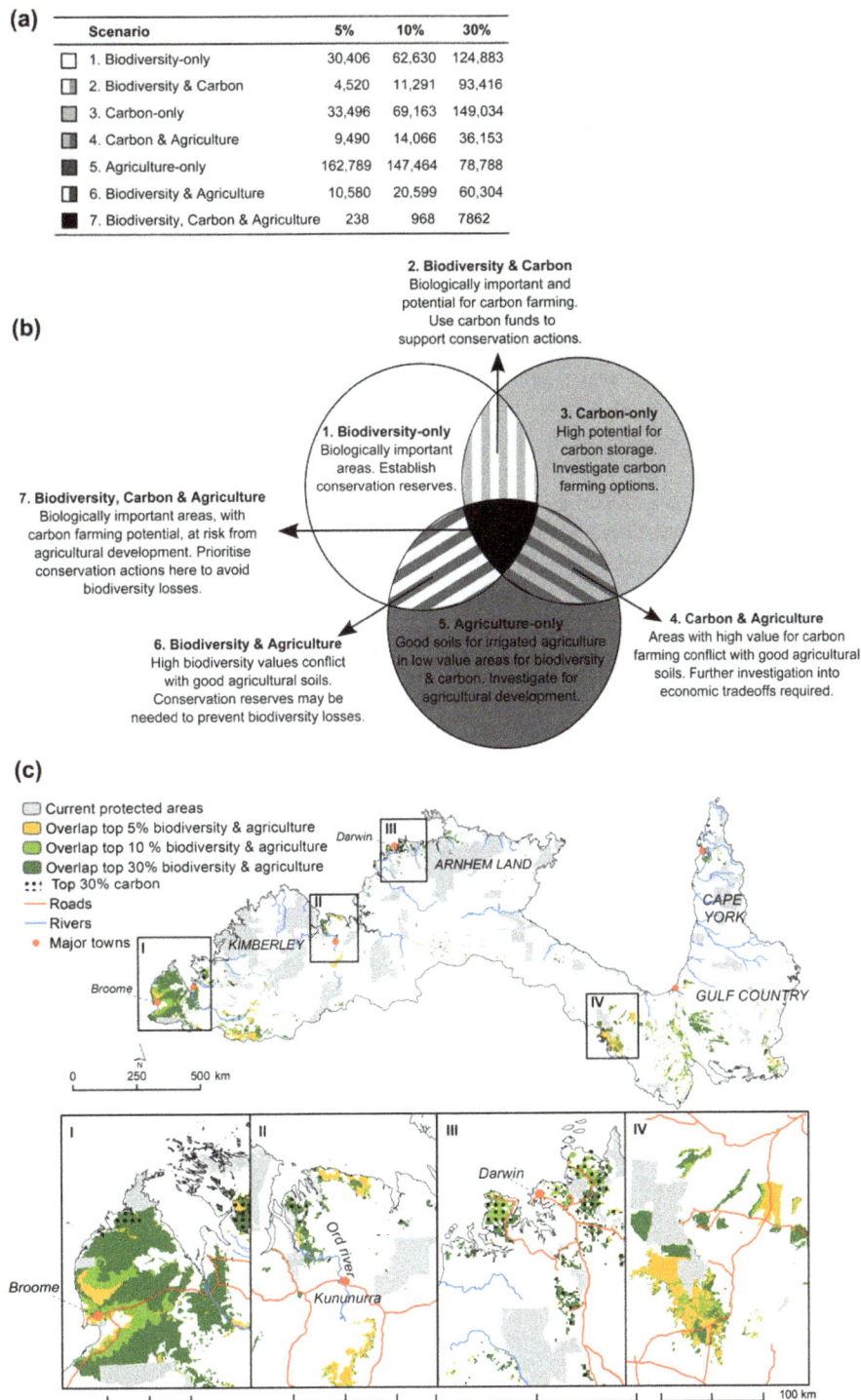

Figure 3 (a) Degree of overlap between any area suitable for agriculture and high priority areas (best 5%, 10%, and 30%) for biodiversity conservation only and carbon storage only (area in squared km). For example, whereas 30,406 km² of northern Australia has been identified as high priority for biodiversity (within the top 5 % of the *biodiversity-only* scenario landscape ranking), only 4,520 km² overlaps with high priority areas for carbon storage (within the top 5% of the *carbon-only* scenario). (b) Venn diagram showing the areas of potential conflict (trade-offs) or synergies between the three land-uses as well as their implications for policy making. (c) Location of sites where there is spatial overlap between the high priority areas (best 5%, 10%, and 30%) for biodiversity conservation only and any area suitable for agriculture in northern Australia (i.e., areas of potential conflict between biodiversity and agriculture). The map also shows the sites where these areas of potential conflict between biodiversity and agriculture overlap with high values for carbon storage (top 30% of the *carbon-only* scenario). Panels I, II, III, and IV show these overlaps in detail for four different areas of the study region.

increase the representation of biodiversity features under some form of protection from 29% to 57%.

Trade-offs and synergies between land-uses

Eighty-eight percent of the best soils for agriculture occur outside the current protected area network. However, there is considerable overlap between priority areas for biodiversity conservation outside current protected areas and locations most suitable for agriculture (60,304 km^2, Figure 3a, scenario 6). If agricultural development is expanded in northern Australia, there is likely to be future conflict between these two land-uses (Figure 3b, scenario 6). The largest areas of overlap between biodiversity and agriculture occur in the southern parts of the study area (Figure 3c, panels I and IV). Conversely, 56,441 km^2 (~30%) of the best agricultural soils occur within areas of relatively low conservation priority based on our criteria (the bottom 30% of the *biodiversity-only* scenario). Areas that are most important for all three land-uses represent a small fraction of the overall landscape (<0.025 %; Figures 3a–b, scenario 7) and are mainly located in the north of the study area (Figure 3c, panels II and III).

Prioritizing land-use based only on opportunities for high intensity irrigated agriculture or carbon storage is predicted to lead to total habitat loss for at least one species, even when only a small proportion of the landscape is converted (Figures 4b–c). By explicitly including species distributions in prioritizations of agricultural area development, even the most heavily impacted species retain a small proportion of their current distribution with relatively high rates of land-use conversion toward agriculture (Figures 4d–e). The performance of the worst-off 10% of species and communities (average performance of biodiversity features within the bottom 10th percentile of data) was markedly higher under the *all-equal* and (especially) the *biodiversity-weighted* analysis compared with both *carbon*-only and *agriculture-only* scenarios (Figure 5a). The *agriculture-only* and *carbon-only* analyses predict much larger losses in the distributions of biodiversity features than the *all-equal* and *biodiversity-weighted* analyses (Figure 5b; Table 1). For example, when approximately 20% of the landscape is converted to agriculture, all known records of three biodiversity features and the total extent of one vegetation community would likely be totally lost under the *agriculture-only* scenario. A further 36 species and vegetation communities would have more than 50% of their current distribution impacted. In contrast, converting the same area of land under the *all-equal* or *biodiversity-weighted* scenarios would lead to no species losing their last remaining suitable habitat, and only seven or five biodiversity features having 50–

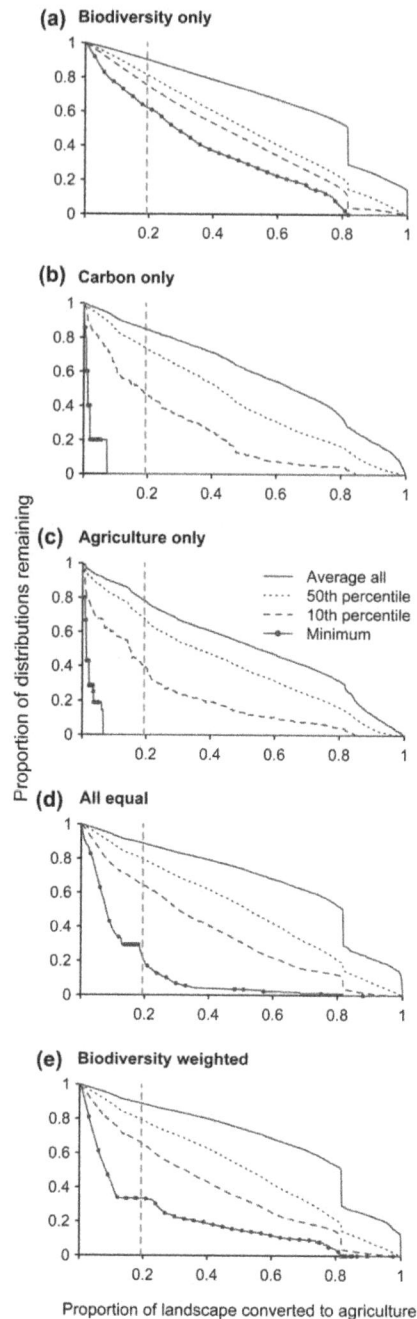

Figure 4 Relationship between the proportion of the landscape converted to agriculture and the performance of the biodiversity features under five prioritization scenarios: (a) *biodiversity-only*, (b) *carbon-only*, (c) *agriculture-only*, (d) and (e) all land-uses (*all-equal* and *biodiversity-weighted*). The grey lines show the average proportion of distributions remaining for all 654 biodiversity features (solid line "average all"), the worst-off 50% and 10% of biodiversity features (dotted -50th percentile- and dashed lines – 10th percentile, respectively), and the feature with the absolute lowest distribution remaining (dotted-solid line, "minimum"). The dashed red line marks the threshold corresponding to the total area covered by the most suitable soils for irrigated agriculture across northern Australia (approximately 20% of the landscape).

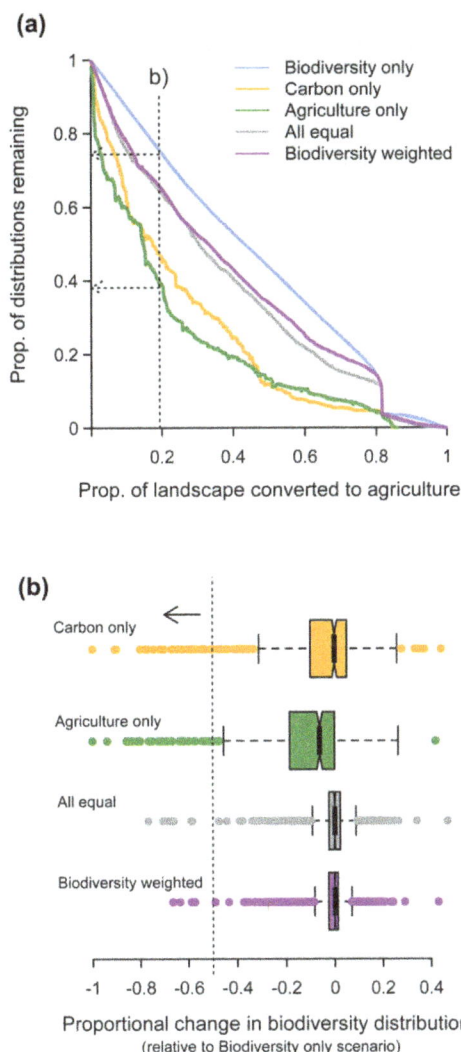

(a)

(b)

Figure 5 Performance of biodiversity features under five prioritization scenarios. (a) Proportion of the biodiversity features' distributions remaining at different levels of landscape lost due to conversion to agriculture. Lines represent the average performance of the worst-off 10% of the biodiversity features for each scenario (*biodiversity-only*, *carbon-only*, *agriculture-only*, *all-equal*, and *biodiversity-weighted*). Comparison between scenarios can be made at any threshold of landscape conversion along the X-axis. For example, a conversion of all suitable soils for agriculture into irrigated crops or pasturelands would imply approximately 20% of landscape loss for other land-uses (dotted black vertical line linking with plot b). At this proportion of landscape loss, the *agriculture-only* scenario predicts that the average distributions remaining for the worst 10% of the biodiversity features is 0.38 versus the 0.75 predicted by the biodiversity only scenario (i.e., a reduction of approximately 50% in predicted distributions between the two scenarios). (b) Relative change in the distributions of each of the 654 biodiversity features predicted under the *carbon-only*, *agriculture-only*, *all-equal* and *biodiversity-weighted* scenarios compared to the *biodiversity-only* scenario when approximately 20% of the landscape is converted into agricultural lands (i.e., when all suitable soils for agriculture are developed). Biodiversity features to the left of the dotted line (following arrow direction) under any of these four scenarios are predicted to lose more than 50% of their distributions.

75% of their current distribution impacted, respectively (Table 1).

Discussion

The policy document *Our North, Our Future: White Paper on Developing Northern Australia* (Australian Government 2015) pays little attention to the potential impact that agricultural development options may have on biodiversity and associated industries (e.g., tourism). Nor does it mention how such impacts would be assessed, risks to biodiversity managed, and appropriately balanced trade-offs between biodiversity, agriculture and other sectors will be achieved. The substantial overlap between agricultural potential and biodiversity value suggests that agricultural development, based solely on considerations about production potential, could have significant negative impacts on biodiversity. Some trade-offs will be necessary if the loss of significant biodiversity values due to agricultural intensification is to be avoided.

In addition to the potential impacts of non-strategic or poorly-regulated agricultural development, our results also highlight the fact that there are up to 56,000 km^2 of high agriculture potential soils within the bottom 30% of the biodiversity values analyzed. That is, there are potentially many opportunities to develop irrigated agriculture in areas that are not high priority for biodiversity (measured as the representativeness of a defined set of features). Our broad-scale prioritization should help guide future finer-scale examination of factors that can further threaten biodiversity conservation including accessibility, existing and likely future irrigation infrastructure, and factors likely to reduce threat from agricultural development, such as flood and cyclone risk or soil erosion (Wilson *et al.* 2013). If specific proposals for agricultural development emerge that identify particular annual or perennial crops or pastures as economically viable in particular places, then the specific impacts of these options on biodiversity, carbon (or other values such as water) can be evaluated using the analytical approach we demonstrate here. However, until specific proposal emerge, the resolution of our analysis seems appropriate for identifying broad areas of potential land-use conflict.

Results of the *all-equal* analysis indicate that the biodiversity features considered in this work could maintain their representation across the study area even with fairly high levels of agricultural development (Figure 5b). However, the actual long-term persistence of those biodiversity features will also depend on ecological processes such as connectivity, dispersal, changing climate and fire regimes, or predation and/or competition from invasive species. The demographic and environmental data

Table 1 Total number of biodiversity features (species and vegetation communities) that would lose >50%, >75% or 100% of their current distribution ranges in northern Australia under the different land-use scenarios and at two different fractions of the landscape converted into agriculture. The number of nationally threatened biodiversity features is indicated in brackets. The development of 50% or 100% of best suitable soils for agriculture corresponds respectively to the conversion of approximately 10% and 20% of the total landscape of northern Australia into irrigated agricultural and pasturelands

Scenarios	Development of 50% of suitable soils for agriculture			Development of 100% of suitable soils for agriculture		
	>50%	>75%	100%	>50%	>75%	>100%
Agriculture-only	15 [1]	3 [0]	1 [0]	40 [1]	15 [1]	4 [0]
Carbon-only	11 [0]	2 [0]	1 [0]	31 [0]	9 [0]	3 [0]
All-equal	4 [1]	0 [0]	0 [0]	7 [1]	1 [0]	0 [0]
Biodiversity-weighted	2 [1]	0 [0]	0 [0]	5 [1]	0 [0]	0 [0]

required to model persistence under threat and land-use change scenarios are typically only available for a small subset of well-studied species. Development of such models for these species can provide further insights into the sustainability of competing land-use, management and impact mitigation options (Sebastián-González et al. 2011). Coupled with the need to include information on key breeding areas, refugia and sites of endemicity for species, this could be an appropriate next step in northern Australia to help decision-makers understand the potential implications of development options and the additional conservation investments needed to secure biodiversity persistence.

Future extensions of our study should account for the trade-offs involving other economic development opportunities such as nature-based tourism or shale gas expansion, or for indigenous cultural values that are important from a social and legal perspective. Moreover, the *indirect* impacts of agricultural development on biodiversity (e.g., the construction of transport networks, dams and pipelines) could outweigh the direct impacts of the land-use changes we have analyzed (Kingsford 2000; Letnic et al. 2014). On the flipside, the cost and practical impediments to infrastructure development needed to support agricultural intensification are likely to change agricultural priority areas in more complex ways than we have analyzed here. Similarly, we assume that any land-use could occur anywhere outside the existing reserve system, though this is clearly not the case in some areas, due to a range of cultural and regulatory constraints. Refining the biodiversity (i.e., accounting for more species) agriculture and carbon potential mapping, and combining those with other land-use options and constraints currently not included in our analyses constitute obvious extensions to the work presented here. For example, innovation in the agricultural sector could lead to new ways of conducting intensified agriculture in the region that secures more carbon and biodiversity at the site level, making the three land-uses more compatible. We have included the best current information about impacts of proposed agricultural activities on carbon and biodiversity, though our method of analysis easily accommodates new information.

Land-use change remains, arguably, the most potent threat to biodiversity conservation globally. We have demonstrated an approach to quantifying impacts of development options across multiple species and ecological communities, and exploring trade-offs and synergies to minimize impacts through judicious positioning of impacts and conservation measures. Such analyses can provide support to complex land-use planning problems because they can encapsulate relatively complex conservation ideas such as irreplaceability, complementarity, connectivity, and cost-effectiveness in relatively simple map outputs. This has immediate relevance for policy-makers and planners considering the development of northern Australia for agriculture, but the approach presented here can conceivably be adapted to any spatial, multi-objective land-use planning challenge.

Acknowledgments

This work was supported by the National Environmental Science Program Environmental Decisions Hub. Wintle was supported by an ARC Future Fellowship (FT100100819). Luck was supported by an ARC Future Fellowship (FT0990436). Adam Liedloff, Linda Gregory, Peter Wilson, and Brett Murphy made significant contributions to this work through the provision of data and expertise. We also thank M.W. Schwartz, R. Lewison and the anonymous reviewers who provided very constructive comments on the manuscript.

Supporting Information

Appendix 1. Description of source data used in the spatial prioritization

Appendix 2. Description of the spatial prioritization methodology

References

Andersen, A.N., Woinarski, J.C.Z. & Parr, C.L. (2012). Savanna burning for biodiversity: fire management for faunal conservation in Australian tropical savannas. *Austral Ecol.*, **37**, 658-667.

Australian Government (2015). Our North, Our Future: White Paper on Developing Northern Australia. Canberra, ACT. http://industry.gov.au/ONA/WhitePaper/Documents/northern_australia_white_paper.pdf

Bradshaw, C.J.A. (2012). Little left to lose: deforestation and forest degradation in Australia since European colonization. *J. Plant Ecol.*, **5**, 109-120.

Bradshaw, C.J.A., Bowman, D.M.J.S., *et al.* (2013). Brave new green world—consequences of a carbon economy for the conservation of Australian biodiversity. *Biol. Conserv.*, **161**, 71-90.

Cabeza, M. & Moilanen, A. (2006). Replacement cost: a practical measure of site value for cost-effective reserve planning. *Biol. Conserv.*, **132**, 336-342.

Commonwealth of Australia. (2015). Budget Paper No. 2, Budget Measures 2015-2016. http://www.budget.gov.au/2015-16/content/bp2/download/BP2_consolidated.pdf

Cook, G.D., Jackson, S. & Williams, R.J. (2012). A revolution in northern Australian fire management: recognition of Indigenous knowledge, practice and management. Pages 293-305 in Bradstock, R.A. & Williams, R.J., editors. *Flammable Australia*. CSIRO Publishing, Collingwood, Australia.

Cook, G.D., Liedloff, A.C., Cuff, N.J., Brocklehurst, P.S. & Wiliams, R.J. (2015). Stocks and dynamics of carbon in trees across a rainfall gradient in a tropical savanna. *Aust. Ecol.*, **40**, 845-856.

Cook, G.D. & Meyer, C.P. (2009). Fire, fuels and greenhouse gases. Pages 313-328 in Russell-Smith, J., Whitehead, P.J. & Cooke, P.M., editors. *Culture, ecology, and economy of fire management in north Australian savannas: rekindling the Wurrk tradition*. CSIRO Publishing, Melbourne.

Cook, G.D., Williams, R.J., Stokes, C.J., Hutley, L.B., Ash, A.J. & Richards, A.E. (2010). Managing sources and sinks of greenhouse gases in Australia's rangelands and tropical savannas. *Rangeland Ecol. Manage.*, **63**, 137-146.

Fischer, J. & Lindenmayer, D.B. (2007). Landscape modification and habitat fragmentation: a synthesis. *Glob. Ecol. Biogeogr.*, **16**, 265-280.

Fitzsimons, J., Russell-Smith, J., James, G., *et al.* (2012). Insights into the biodiversity and social benchmarking components of the Northern Australian fire management and carbon abatement programmes. *Ecol. Manage. Restor.*, **13**, 51-57.

Foley, J.A., Defries, R., Asner, G.P., *et al.* (2005). Global consequences of land use. *Science*, **309**, 570-574.

Guo, L.B. & Gifford, R.M. (2002). Soil carbon stocks and land use change: a meta analysis. *Glob. Chang. Biol.*, **8**, 345-360.

Joint Select Committee on Northern Australia. (2014). Pivot North. Inquiry into the Development of Northern Australia - Final Report. The Parliament of the Commonwealth of Australia, Canberra, ACT.

Kingsford, R.T. (2000). Ecological impacts of dams, water diversions and river management on floodplain wetlands in Australia. *Austral Ecol.*, **25**, 109-127.

Letnic, M., Webb, J.K., Jessop, T.S., Florance, D. & Dempster, T. (2014). Artificial water points facilitate the spread of an invasive vertebrate in arid Australia. *J. Appl. Ecol.*, **51**, 795-803.

Lindenmayer, D.B. & Possingham, H.P. (2013). No excuse for habitat destruction. *Science*, **340**, 680.

Luck, G.W., Hunt, K. & Carter, A. (2015). The species and functional diversity of birds in almond orchards, apple orchards, vineyards and eucalypt woodlots. *Emu*, **115**, 99-109.

Luo, Z., Wang, E. & Sun, O.J. (2010). Soil carbon change and its responses to agricultural practices in Australian agro-ecosystems: a review and synthesis. *Geoderma*, **155**, 211-223.

Maggini, R., Kujala, H., Taylor, M., *et al.* (2013). Protecting and restoring habitat to help Australia's threatened species adapt to climate change. *National Climate Change Adaptation Research Facility*. Gold Coast, Australia.

Moilanen, A., Anderson, B.J., Eigenbrod, F., *et al.* (2011). Balancing alternative land uses in conservation prioritization. *Ecol. Appl.*, **21**, 1419-1426.

Moilanen, A., Franco, A.M.A., Early, R.I., Fox, R., Wintle, B. & Thomas, C.D. (2005). Prioritizing multiple-use landscapes for conservation: methods for large multi-species planning problems. *Proc. Biol. Sci.*, **272**, 1885-1891.

Moilanen, A., Meller, L., Leppänen, J., Pouzols, F.M. & Arponen, A. (2012). Zonation: spatial conservation planning framework and software. Version 3.1 User Manual, 288.

Russell-Smith, J., Cook, G.D., Cooke, P.M., *et al.* (2013). Managing fire regimes in north Australian savannas: applying aboriginal approaches to contemporary global problems. *Front. Ecol. Environ.*, **11**, e55-e63.

Sala, O.E., Chapin, F.S., Armesto, J.J., *et al.* (2000). Global biodiversity scenarios for the year 2100. *Science*, **287**, 1770-1774.

Sebastián-González, E., Sánchez-Zapata, J.A., Botella, F., Figuerola, J., Hiraldo, F. & Wintle, B.A. (2011). Linking cost efficiency evaluation with population viability analysis to prioritize wetland bird conservation actions. *Biol. Conserv.*, **144**, 2354-2361.

Thomas, C.D., Anderson, B.J., Moilanen, A., *et al.* (2013). Reconciling biodiversity and carbon conservation. *Ecol. Lett.*, **16**, 39-47.

Venter, O., Laurance, W.F., Iwamura, T., Wilson, K.A., Fuller, R.A. & Possingham, H.P. (2009). Harnessing carbon payments to protect biodiversity. *Science*, **326**, 1368.

Walton, N. & Fitzsimons, J. (2015). Payment for ecosystem services in practice – savanna burning and carbon abatement at Fish River, northern Australia. Pages 78-83 in Figgis, P., Mackey, B., Fitzsimons, J., Irving, J. & Clarke, P., editors. *Valuing nature: protected areas and ecosystem services*. Australian Committee for IUCN, Sydney, Australia.

Wilson, P., Gregory, L. & Watson, I. (2013). Land and soil resources. Pages 76-111 in Grice, A.C., Watson, I. & Stone, P., editors. *Mosaic irrigation for the northern Australian beef industry. An assessment of sustainability and potential*. Synthesis Report prepared for the Office of Northern Australia, CSIRO, Brisbane, Australia.

Wilson, P., Ringrose-Voase, A., Jacquier, D., *et al.* (2009). *Land and soil resources in northern Australia*. Northern Australian Land and Water Science Review.

Woinarski, J., Mackey, B., Nix, H. & Traill, B. (2007). *The nature of northern Australia: natural values, ecological processes and future prospects*. ANU e-press, Canberra, Australia. http://press.anu.edu.au?p=34501.

Woinarski, J.C.Z., Burbidge, A. A. & Harrison, P.L. (2015). Ongoing unraveling of a continental fauna: decline and extinction of Australian mammals since European settlement. *Proc. Natl. Acad. Sci.*, **112**, 4531-4540.

Woinarski, J.C.Z., Hempel, C., Cowie, I., *et al.* (2006). Distributional pattern of plant species endemic to the Northern Territory, Australia. *Aust. J. Bot.*, **54**, 627-640.

Woinarski, J.C.Z., Legge, S., Fitzsimons, J.A., *et al.* (2011). The disappearing mammal fauna of northern Australia: context, cause, and response. *Conserv. Lett.*, **4**, 192-201.

11

Adding Some Green to the Greening: Improving the EU's Ecological Focus Areas for Biodiversity and Farmers

Guy Pe'er[1,2], Yves Zinngrebe[1,3], Jennifer Hauck[4,5], Stefan Schindler[6,7], Andreas Dittrich[8], Silvia Zingg[9,10], Teja Tscharntke[11], Rainer Oppermann[12], Laura M.E. Sutcliffe[12,13], Clélia Sirami[14], Jenny Schmidt[15], Christian Hoyer[8], Christian Schleyer[16], & Sebastian Lakner[3]

[1] Department of Conservation Biology, UFZ – Helmholtz Centre for Environmental Research, Permoserstr. 15, 04318 Leipzig, Germany
[2] German Centre for Integrative Biodiversity Research (iDiv) Halle-Jena-Leipzig, Deutscher Platz 5e, 04103 Leipzig, Germany
[3] Georg-August-University Göttingen,Department for Agricultural Economics and Rural Development, Platz der Göttinger Sieben 5, 37073 Göttingen, Germany
[4] CoKnow Consulting – Coproducing Knowledge for Sustainability, Mühlweg 3, 04838 Jesewitz, Germany
[5] Department of Environmental Politics, UFZ – Helmholtz Centre for Environmental Research, Permoserstr. 15, 04318 Leipzig, Germany
[6] Environment Agency Austria, Spittelauer Lände 5 (A-1090) Vienna, Austria
[7] Department of Conservation Biology, Vegetation & Landscape Ecology, University of Vienna, Rennweg 14 (A-1030) Vienna, Austria
[8] Department of Computational Landscape Ecology, UFZ – Helmholtz Centre for Environmental Research, Permoserstr. 15, 04318 Leipzig, Germany
[9] Division of Conservation Biology, Institute of Ecology and Evolution, University of Bern, 3013 Bern, Switzerland
[10] Bern University of Applied Sciences,School of Agricultural, Forest and Food Sciences, 3052 Zollikofen, Switzerland
[11] Agroecology, Department of Crop Sciences, University of Göttingen, Grisebachstraße 6, 37077 Göttingen, Germany
[12] Institute for Agro-ecology and Biodiversity (IFAB), Böcklinstr. 27, 68163 Mannheim, Germany
[13] Plant Ecology and Ecosystems Research, University of Göttingen, Untere Karspüle 2, 37073 Göttingen, Germany
[14] Dynafor, Université de Toulouse, INRA, INPT, INP-EI Purpan, Castanet Tolosan, France
[15] Department of Environmental Politics, UFZ – Helmholtz Centre for Environmental Research, Permoserstr. 15, 04318 Leipzig, Germany
[16] Institute of Social Ecology, Alpen-Adria University Klagenfurt, Schottenfeldgasse 29, 1070 Vienna, Austria

Keywords

Agriculture; biodiversity; Common Agricultural Policy; Ecological Focus Areas; farmers' choices; greening measures; policy implementation; policy simplification.

Correspondence

Guy Pe'er, Department of Conservation Biology, UFZ – Helmholtz Centre for Environmental Research, Permoserstr. 15 04318 Leipzig, Germany. E-mail: Guy.peer@ufz.de

Abstract

Ecological Focus Areas (EFAs) are one of the three new greening measures of the European Common Agricultural Policy (CAP). We used an interdisciplinary and European-scale approach to evaluate ecological effectiveness and farmers' perception of the different EFA options. We assessed potential benefits of EFA options for biodiversity using a survey among 88 ecologists from 17 European countries. We further analyzed data on EFA uptake at the EU level and in eight EU Member States, and reviewed socio-economic factors influencing farmers' decisions. We then identified possible ways to improve EFAs. Ecologists scored field margins, buffer strips, fallow land, and landscape features as most beneficial whereas farmers mostly implemented "catch crops and green cover," nitrogen-fixing crops, and fallow land. Based on the expert inputs and a review of the factors influencing farmers' decisions, we suggest that EFA implementation could be improved by (a) prioritizing EFA options that promote biodiversity (e.g., reducing the weight or even excluding ineffective options); (b) reducing administrative constraints; (c) setting stricter management requirements (e.g., limiting agrochemical use); and (d) offering further incentives for expanding options like landscape features and buffer strips. We finally propose further improvements at the next CAP reform, to improve ecological effectiveness and cost-effectiveness.

Introduction

Agricultural intensification and land abandonment exert major pressures on farmland biodiversity and diminish ecosystem functions and services. The ongoing decline in biodiversity in and around farmland is a source of major concern both in Europe (EEA 2015) and globally (Maxwell *et al.* 2016). In the European Union (EU), a key instrument that could help mitigate these trends is the Common Agricultural Policy (CAP). The CAP provides payments under two "Pillars": "direct payments and market-related expenditures" (Pillar 1, circa €37 Billion/yr) and "Rural Development" (Pillar 2, circa €14 Billion/yr).

Since its first implementation in 1962, the CAP has been repeatedly reformed to reflect changes in societal demands. In response to the increasing demand for biodiversity conservation (Hodge *et al.* 2015), the latest reform of 2015 introduced a "greening" of Pillar 1. Consequently, 30% of the payment are now linked to one or more of three new greening measures (EC 2013a, Article 43): (a) crop diversification, requiring farms with arable land exceeding 20 or 30 hectares to grow at least two or three crops, respectively; (b) maintenance of permanent pastures, allowing only for a maximum loss of 5% by 2020; and (c) promotion of Ecological Focus Areas (EFAs), requiring farms with arable areas exceeding 15 hectares to dedicate 5% of such areas to ecologically beneficial elements as defined by the European Commission (EC). Such elements include landscape features such as terraces, hedges, or ponds, but also fallow land, nitrogen-fixing crops, and "catch crops and green cover" (EC 2013a, 2014; see Table 1). As the ecological value and implementation costs of different EFA options vary, the EC introduced weighting factors (see Table 2): for example, one hectare of landscape features is counted as 1.5 ha, whereas the same area of nitrogen-fixing crops is counted as 0.3 hectare. Each Member State (MS) had the opportunity to select which of the ten EFA options defined by the EC are eligible for their national direct payments. In addition, MSs may support other "equivalent measures" that offer a similar or greater benefit for the environment, as long as they are approved by the EC prior to implementation. Each farmer may then choose which EFA options and/or national equivalent measures to implement (Oppermann 2015).

Challenges for biodiversity and farmers

The new greening measures have been criticized by ecologists and environmental organizations for setting requirements that are too low to halt the loss of farmland biodiversity (Pe'er *et al.* 2014a), and for not selecting the most effective measures for conserving biodiversity under current financial constraints (Dicks *et al.* 2014; Sutherland *et al.* 2015). In addition, the introduction of new greening measures resulted in increased administrative burdens for farmers and authorities, and therefore simplification of CAP's implementation will play an important role in its upcoming mid-term review, scheduled for March 2017 (European Council 2015).

Consequently, while the general public largely supports the CAP greening (EC 2016c), it is essential to assess whether its current design and implementation can yield significant positive impacts on biodiversity while being practicable for farmers. The recent release of reports by each MS on the implementation of CAP greening measures in 2015 provides a unique opportunity to do so.

Objectives of this article

This article examines how greening measures are currently designed and implemented, and how they could be improved to the benefit of both biodiversity and farmers. We focus here on Ecological Focus Areas (EFAs) because they represent a new element of the CAP whose effects on biodiversity are poorly documented. Furthermore, EFAs are likely to be subjected to reforms during the 2017 mid-term review, including the expansion of EFAs from 5% to 7% of arable land, and will likely remain part of the policy mix for the CAP beyond 2020. It is therefore critical and timely to assess the current design and implementation of EFAs and propose recommendations to improve their effectiveness for biodiversity while overcoming possible implementation barriers for farmers.

This article presents an interdisciplinary evaluation of the various EFA options, combining ecological experts' assessments on their potential effects on biodiversity with social scientists' review and evaluations of the factors influencing farmers' implementation decisions. To this end, we (1) conducted a European-scale survey among ecologists to assess potential biodiversity effects of EFA options; (2) collected data on farmers' uptake to examine on-the-ground EFA implementation; (3) synthesized expert opinions and a review of the factors influencing farmers' decisions; and (4) compared EFA options according to their impacts on biodiversity and their relevance for farmers to identify possible improvements of EFA design and implementation. Through this interdisciplinary and European-scale approach, we develop recommendations aiming to increase the uptake and best management of biodiversity-friendly options by farmers; reduce administrative burdens; and promote coherence between CAP and EU's nature conservation goals.

Methods

The conceptual framework of this study was developed over three interdisciplinary workshops between June and September 2015 (see Supporting Information [SI] 1). The spectrum of methods chosen included:

(1) **Ecologists' evaluation of EFA impacts on biodiversity**: We conducted a survey among

Table 1 Overview of the options defined by the EU as eligible for EFAs, alongside the weighting factor defined by the EC for a given area taken up for each option, and the number of MSs implementing each EFA option (= "Num. MSs")

Category	Description	Weighting factor	Num. MSs
(a) Fallow land	Land without any crop production or grazing, but maintained for production in the following years.	1.0	26
(b) Terraces	Terraces without use of pesticides.	1.0	8
(c) Landscape features	Elements subject to cross-compliance like hedgerows, single trees, rows or groups of trees, boundary ridge, ditches, other landscape elements.	See below	24
(d) Buffer strips	Strips without productive use alongside a watercourse adjacent to a field or within a field higher upon a slope.	1.5	17
(e) Agro-forestry	Land-use systems in which trees are grown in combination with agriculture on the same land with a maximum number of trees per hectare.	–	11
(f1) Strips along forest edges – with production	Strips of arable land adjacent to forest, with production but limited agrochemical inputs; with a width between 1 and 10 meters.	0.3	5
(f2) Strips along forest edges – without production	Strips on arable land adjacent to forest, without production; with a width between 1 and 10 meters.	1.5	9
(g) Short rotation coppice	Production of wood with specific, fast growing tree species.	0.3	20
(h) Afforested areas	Areas with afforestation on former arable land (in most cases supported by Pillar 2 measures).	1.0	14
(i) Catch crops, or green cover	Catch crops are a mixture of productive crops and/or grass following a productive crop to protect soils and use available nutrients during the winter.	0.3	19
(j) Nitrogen-fixing crops	A list of productive leguminous plants	0.3 (DE: 0.7)[a]	27
Specific landscape features:			
Hedges	Hedges or wooded strips (width up to 10 meters).	2.0	
Isolated trees	Isolated trees with a crown diameter of minimum 4 meters.	1.5	
Trees in lines	Trees in line with a crown diameter of minimum 4 meters. Space between crowns shall not exceed 5 meters.	2.0	
Trees in groups	Trees in group, where trees are connected by overlapping crown cover, and field copses of maximum 0.3 ha in both cases.	1.5	
Traditional stone walls	Wall with a length of minimum 5 meters that are not part of a terrace.	1.0	
Ditches	Ditches with a maximum width of 6 meters, including open watercourses for irrigation or drainage (excluding channels with concrete walls).	2.0	
Ponds	Ponds of up to 0.1 ha (excluding reservoirs made of concrete or plastic.	1.5	
Field margins	Field margins with a width between 1 and 20 meters, with no agricultural production.	1.5	

[a] In Germany, the weighting factor for nitrogen-fixing crops is 0.7, using the flexibility allowed by EC delegated regulation (EU) No 639/2014.
Source: EC 2014.

ecologists in the EU and Switzerland working on biodiversity in agro-ecosystems. As potential experts we considered persons who perform ecological research, monitoring or conservation management in agricultural landscapes or farmland areas. Familiarity with at least some of the features eligible for EFAs was required, while policy knowledge was not. Experts were identified as such by workshop partici-

pants or suggested by other respondents to our survey (i.e., a snowball approach). In total, invitations to complete the survey were sent to circa 310 experts, asking them to only fill out the survey if they felt they had sufficient expertise in the subject area. Respondents were asked to state their area of expertise (geographic, methodological, and taxonomic) and assess the impacts of EFA options for up to three

Table 2 Share of the different Ecological Focus Area (EFA) options taken up (a) at 8 MSs and the EU level, and, (b) in the different federal states of Germany (shares of area are in % and before applying weighting factors)

a) Member State

	Fallow land	Buffer strips	Landscape features	Catch crops and green cover	Nitrogen-fixing crops	Short rotation coppice	Afforested areas	% EFA (absolute)	% EFA (weighted)
				Share of Ecological Focus Area (in per cent)[f]					
Germany	16.2	1.2	2.4	68.0	11.8	0.2	0.1	11.5	5.8
Austria	19.1	n.a.[d]	0.03[c]	32.3	47.9	0.7	n.a.	–	–
Czech Republic	5.4	0.5	0.08	33.4	60.5	0.03	0.06	–	–
Denmark	7.6	6.0	0.3[b]	84.6	n.a.	1.5	n.a.	–	–
England	33.1	1.0	4.7[a]	5.5	55.8	n.a.	n.a.	–	–
Estonia	25.5	n.a.	2.0	n.a.	72.5	n.a.	n.a.	–	–
The Netherlands[g]	n.a.	1.8	n.a.	95.1	3.1	0.01	n.a.	–	–
Poland[h]	4.7	0[h]	0.4[h]	57.4	36.6	0.2	0.8	16.8	8.8
European Union (EU)[e]	**21.2**	**0.7**	**4.3**	**27.7**	**45.4**	**0.2**	**0.6**	**14.0**	**9.0**

b) German federal state

	Fallow land	Buffer strips	Landscape features	Catch crops and green cover	Nitrogen-fixing crops	Short rotation coppice	Afforested areas	% EFA (absolute)**	% EFA (weighted)**
				Share of Ecological Focus Area (in % before applying weighting factors)					
Baden-Württemberg	11.4	0.6	0.3	70.6	17.0	0.1	0.0	**12.1**	**5.6**
Bavaria	12.9	0.9	0.4	72.1	13.4	0.1	0.0	**11.5**	**5.2**
Brandenburg*	29.7	0.6	1.8	48.0	19.0	1.0	0.0	**11.0**	**6.7**
Hessen	28.6	1.2	0.5	60.8	8.9	0.0	0.0	**9.6**	**5.3**
Lower Saxony*	8.7	0.6	0.6	87.5	2.6	0.1	0.0	**15.4**	**5.8**
Mecklenburg-Vorpommern	29.2	3.2	3.3	57.0	5.9	0.0	1.4	**10.0**	**6.2**
North Rhine-Westphalia	6.8	1.6	1.2	87.1	3.4	0.1	0.0	**15.0**	**5.9**
Rhineland-Palatinate	33.0	0.9	1.1	55.4	9.6	0.1	0.0	**10.4**	**6.1**
Saarland	46.3	1.9	5.3	37.0	9.4	0.1	0.0	**6.0**	**4.5**
Saxony	13.9	1.2	1.0	64.1	19.5	0.1	0.3	**11.1**	**5.6**
Saxony-Anhalt	26.1	0.8	1.1	47.7	24.2	0.1	0.1	**10.0**	**6.0**
Schleswig-Holstein*	9.4	3.3	47.7	35.7	3.7	0.1	0.0	**5.9**	**6.0**
Thuringia	18.5	2.3	1.4	35.9	41.9	0.0	0.0	**9.0**	**5.8**
Germany	**16.2**	**1.2**	**2.4**	**68.0**	**11.8**	**0.2**	**0.1**	**11.5**	**5.8**

Source: Results for MSs were reported by national Ministries for Agriculture from October 2015 until February 2016. EU-wide results were reported by the EU Commission in June 2016.

[a] England: Hedges.

[b] Denmark: Ponds and archaeological sites between 0.01 and 0.2 ha.

[c] Austria: Ponds and ditches are offered as landscape features.

[d] "n.a.": measure not approved in that MS.

[e] Preliminary Data for all MSs except France.

[f] Share of area before applying weighting factors.

[g] The "collective approaches" to EFA and the "Skylark-Program" are not included in the figures.

[h] In Poland, the figures for some of the landscape features and buffer strips were only available in "linear meters," not in hectares. So for these two options, the presented EFA share is probably slightly underestimated.

Source: German Ministry for Food and Agriculture (BMEL) 2015b.

*The city-states Berlin (BE), Bremen (HB), and Hamburg (HH) were added to Brandenburg, Lower Saxony, and Schleswig-Holstein, respectively.

**Note that because of the weighting factors employed, the total area of EFAs cannot be interpreted based on the presented shares (in percentage). For raw numbers (in area) see Table S5 in SI 4.

"groups" defined by taxon and habitat affiliation (e.g., forest birds, grassland butterflies). Experts then scored the effects of each EFA option on each "group," in their view, from +5 (very positive) to -5 (very negative) or "mixed effects." Experts were also asked to identify conditions under which each EFA option could most benefit biodiversity (e.g., agricultural management, spatial design, vegetation composition, implementation duration, structural maintenance, and other). For further details on the methods and profile of the respondents, see SI 1. For an overview of the recommendations, see SI 2.

The answers of the experts were analyzed using descriptive statistics (mean, median, quantiles, and 95% CIs). We used the average score to define an ecological "win" (average > 1), "lose" (< -1), or "mixed" (-1 < average < 1). This assignment was verified against the three quartiles of score values, identifying the number of cases above, at, or under 0, as well as by inspecting the frequency distribution of scores. We refrained from conducting significance tests due to the high variance in score values (see Figure 1), related to regional and taxonomic differences which are beyond the scope of this study.

2 **Statistics on the implementation of EFAs in 2015**: We asked agricultural ministries across MSs for data on the number of farmers and total area registered under each EFA option in 2015. We considered that these uptake levels represent good indicators of farmers' preferences. We received data from eight MSs: Germany, Estonia, The Netherlands, Denmark, Austria, Czech Republic, England, and Poland as well as a preliminary data at the EU level based on all MSs except France (EC 2016b). We defined EFAs as "win" (+) and "lose" (-) based on high or low uptakes at the EU level, assuming that these indicate attractiveness from the farmers' perspectives. This assignment was confirmed through interviews with farmer representatives in a complementary study (Zinngrebe et al. submitted).

We obtained additional data on variations in farmers' uptake levels across German federal states, including (a) the total cover of each EFA option declared in 2015 compared to the total Utilized Agricultural Area (UAA), and (b) the total cover of nitrogen-fixing crops and fallow land over time since 2004.

3 **Expert knowledge on factors influencing farmers' uptake**: We collected inputs during the three workshops on the determinants that may influence farmer's EFA decisions. At the third workshop, we structured these inputs, and divided these determi-

nants into three categories: economic determinants, administrative conditions, and farm-level management. These categories were used to guide a literature review of >30 publications in English and German (both peer reviewed and gray literature), to aid interpreting the implementation statistics, and thereby, gaining a better understanding of farmer's preferences and constraints.

4 **Synthesis and collation of recommendations**: We compared EFA options according to both their impacts on biodiversity and their uptake by farmers, to identify different categories of EFA options ("win-win," "win-lose," etc.). Based on this simple categorization, combined with the expert opinions provided during the workshops and survey, we then developed recommendations on the ways to mitigate potential trade-offs and conflicts.

We note that ecologists participating in our surveys focused on above-ground biodiversity, and particularly farmland biodiversity. Less attention was also given to biodiversity in forested areas. We did not analyze geographical differentiation in scoring values, nor impacts of EFAs on ecosystem services, as these aspects were beyond the scope of this study.

Results

EFA impacts on biodiversity

We received 88 expert responses to our survey, from 16 MSs and Switzerland. The number of responses varied among EFA options, ranging from 67 for catch crops to 87 for landscape features. Taxonomic expertise included birds, plants, arthropods, mammals, amphibians, and reptiles. Scores were associated with high variance for some EFA options (e.g., afforested areas, agroforestry, or terraces), probably as a result of taxonomic and geographic differences, as well as differences on how envisioned practices were implemented in detail (Figure 1). Overall, buffer strips, fallow land, and landscape features received mostly positive scores, while agroforestry, afforestation, and short-rotation coppice received generally negative scores in terms of their perceived impact on biodiversity (Figure 1). Among landscape features, hedges, field margins, and traditional stone walls received the highest scores (Figure 1). All options apart from catch crops and short-rotation coppice were considered to have an overall positive effect on generalist species, whereas specialist species were considered to benefit primarily from fallow land, buffer strips, and landscape features (Figure 1C). Fallow land and buffer strips were considered to benefit farmland species, while landscape features and nitrogen-fixing crops had more variable scores (Figure 1D).

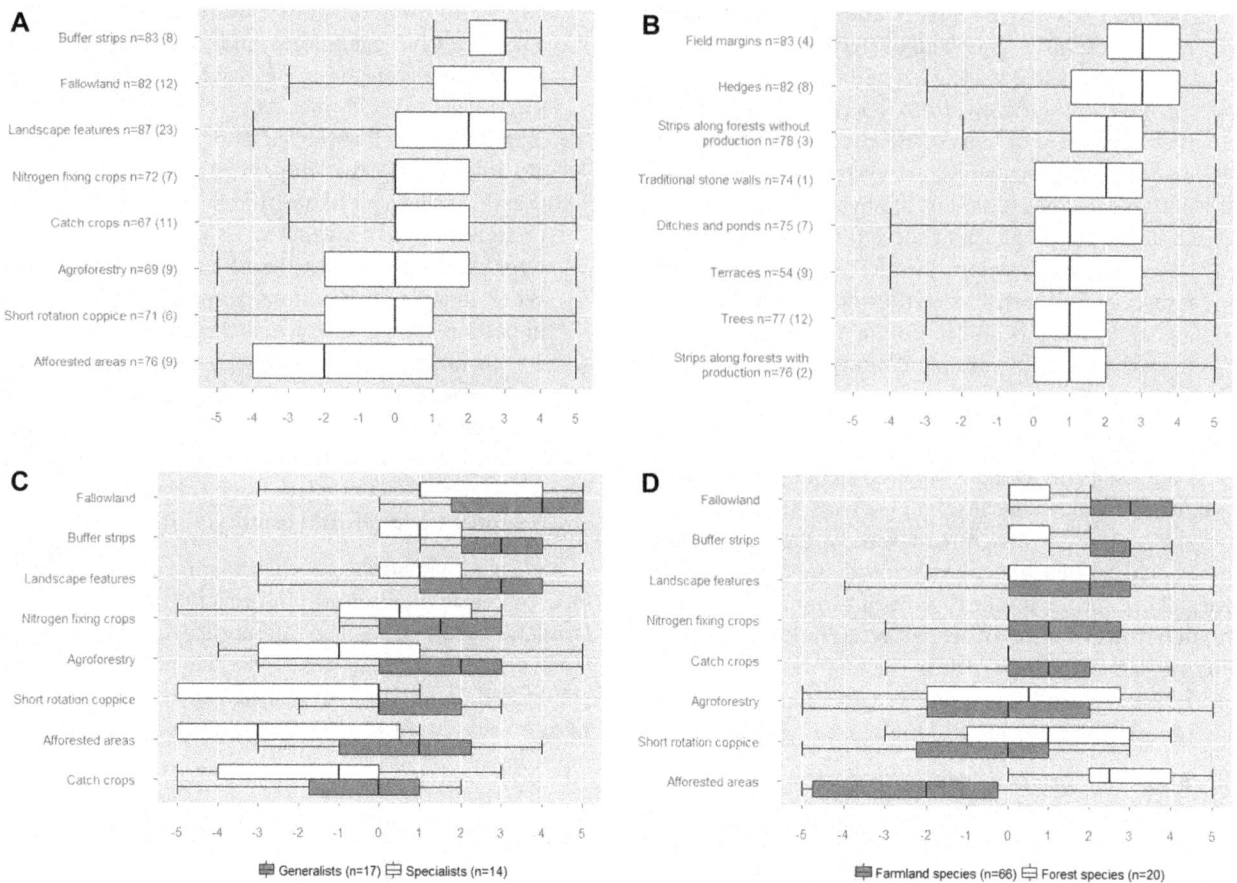

Figure 1 Outcome of EFA scoring by ecologists (A) for all EFAs, (B) for every type of landscape feature, (C) categorized based on specialization of assessed groups into generalists versus specialists, and (D) categorized based on habitat affiliation (forest versus farmland species). Box plots depict the medians, quantiles, and standard deviation. N is the number of experts assessing a given EFA or species' group. The number of experts reporting "mixed" effects is given in brackets for (A) and (B).

Despite an overall negative score, afforested areas were scored as benefitting forest species (Figure 1D).

EFA implementation in 2015

The number of EFA options eligible for implementation varied between MSs: 14 MSs approved >10 EFA options, 9 MSs approved between 5 and 9 options, and 5 MSs approved ≤ 4 options. Nitrogen-fixing crops, fallow land, and landscape features were taken up by the largest number of MSs (Table 1). Overall in the EU, farmers registered 16% of the arable land as EFAs, equivalent to 10% after applying weighting factors (Table 2). Three EFA options accounted for the vast majority of EFA cover: nitrogen-fixing crops, "catch crops and green cover," and fallow land. In most MSs assessed, landscape features had a very low uptake, as did buffer strips. Implementation levels varied both among and within MSs.

Nitrogen-fixing crops had a share of 46% across the EU but ranged in the assessed MSs from 3% (The Netherlands) to 73% (Estonia). Catch crops (27% at EU level) ranged from 6% to 95%, while fallow land (21%) ranged from 5% to 33% (Table 2). In Germany, "catch crops and green cover" represented the main EFA option (68% of total EFA), but uptake levels were particularly high in the federal states of Lower Saxony, North Rhine-Westphalia, Bavaria, and Baden-Württemberg (Table 2). Landscape features made up only 2.4% of the EFA area on average in Germany, but 48% in Schleswig-Holstein. The area of nitrogen-fixing crops and fallow land in Germany increased by 74% and 62%, respectively, between 2014 and 2015 (Figure 2), indicating a direct impact of EFA implementation. Nevertheless, the area of fallow land and nitrogen-fixing crops remains considerably smaller than during the obligatory set-aside policy of the CAP prior to 2008 and the support of leguminous plants in the 2000s

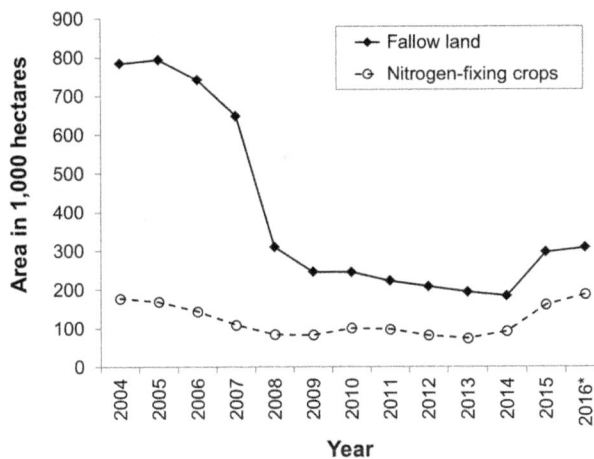

Figure 2 Cover of fallow land and nitrogen-fixing crops in Germany through the years 2004–2016. Source: Destatis 2005–2016. * Note that the data for 2016 are preliminary.

(Figure 2). Also, it is estimated that only 15.1–41.5% of the existing landscape features and buffer strips in Germany were registered as EFAs in 2015 (Isermeyer et al. 2014, p. 15).

Determinants of uptake among farmers

Our literature review identified multiple theories and approaches which can be used to explain farmers' decision-making (Ajzen 1991; Van der Ploeg 1994; de Snoo et al. 2013; Home et al. 2014). Following Lange et al. (2015), we clustered determinants into three key categories: economic determinants, administrative restrictions, and farmers' perceptions and knowledge (see also Table 3).

(a) Economic determinants

Economic considerations, including minimizing the production, opportunity, and/or transaction costs, are central for farmers when evaluating policies. Farmers perceive greening restrictions as costly (Schulz et al. 2014) and therefore tend to choose the most productive and cheapest options (Lakner & Holst 2015). Consequently, 73.1% of the total EFA-area in the EU is covered by "productive options" (EC 2016b; see also Table 2). Economic considerations also tend to favor existing features (e.g., the high registration of landscape features in the federal state of Schleswig-Holstein, Table 2), as well as practices that are easy to implement. This is reflected by the large proportion of "catch crops and green cover" in some MSs (Table 2). Land productivity, expressed in costs of land rent, affects EFA choices too. For instance, higher uptake of fallow land in Germany occurred in

federal states with lower land-rental prices and usually lower productivity, indicating low opportunity costs (see **Figure S1** in SI 3). Production costs and opportunity costs for land use, but also transaction costs and property rights, limit the establishment of new landscape elements such as hedges (Schleyer & Plieninger 2011). Finally, for economic reasons the expectation is that most farmers will not abandon the system of direct payments, since the implementation costs are clearly below the received greening payments (Heinrich 2012; de Witte & Latacz-Lohmann 2014; Schmidt et al. 2014; Lakner & Bosse 2016; Lakner et al. 2016).

b) Administrative restrictions

The implementation of EFAs by farmers is controlled by authorities, and might lead to sanctions if rules are broken. For example, there are strict minimum and maximum widths allowed for buffer strips in Germany (BMEL 2015a). This creates a relatively high risk of repayment or fines due to inaccuracies in the measurement of strip size or width by either farmer or authorities. Thus, although this option is generally economically viable (de Witte & Latacz-Lohmann 2014; Schmidt et al. 2014; Lakner & Bosse 2016; Lakner et al. 2016), the legal restrictions translate into risk-related costs or transaction costs, which partly explain the low uptake of buffer strips and landscape features.

The low uptake of landscape features can also be explained by property rights. Often, ownership and hence the right to register a feature is unclear, or there can be multiple owners (e.g., Schleyer & Plieninger 2011; Hauck et al. 2016). The exceptionally high uptake of landscape elements in Schleswig-Holstein may be explained not only by their unusual prevalence, but also by clearly defined land-property rights compared to other German regions. Landscape features in Germany and other countries are also subject to overlapping regulation, such that in many cases landscape features require safeguarding under the cross compliance (CC) rules of Pillar 1 (EC 2009, article 6; 2016a) or the EU's Habitats Directive. This increases the control risks for farmers, making them reluctant to register, restore, or establish new landscape features.

c) Farmers' perceptions and knowledge

Traditional land-use and established farming practices influence EFA decisions. For example, hedges in Schleswig-Holstein ("Knicks") are part of the traditional farming system and landscape going back into the 18th century (Beyer & Schleuß 1991). Their protection is consequently both pragmatic (as boundary structures) and culturally beneficial (Piorr & Reutter 2002). **Existing practices**, farm structures, available technologies,

Table 3 **Synthesis of expert knowledge on factors influencing farmers' uptake,** with examples for implications and outcomes of these factors

	Determinants	Implications and examples
Economic	Productivity	Higher uptake of productive EFA options (e.g., catch crops)
	Opportunity cost	Higher uptake of fallow land where rental prices are lower
Admin	Repayment risk	Width of buffer strips hard to measure → low uptake
	Property rights	Unclear ownership of landscape features → low uptake
Perception	Traditions	High coverage of hedges in Schleswig-Holstein
	Experience	Higher uptake of nitrogen-fixing crops in Eastern Germany
	Personal attitude	Strong variations in EFA preferences between farmers

and established management experiences can influence farmers' EFA choices too (Schulz et al. 2014). For instance, growing nitrogen-fixing crops such as beans and peas requires specific knowledge and particular harvesting equipment. This translates into a higher uptake of these EFA options in Eastern Germany where the climate is suitable for these crops and farmers have more experience and better equipment for implementation.

The perception of "productivity" in land management is another important criterion for EFA decisions, going beyond pure economic reasoning, since many farmers maintain a **self-perception** as "producers" (Burton et al. 2008; de Snoo et al. 2013; Home et al. 2014), whose primary role is to ensure the provision of food for society rather than protecting biodiversity. Accordingly, EFA decisions show a clear preference for "productive" EFA options.

Literature concerning the uptake of Agri-Environmental Measures (AEM) and other programs aiming to enhance farmland biodiversity suggests that their uptake, as well as the sense of ownership, are influenced by personal attitudes, subjective norms, and social interaction and control (Burton & Wilson 2006; Defrancesco et al. 2006; Burton et al. 2008; Ahnström et al. 2009; Ahnström et al. 2013; Home et al. 2014; Sulemana & James 2014). For example, Lokhorst et al. (2014) found that a stronger ownership of conservation activities among farmers related to personal connectedness to nature and a self-perception as conservationists. Additionally, social pressure can influence on decision-making since farmers have an interest in maintaining their fields in a productive and "tidy" status as perceived by their peers and neighbors (Hauck et al. 2016).

Comparison of EFA options and recommendations for future improvements

We identified one EFA option that is a "win-win" for farmland biodiversity and farmers (fallow land), two "win-lose" options, beneficial for farmland biodiversity but unattractive for farmers (buffer-strips and landscape features), two "mixed-win" options with limited or unclear benefits for biodiversity but favored by farmers (nitrogen-fixing crops and "catch crops and green cover"), two "mixed-lose" options (agroforestry and short-rotation coppice), and one "lose-lose" option (afforested areas) from a perspective of farmland biodiversity (Table 4). We did not find "lose-win" options (receiving strongly negative scores by ecologists but favored by farmers).

We obtained a total of 895 recommendations to improve the effectiveness of EFA (for a full overview see SI 2). Options with more positive scores received also a greater number of recommendations ($n = 70–82$ for buffer strips, fallow land, and field margins), while EFA options with lower scores received fewer recommendations ($n = 31–42$ for short-rotation coppice, afforested areas, and agroforestry). Most recommendations were related to the category "agricultural management" ($n = 286$), with a general call for setting management specifications ($n = 107$), limiting the use of agrochemicals ($n = 72$), and defining specific harvesting and mowing regimes ($n = 52$). Within the category "spatial design" ($n = 264$), the most frequent recommendations were a general call for design and location properties ($n = 88$), the importance of defining a size or area ($n = 63$), and the potential benefits of combining buffer strips or landscape elements with other EFA options ($n = 42$). For the category "vegetation structure and composition" ($n = 262$), the majority of recommendations called for plant composition ($n = 219$), specifically mentioning the importance of plant diversity ($n = 69$), native plants ($n = 47$), and the support of flowering species ($n = 38$).

The need to consider the duration of EFA implementation was also mentioned, in particular for fallow land ($n = 11$).

Discussion

Our study indicates a mismatch between EFA design and implementation, where most EFA options that were

Table 4 Comparison of EFA options according to their score by ecologists (Figure 1) compared to their uptake by farmers (Table 2) as a measure of attractiveness. We defined an ecological win (+), lose (-) or mixed effects (±) based on the average and 3 quartiles (25, 50, 75%) of their score; while defining win or lose for farmers' implementation based on the share of the different EFA in the EU

	Ecologists scores		Farmers uptake		Biodiversity vs. Farmers perspective
	Average score		EFA in the EU [%]		
Fallow land	2.4	+	20.8	+	**win - win**
Buffer strips	2.5	+	0.6	−	**win - lose**
Landscape features	1.6	+	4.3	−	**win - lose**
Nitrogen-fixing Crops	0.7	±	46.2	+	**mixed - win**
Catch crops & green cover	0.4	±	27.2	+	**mixed - win**
Agroforestry	-0.1	−	n.a.	−	**mixed - lose**
Short-rotation coppice	-0.4	−	0.8	−	**mixed - lose**
Afforested areas	-1.4	−	n.a.	−	**lose - lose**

considered beneficial to biodiversity had low uptake among farmers. Moreover, we observed that the proportion of EFA surfaces registered were higher than the 5% currently required, with an EU-average of > 10%. Consequently, our study suggests that increasing the required EFA surface from 5% to 7%, as currently discussed in the context of the mid-term review, is unlikely to yield significant improvements in terms of EFAs' contribution to biodiversity conservation. Instead, efforts should rather focus on improving EFA option design and implementation, considering biodiversity, the determinants of farmers' decisions, and current obstacles to EFA implementation.

Improving EFA implementation in the current framework: recommendations for the 2017 mid-term review

The upcoming mid-term review of the CAP, scheduled for 2017, can be used to address implementation issues within the current legal framework. Accordingly, we first provide five major recommendations for improving EFA effectiveness for biodiversity within the current regulatory framework.

(1) Prioritize EFA options with clear benefits for biodiversity and reduce incentives for less effective ones. This could be achieved by

- **Ensuring that MSs approve effective EFAs**: Some EFA options that were scored highly by ecologists were not approved by all MSs (EC 2015a, p. 13). The MSs that did not include fallow land (the only "win-win" option), buffer strips, and especially the four MSs that did not approve landscape features, should be encouraged to revise their decisions.

- **Reconsidering ineffective EFA options**: Five EFA options were found to have mixed or even negative effects on farmland biodiversity. While they may support other environmental objectives beyond biodiversity (such as soil retention or carbon sequestration), they need to be employed under careful management criteria, limited to a maximum area at the regional or national level (see, e.g., The Netherlands with 95% catch crops), or even considered for removal to reduce competition with more effective EFA options. This may also help simplifying implementation. We also note that short-rotation coppice, agro-forestry, and afforestation are already promoted through other policies including AEMs.

- **Expanding the use of equivalent measures**: Equivalent practices have been developed by only five MSs to adapt greening to specific farming systems and environmental priorities, and none has taken a regional approach (Hart 2015). Implementation barriers could be reduced through local governance fora, integrating regional knowledge, stakeholder interests, and scientific expertise (Dosch & Schleyer 2005).

- **Adapting weighting factors** to the ecological value and implementation costs and benefits of each EFA option: Particularly, the weighting factors for catch crops and green cover and for nitrogen-fixing crops should be reduced as these options are easy to implement with little or no costs, while having unclear or small benefits for biodiversity.

(2) Reduce farmers' administrative burdens
Reducing administrative burdens could enhance the uptake of EFA options that are otherwise avoided,

such as buffer strips and landscape features ("win-lose"). This could be achieved by:

- **Simplifying technical requirements and relaxing sanctions**: Reducing and simplifying some technical requirements of biodiversity supporting EFAs, such as buffer strip widths and landscape feature area calculations, could significantly incentivize their uptake. Furthermore, sanctions can be relaxed if they emerge from innocent errors, such as mapping mistakes or area miscalculations that may not impede EFA aims.

- **Increasing capacities for administrative support and ecological advice**: Stronger support provided by authorities would contribute to increasing farmers' confidence in effective EFA options. Particularly, funds are now available for Farm Advisory Systems (FAS) in Pillar 1, but they should be further developed in terms of their capacity to provide ecological knowledge and to communicate the benefits of some EFAs in terms of ecosystem services like erosion control, water quality, pollination, and pest control.

- **Extending the eligible implementation duration of selected EFA options**: Under the current legislation, the status of buffer strips and fallow land (if continuously covered by grass-dominated vegetation) may change if implemented for more than 5 years, turning from arable to "permanent grassland." This change of status lowers land prices substantially and is legally difficult to reverse. To avoid these negative consequences, farmers usually convert buffer strips and fallow land back to arable land after 5 years, with potential loss of various benefits for biodiversity. Excluding EFA areas from this rule could enhance implementation duration, promote habitat stability over time, and improve the potential of these EFAs to contribute to landscape connectivity (SI 2; see also Henderson *et al.* 2000).

(3) Set targeted and clear management requirements
Ecological experts provided a wide range of specific recommendations to improve the effectiveness of EFA options (for an overview, see SI 2). Notable recommendations are to **restrict the use of agro-chemicals**, and ensure a **high diversity of (eligible) plant species** while particularly supporting flowering plants.

(4) Combine policy instruments
The protection, restoration, or creation of landscape features and buffer strips can be promoted by **offering top-up payments**, for example, through AEMs. Some German federal states already use additional EFA-top-up payments with AEMs (see Lakner *et al.* 2016), albeit still with differing impacts.

(5) Improve transparency of the implementation process

The European Commission so far has not published any full or comprehensible dataset on EFA-uptake at the MS level (for example, we used data primarily from cooperating ministries). Greater transparency, by publishing the implementation data yearly, can promote learning and improvements, as well as cooperation among stakeholders.

A vision beyond 2020

Our workshop discussions brought up five major points which should be considered in the next CAP reform in order to improve both ecological effectiveness and cost-effectiveness (see also **Table S4**).

(1) Revise exemptions
Currently, farms with less than 15 ha arable land or farms with permanent crops are exempt from EFAs. This exemption should be revisited to improve the ecological effectiveness of EFAs (see **Table S4**).

(2) Reduce windfall-gains to improve cost-effectiveness
Currently, 54% of EU farmers face minimal or no additional costs by these measures (EC 2011, pp. 9, 17), thus functioning as so-called "windfall-gains," i.e., payments for which no additional effort is taken toward the provision of a related service (in this case, biodiversity conservation). In some cases the premium level is well above the real production costs (or costs incurred), especially for catch crops (de Witte & Latacz-Lohmann 2014; Lakner & Holst 2015; Lakner & Bosse 2016). Consequently, there is much room for better differentiating payments based on actual costs and benefits.

(3) Regionalize EFA design and implementation
Our expert survey highlighted an important need to adapt to local EFA settings from both ecological and socio-economic perspectives, by accounting for geographical, societal, and socio-economic specificities, as well as administrative scales (Lehmann *et al.* 2009; Prager 2015b). Recommendations included the following points:

- **Adopt EFA requirements to socio-economic and ecological conditions**: Cost-effectiveness is reduced by ignoring the very marginal production and opportunity costs not only among EFA options but also, for each option, in terms of the immense heterogeneity among farms and farmers across the EU alongside their diversity of interests, motivations, and attitudes toward biodiversity protection (see, e.g., Schmitzberger et al. 2005; Wätzold & Drechsler 2014).

- **Support landscape-targeted and collaborative implementation**: Experiences from AEMs demonstrated that landscape-targeted AEMs are more effective (Tscharntke et al. 2005; Wätzold & Schwerdtner 2005; Wrbka et al. 2008; Merckx et al. 2009; Batáry et al. 2011; Prager 2015a; Tscharntke et al. 2015). Similar spatial targeting should be applied to EFAs. This could be facilitated by local and regional authorities, and supported through Farm Advisory Systems and local governance fora.

- **Reconsider EFA requirement levels at the regional level**: The current option for MSs to reduce the requirement for EFA from 5% to 2.5% in regions with high proportion of forests or protected areas entails that less strict requirements are set for areas where (semi-)natural elements are more abundant and hence, at least from a landscape-ecology perspective, EFAs could be particularly beneficial for biodiversity, for instance, by improving landscape permeability and connectivity.

(4) Enhance cooperation between administration, extension services, and farmers

Cooperation between stakeholders can act to increase uptake of biodiversity-friendly EFA options and improve their practical implementation. It could be achieved by

- **Adapting administrative structures to enable collective implementation**: Collective implementation might provide a range of advantages as farmers can jointly achieve more ambitious targets across spatial and temporal scales and reach critical ecological thresholds such as habitat size (Pe'er et al. 2014b) and connectivity. Yet only two MSs (The Netherlands and Poland) allowed farmers to implement EFAs collectively, providing incentives for contiguous EFAs (Hart 2015); and in 2015, only 45 farmers used this option (EC 2016b, Part 3/6, Annex II, p. 43). These examples demonstrate an underused potential to improve the environmental benefits of EFAs, as well as a need to assess whether collective implementation can indeed achieve its desired impacts.

- **Promoting integrative and participatory approaches**: By supporting integrative and participatory conservation planning, platforms for knowledge exchange and local governance fora, the EU and MSs can promote synergies between sectoral interests, cooperation among stakehold-

ers, and more (cost-)effective use of knowledge (Dosch & Schleyer 2005). Furthermore, bottom-up initiatives can help achieve desired ecological and socio-economic goals that may otherwise seem administratively unfeasible (Prager 2015a,b).

(5) Enhance policy integration

Policy integration (sensu Runhaar et al. 2014) of all three greening measures will need to be carefully inspected in the next negotiations of the CAP regarding the EC's commitment to Policy Coherence for Development (PCD, EC 2015b), as well as CBD Aichi target 3 which requires signatory bodies to eliminate incentives harmful to biodiversity. Particularly, we recommend to:

- **Integrate the CAP with existing policies for farmland biodiversity conservation**: Recent assessments demonstrate limited coherence of the CAP with the Habitats and Birds Directives (Milieu et al. 2015). We further recommend inspecting its coherence with the EU's Green Infrastructure Strategy (European Parliament 2013).

- **Ensure that none of the greening measures supports biodiversity deterioration**: The permanent-grassland greening measure focuses on quantity while lacking quality requirements. The crop-diversification measure allows reducing the number of crops without meeting the threshold requirement of two or three crops (Pe'er et al. 2014a). Moreover, closer monitoring should be employed to ensure that (semi-)natural habitats do not deteriorate under the greening measures.

- **Use experiences from AEM implementation**: Overall, there is much room to use the experiences acquired in the application of AEMs within Pillar 2. Voluntary financial incentives, such as those provided by AEMs, often work better than regulatory mechanisms (Henle et al. 2008) and can be linked to result-based mechanisms that have been tested with some success (de Snoo et al. 2013). Overall, given the experiences and tools already existing in Pillar 2 AEMs, and considering the budget decline for Pillar 2 in the last CAP reform, the next reform should focus on restoring and even expanding Pillar 2, and within it the share of budgets earmarked for protecting biodiversity.

Acknowledgments

We are grateful to the more than 100 ecologists, agronomists, and other experts who participated in our workshops and responded to our survey. This project developed through the support of "Synthesis projects" at the UFZ. Guy Pe'er also acknowledges financial support from the FP7 project EU BON. Clélia Sirami acknowledges funding by the FarmLand project, funded by the ERA-Net BiodivERsA under the French National Research Agency (ANR-11-EBID-0004), the German BMBF & DFG, and the Spanish Ministry of Economy and Competitiveness. Stefan Schindler was partly supported by the grant FPA EEA/NSV/14/001_ETC/ULS. We thank Tibor Hartel and Amanda Sahrbacher for constructive comments on this manuscript.

Supporting Information

Table S1: Respondents' profile according to (a**)** country, taxonomic group of expertise, years of experience and (b) source of expertise.

Table S2: Type of recommendations and the number of experts making them, accompanied by an explanation of the content and/or the recommendation itself.

Table S3: Type of recommendations, and number of experts making them, (**a**) for each EFA option and (**b**) for each type of landscape feature.

Table S4: Synthesis of recommendations provided, by participants at the Round Table Discussion at the ICCB-ECCB conference in Montpellier.

Table S5: Area and share of registered ecological focus area across federal states in Germany, weighted according the legal weighting factors (in hectares).

References

Ahnström, J., Bengtsson, J., Berg, Å., Hallgren, L., Boonstra, W.J. & Björklund, J. (2013). Farmers' interest in nature and its relation to biodiversity in arable fields. *Int. J. Ecol.* 2013, Article ID 617352.

Ahnström, J., Höckert, J., Bergeå, H.L., Francis, C.A., Skelton, P. & Hallgren, L. (2009). Farmers and nature conservation: what is known about attitudes, context factors and actions affecting conservation? *Renew. Agric. Food Syst.*, **24**, 38.

Ajzen, I. (1991). The theory of planned behavior. *Organ. Behav. Hum. Decis. Process.*, **50**, 179-211.

Batáry, P., Andras, B., Kleijn, D. & Tscharntke, T. (2011). Landscape-moderated biodiversity effects of agri-environmental management: a meta-analysis. *Proc. Biol. Sci.*, **278**, 1894-1902.

Beyer, L. & Schleuß, U. (1991). The soils of wall-hedges in Schleswig-Holstein – classification and genesis (in German). *Z. Pflanzenernährung Bodenkunden*, **154**, 431-436.

BMEL. (2015a). *Implementation of the agricultural reform in Germany 2015 (Government Brochure, in German)*. German Ministry for Food and Agriculture (BMEL), Berlin, Germany.

BMEL (2015b). Reply of the German Ministry for Food and Agriculture to the German Parliament to a formal request of Parliament Member Dr. K. Tackmann, October 2015, Bundestags-Document No. 18/6529. German Ministry for Food and Agriculture (BMEL), Berlin, Germany.

Burton, R.J.F., Kuczera, C. & Schwarz, G. (2008). Exploring farmers' cultural resistance to voluntary agri-environmental schemes. *Sociologia Ruralis*, **48**, 16-37.

Burton, R.J.F. & Wilson, G.A. (2006). Injecting social psychology theory into conceptualisations of agricultural agency: towards a post-productivist farmer self-identity? *Journal of Rural Studies*, **22**, 95-115.

de Snoo, G.R., Herzon, I., Staats, H., *et al.* (2013). Toward effective nature conservation on farmland: making farmers matter. *Conservation Letters*, **6**, 66-72.

de Witte, T. & Latacz-Lohmann, U. (2014). Was kostet das Greening? *Top Agrar*, **4**, 36-41.

Defrancesco, E., Gatto, P., Runge, F. & Trestini, S. (2006). Factors affecting farmers' participation in agri-environmental measures: evidence from a case study. *10th Joint Conference on Food, Agriculture and the Environment.* Duluth, Minnesota.

Destatis (div. years): Arable production in Germany (in German), Series 3.1.2, Federal Statistical Office (Destatis), Wiesbaden, Germany, Volumes 2005, 2007-2016.

Dicks, L.V., Hodge, I., Randall, N., *et al.* (2014). A transparent process for 'evidence-informed' policy making. *Conserv. Lett.*, **7**, 119-125.

Dosch, A. & Schleyer, C. (2005). Transdisciplinary approaches in natural resource management: the case of an agri-environmental forum in Brandenburg (Germany). *Zeitschrift für angewandte Umweltforschung (ZAU)*, **17**, 65-79.

EEA. (2015). State of nature in the EU. *Results from reporting under the nature directives 2007–2012*, Technical report No 02/2015. p. 178 pp. European Environment Agency, Copenhagen.

EC (European Commission). (2009). *Council regulation (EC) No. 73/2009 of 19 January 2009 establishing common rules for direct support schemes for farmers under the common agricultural policy and establishing certain support schemes for farmers.* European Union, Brussels.

EC (European Commission). (2011). *Impact assessment: Common Agricultural Policy towards 2020 ANNEX 2D*, Commission Staff Working Paper, 20.10.2011 SEC (2011) 1153 final/2. European Union, Brussels, Belgium.

EC (European Commission). (2013a). *CAP reform - an explanation of the main elements.* European Commission, Brussels, Belgium.

EC (European Commission). (2013b). *Regulation (EU) No 1307/2013 of the European Parliament and of the Council of 17*

December 2013 establishing rules for direct payments to farmers under support schemes within the framework of the common agricultural policy and repealing Council Regulation (EC) No 637/2008 and Council Regulation (EC) No 73/2009. European Commission, Brussels, Belgium.

EC (European Commission). (2014). *Commission Delegated Regulation (EU) No 639/2014 of 11 March 2014, supplementing Regulation (EU) No 1307/2013 of the European Parliament and of the Council establishing rules for direct payments to farmers under support schemes within the framework of the common agricultural policy and amending Annex X to that Regulation.* European Union, Brussels, Belgium.

EC (European Commission). (2015a). *Direct payments post 2014 Decisions taken by Member States by 1 August 2014 - State of play on 07.05.2015* - Information note. European Commission, Brussels, Belgium.

EC (European Commission). (2015b). *Commision staff working document: Policy Coherence for Development.* 2015 EU Report, SWD(2015) 159 final. European Commission, Brussels, Belgium.

EC (European Commission). (2016a). *Glossary of terms related to the Common Agricultural Policy.* European Union, Brussels, Belgium.

EC (European Commission). (2016b). *Review of Greening after one year, Commission staff working document from June 22, 2016, SWD (2016) 218 final.* European Commission, Brussels, Belgium.

EC (European Commission). (2016c). Special Eurobarometer 440: Europeans, Agriculture and the CAP. *Survey requested by the European Commission,* Directorate-General for Agriculture and Rural Development and co-ordinated by the Directorate-General for Communication, Brussels, Belgium.

European Council. (2015). *Draft Council conclusions on the Simplification of the Common Agriculture Policy, Document 7524/2/15 REV 2.* European Council, Brussels, Belgium.

Hart, K. (2015). *Green direct payments: implementation choices of nine Member States and their environmental implications.* IEEP, London, UK.

Hauck, J., Schmidt, J. & Werner, A. (2016). Using social network analysis to identify key stakeholders in agricultural biodiversity governance and related land-use decisions at regional and local level. *Ecology and Society,* 21(2):49, http://dx.doi.org/10.5751/ES-08596-210249. Accessed November 1, 2016.

Heinrich, B. (2012). Calculating the 'Greening'-effect – a case study approach to predict the gross margin losses in different farm types in Germany due to the reform of the CAP. *Discussion paper 1205 of the Department of Agricultural Economics and Rural Development.* Georg-August-University Göttingen, Göttingen, Germany.

Henderson, I.G., Cooper, J., Fuller, R.J. & Vickery, J. (2000). The relative abundance of birds on set-aside and neighbouring fields in summer. *J. Appl. Ecol.,* **37,** 335-347.

Henle, K., Alard, D., Clitherow, J. *et al.* (2008). Identifying and managing the conflicts between agriculture and biodiversity conservation in Europe - A review. *Agric. Ecosyst. Environ.,* **124,** 60-71.

Hodge, I., Hauck, J. & Bonn, A. (2015). The alignment of agricultural and nature conservation policies in the European Union. *Conserv. Biol.,* **29,** 996-1005.

Home, R., Balmer, O., Jahrl, I., Stolze, M. & Pfiffner, L. (2014) Motivations for implementation of ecological compensation areas on Swiss lowland farms. *J. Rural Stud.,* **34,** 26-36.

Isermeyer, F., Forstner, B., Nieberg, H. *et al.* (2014). Stellungnahme im Rahmen einer öffentlichen Anhörung des Ausschusses für Ernährung und Landwirtschaft des Deutschen Bundestages am 7. April 2014.

Lakner, S. & Bosse, A. (2016). Mühsames Abwägen (Zur ökologische Vorrangfläche in Sachsen-Anhalt in German). *Bauernzeitung,* **10,** 50-51.

Lakner, S. & Holst, C. (2015). Farm implementation of greening requirement: economic determinants (in German). *Natur und Landschaft,* **90,** 271-277.

Lakner, S., Schmitt, J., Schuler, S. & Zinngrebe, Y. (2016). Naturschutzpolitik in der Landwirtschaft: Erfahrungen aus er Umsetzung von Greening und der ökologischen Vorrangfläche 2015. *Conference of the German Association of Agricultural Economists (Gewisola),* Bonn.

Lange, A., Siebert, R. & Barkmann, T. (2015). Sustainability in land management: an analysis of stakeholder perceptions in rural northern Germany. *Sustainability,* **7,** 683-704.

Lehmann, P., Schleyer, C., Wätzold, F. & Wüstemann, H. (2009). Promoting multifunctionality of agriculture: an economic analysis of new approaches in Germany. *Journal of Environmental Policy & Planning,* **11,** 315-332.

Lokhorst, A.M., Hoon, C., le Rutte, R. & de Snoo, G. (2014). There is an I in nature: the crucial role of the self in nature conservation. *Land Use Policy,* **39,** 121-126.

Maxwell, S., Fuller, R., Brooks, T. & Watson, J. (2016). Biodiversity: the ravages of guns, nets and bulldozers. *Nature,* **536,** 143.

Merckx, T., Feber, R.E., Riordan, P., *et al.* (2009). Optimizing the biodiversity gain from agri-environment schemes. *Agriculture Ecosystems & Environment,* **130,** 177-182.

Milieu, L., IEEP, ICF International & Ecosystems, L. (2015). *Evaluation Study to support the Fitness Check of the Birds and Habitats Directives.* Draft - Emerging Findings. p. 68. Milieu Ltd., Brussels, Belgium.

Oppermann, R. (2015). Ökologische Vorrangflächen – Optionen der praktischen Umsetzung aus Sicht von Biodiversität und Landwirtschaft. *Natur und Landschaft,* **90,** 263-270.

Pe'er, G., Dicks, L.V., Visconti, P., *et al.* (2014a). EU agricultural reform fails on biodiversity. *Science,* **344,** 1090-1092.

Pe'er, G., Tsianou, M.A., Franz, K.W. *et al.* (2014b). Toward better application of minimum area requirements in conservation planning. *Biological Conservation*, **170**, 92-102.

Piorr, H.-P. & Reutter, M. (2002). *Linear Landscape Elements as Agricultural Environmental Indicators (in German)*. Fachhochschule Eberswalde, Eberswalde, Germany.

Prager, K. (2015a). Agri-environmental collaboratives as bridging organisations in landscape management. *J. Environ. Manage.*, **161**, 375-384.

Prager, K. (2015b). Agri-environmental collaboratives for landscape management in Europe. *Curr. Opin. Environ. Sustainability*, **12**, 59-66.

Runhaar, H., P. Driessen, and C. Uittenbroek. 2014. Towards a systematic framework for the analysis of environmental policy integration. *Environmental Policy and Governance* **24**, 233-246.

Schleyer, C. & Plieninger, T. (2011). Obstacles and options for the design and implementation of payment schemes for ecosystem services provided through farm trees in Saxony, Germany. *Environ. Conserv.*, **38**, 454-463.

Schmidt, T., Röder, N., Dauber, J., *et al.* (2014) *Biodiversitätsrelevante Regelungen zur nationalen Umsetzung des Greenings der Gemeinsamen Agrarpolitik der EU nach 2013*. Thünen Istitute for Agricultural Research, Working, Braunschweig, Germany.

Schmitzberger, I., Wrbka, T., Steurer, B., Aschenbrenner, G., Peterseil, J. & Zechmeister, H.G. (2005). How farming styles influence biodiversity maintenance in Austrian agricultural landscapes. *Agric. Ecosyst. Environ.*, **108**, 274-290.

Schulz, N., Breustedt, G. & Latacz-Lohmann, U. (2014). Assessing farmers' willingness to accept "greening": insights from a discrete choice experiment in Germany. *J. Agric. Econ.*, **65**, 26-48.

Sulemana, I. & James, H.S. (2014). Farmer identity, ethical attitudes and environmental practices. *Ecol. Econ.*, **98**, 49-61.

Sutherland, W.J., Dicks, L.V., Ockendon, N. & Smith, R.K., editors. (2015). *What works in conservation*. Open Book Publishers, Cambridge, UK.

Tscharntke, T., Klein, A.M., Kruess, A., Steffan-Dewenter, I. & Thies, C. (2005). Landscape perspectives on agricultural intensification and biodiversity - ecosystem service management. *Ecol. Lett.*, **8**, 857-874.

Tscharntke, T., Milder, J.C., Schroth, G., *et al.* (2015). Conserving biodiversity through certification of tropical agroforestry crops at local and landscape scales. *Conserv. Lett.*, **8**, 14-23.

Van der Ploeg, J.D. (1994). Styles of farming: an introductory note on concepts and methodology. Pages 7-30 in J.D. van der Ploeg & A. Long, editors. *Born from within: practice and perspectives of endogenous rural development*. Van Gorcum, Assen, the Netherlands.

Wätzold, F. & Drechsler, M. (2014). Agglomeration payment, agglomeration bonus or homogeneous payment? *Resour. Energy Econ.*, **37**, 85-101.

Wätzold, F. & Schwerdtner, K. (2005). Why be wasteful when preserving a valuable resource? A review article on the cost-effectiveness of European biodiversity conservation policy. *Biol. Conserv.*, **123**, 327-338.

Wrbka, T., Schindler, S., Pollheimer, M., Schmitzberger, I. & Peterseil, J. (2008) Impact of the Austrian Agri-Environmental Scheme on diversity of landscape, plants and birds. *Community Ecol.*, **9**, 217-227.

Zinngrebe, Y., Pe'er, G., Schueler, S., Schmitt, J., Schmidt, J. & Lakner, S. (submitted) The EU's Ecological Focus Areas – explaining farmers' choices in Germany.

12

Where are Ecology and Biodiversity in Social–Ecological Systems Research? A Review of Research Methods and Applied Recommendations

Adena R. Rissman[1] & Sean Gillon[1,2]

[1] Forest and Wildlife Ecology, University of Wisconsin-Madison, 1630 Linden Drive, Madison, WI 53706, USA
[2] Food Systems and Society, Marylhurst University, 17600 Pacific Highway, Portland, OR, 97036, USA

Keywords

Applied research; interdisciplinary; social–ecological systems; sustainability science; systematic literature review.

Correspondence

Adena Rissman, Forest and Wildlife Ecology, University of Wisconsin-Madison, 1630 Linden Drive, Madison, WI 53706, USA.
E-mail: arrissman@wisc.edu

Editor

Rudolf de Groot

Abstract

Understanding social–ecological systems (SES) is critical for effective sustainability and biodiversity conservation initiatives. We systematically reviewed SES research to examine whether and how it integrates ecological and social domains and generates decision-relevant recommendations. We aim to inform SES research methods and improve the relevance of SES research. Of 120 SES articles, two-thirds included an ecological variable while all but one included a social variable. Biodiversity was a less common ecological variable than resource productivity, land cover, and abiotic measures. We found six diverse social–ecological linking methods: modeling (9%), causal loop diagrams (18%), quantitative correlations (8%), separate quantitative measures (13%), indicators (14%), and rich description (37%). Policy recommendations addressing social–ecological dynamics were more likely in articles including both ecological and social variables, suggesting the importance of research approach for policy and practice application. Further integration of ecology and biodiversity is needed to support governance, policy, and management for SES sustainability.

Introduction

Understanding the dynamic connections between ecological and social systems is critical for the design of effective sustainability and biodiversity conservation initiatives (Liu *et al.* 2007; McClanahan *et al.* 2008). Research on social–ecological systems (SES) is rapidly advancing to understand relationships between social and ecological conditions, interactions, and outcomes (Ostrom 2009). In response to calls for interdisciplinary and solutions-oriented research (Metzger & Zare 1999; Palmer *et al.* 2005), governments and foundations have increased their investments in SES research (MA 2005). To make strong recommendations for SES sustainability, researchers must know how ecological and social dynamics relate. We conducted a systematic review to examine whether and how SES research is achieving interdisciplinarity across ecological and social domains to generate decision-relevant recommendations, with a focus on the roles of ecology and biodiversity.

SES research focus and methodologies are just beginning to be assessed (Binder *et al.* 2013), leading to an emerging concern about whether the "E" is sufficiently represented in SES research (Vogt *et al.* 2015). Following SES terminology, we understand ecological to mean ecological or environmental variables. Folke *et al.* (2005) indicate the importance of integrating social and ecological analyses, suggesting that social-only research "will not be sufficient to guide society toward sustainable outcomes," while ecological-only research "as a basis for decision making for sustainability may lead to too narrow conclusions." The SES field represents an important effort to develop innovative research methodologies to operationalize transdisciplinary research (Ostrom 2009) and provide more comprehensive analyses and implications than research focused on social or ecological dynamics alone (Folke *et al.* 2005). Researchers face numerous choices in framing, design, and analysis that influence their recommendations and the visibility of certain topics. We focus on four key issues in SES research: the choice of social and ecological variables, methods for linking social and ecological measures, recommendations for policy and management, and framing of human–environment relations.

The first two key issues – the choice of social and ecological variables and methods for linking them – reflect diverse disciplines, epistemologies, and applications (Binder *et al.* 2013). Many of the first SES researchers were modelers, but a wide range of methodologies are now used to understand relationships between diverse quantitative and qualitative variables (Miller *et al.* 2008). Evolving disciplinary and epistemological communities influence researcher choice of variables. For example, the SES framework, developed through the Ostrom Workshop, originated with social scientists conducting institutional analyses of the commons (Anderies *et al.* 2004; McGinnis & Ostrom 2014). Other SES research emerged from natural scientific inquiry, such as resilience analysis (Berkes *et al.* 2003). Each approach suggests different variables and methods for analysis, influencing findings and applications.

A third key area of methodological interest lies in how SES researchers translate findings into relevant recommendations. SES research promises to enhance outcomes by providing guidance for policy on improving SES sustainability or resilience (Berkes *et al.* 2003; Folke 2006). Less is known, however, about how frequently policy recommendations are made or which SES aspects they address. Although some SES frameworks are designed to diagnose opportunities and obstacles for increasing sustainability (Ostrom 2007), it is not clear whether SES articles with both ecological and social variables are more likely to make recommendations. Diverse SES research approaches are likely to differ in how they inform policy and management, and how policymakers might draw on SES research to improve outcomes.

Finally, the framing of human–environment relationships is another important methodological issue (Binder *et al.* 2013). Ecosystems both provision and present risk for humans (Turner *et al.* 2003; Ostrom 2009). Common pool resources research has historically focused on services or benefits to humans from fisheries, forests, and water. Vulnerability research often focuses on people's exposure and sensitivity to environmental harms. We engage these different conceptualizations of human–environment relations as an opportunity to reflect on SES research's historical legacies, trajectories, and applications for policy and management.

The aim of this research is to better understand emerging SES research and help researchers anticipate how their methodological choices may impact research

relevance for policy and management. We reviewed 120 SES research articles and posed four questions:

(1) How are ecological variables incorporated in SES research?
(2) How do researchers analyze connections between social and ecological variables?
(3) Does SES literature make applied recommendations for SES sustainability, and do these recommendations differ for articles with ecological variables compared to those without?
(4) How does SES literature frame human–environment relationships?

Review Methodology

We retrieved 425 articles with a title, abstract, or keyword phrase of "socio-ecological" or "social-ecological" in Web of Science before August 2012, excluding health sciences fields, in the 10 English-language journals with the most SES publications (see Supplement 1). We restricted this sample to 290 articles reporting empirical results and then randomly selected 120 articles due to researcher capacity.

We coded articles for variables, connections between independent variables (IVs) and dependent variables (DVs), methods, recommendations, and framing of human–environment relationships (Table S2). We identified articles as social–ecological if they included both a social variable (socioeconomics OR governance OR resource management) and an ecological variable (resource productivity OR land cover OR biodiversity OR abiotic). We then categorized the primary methods for linking social and ecological variables. Most articles had identifiable IV–DV connections (101 of 120); we categorized 985 IV–DV connections in these 101 articles. We asked if articles with social and ecological variables were more likely than social-only articles to have different methods, recommendations, and human–environment framings. Differences between social–ecological and social-only articles were examined with two-tailed chi-squared tests. Among the social–ecological articles, we also analyzed whether articles with different social–ecological linking methods were more or less likely to make SES recommendations. Five tests of intercoder reliability were conducted during the coding period in which researchers read and coded the same article and compared results in order to identify and norm variations in coder interpretation.

Results

Ecology and biodiversity in SES research

The SES literature is diverse and emphasizes social variables. Two-thirds of articles (66%, 80 of 120) included

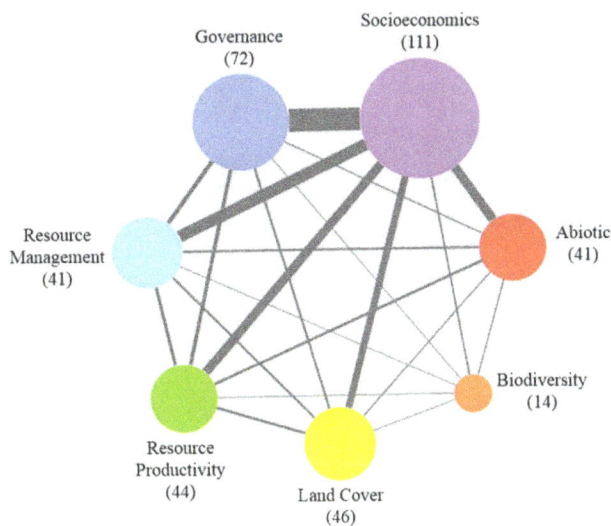

Figure 1 Bubble size and number in parentheses indicate the number of articles that include each type of variable ($n = 120$). Line width indicates the number of articles with each type of IV–DV connection ($n = 101$). Line lengths do not signify meaning.

Table 1 Percent of the 985 independent variable–dependent variable (IV–DV) connections in which each variable was an IV or a DV

		IVs (percent of IVs)	DVs (percent of DVs)
Social	Socioeconomics	53.8%	47.5%
Social	Resource management	8.5%	14.5%
Social	Governance	14.9%	12.5%
Ecological	Resource productivity	4.2%	9.3%
Ecological	Land cover	6.2%	8.0%
Ecological	Abiotic	11.5%	4.5%
Ecological	Biodiversity	0.9%	3.7%

both social and ecological variables. One-third (33%, 39 of 120) had social variables but no ecological variables, and one article had ecological but no social variables. Socioeconomic variables were most likely to be examined, and were commonly associated with governance, followed by resource management and then resource productivity (Figure 1). Connections between biodiversity and other variables were least commonly researched.

Of the 101 articles with identifiable IV–DV connections, 86 examined if a social IV influenced ($\rightarrow$) a social DV, 42 ecological IV $\rightarrow$ social DV, 36 social IV $\rightarrow$ ecological DV, and 27 ecological IV $\rightarrow$ ecological DV. This illustrates that the SES literature is most focused on interactions among social variables, followed by the influence of ecological variables on society. When considering all seven variable categories across the 985 IV–DV connections, we found that abiotic, governance, and socioeconomic variables were more often IVs than DVs, while land cover, resource productivity, resource management, and biodiversity were more often DVs (Table 1).

Linking social and ecological variables in analysis

In response to our methods question about how researchers make connections among social and ecological variables, we found that SES literature includes both qualitative and quantitative methods, with somewhat more emphasis on qualitative methods. Half the arti-

cles (51%) were qualitative-only, one-third (32%) were quantitative-only, and 17% included both qualitative and quantitative methods. Most (62%) of the social-only articles were qualitative, while only 41% of social–ecological articles were qualitative, a statistically significant difference ($\chi^2 = 3.910$, $P = 0.048$). The majority (55%) involved a single case study, 41% had multiple case studies, and 21% were large-n studies ($n > 30$). The time span of phenomena analyzed varied widely from under 1 year to 2,500 years, with a median of 7 years. Nearly half (44%) examined a span of 10 or more years.

We identified six primary methods for linking social and ecological variables: modeling, causal loop diagrams, quantitative correlations, separate quantitative measures, indicators, and rich description (Figure 2 illustrates five of these approaches). We provide a definition, one strong example from our sample, and strengths and weaknesses for each linking method.

Social–ecological modeling

Mechanistic modeling was deployed in 9% of articles (11 of 120) to explain or predict causal relationships among social and ecological system components. In one article, a multiagent model of lobster fisher behavior is linked to a biophysical model of a patchy natural environment, which reveals when individual incentives are aligned with collective action (Wilson *et al.* 2007). SES modeling is "maturing as a discipline in its own right" drawing on natural resource modeling and complex systems research to tackle nonlinear behavior, cross-scale and interdependent dynamics, and uncertainty (Schlüter *et al.* 2012). Models allow for the comparison of scenarios and policy options. Disadvantages include extensive and expensive data requirements, difficulty of modeling all system components, lack of transparency, and difficulty of communicating methods and results to nonexperts.

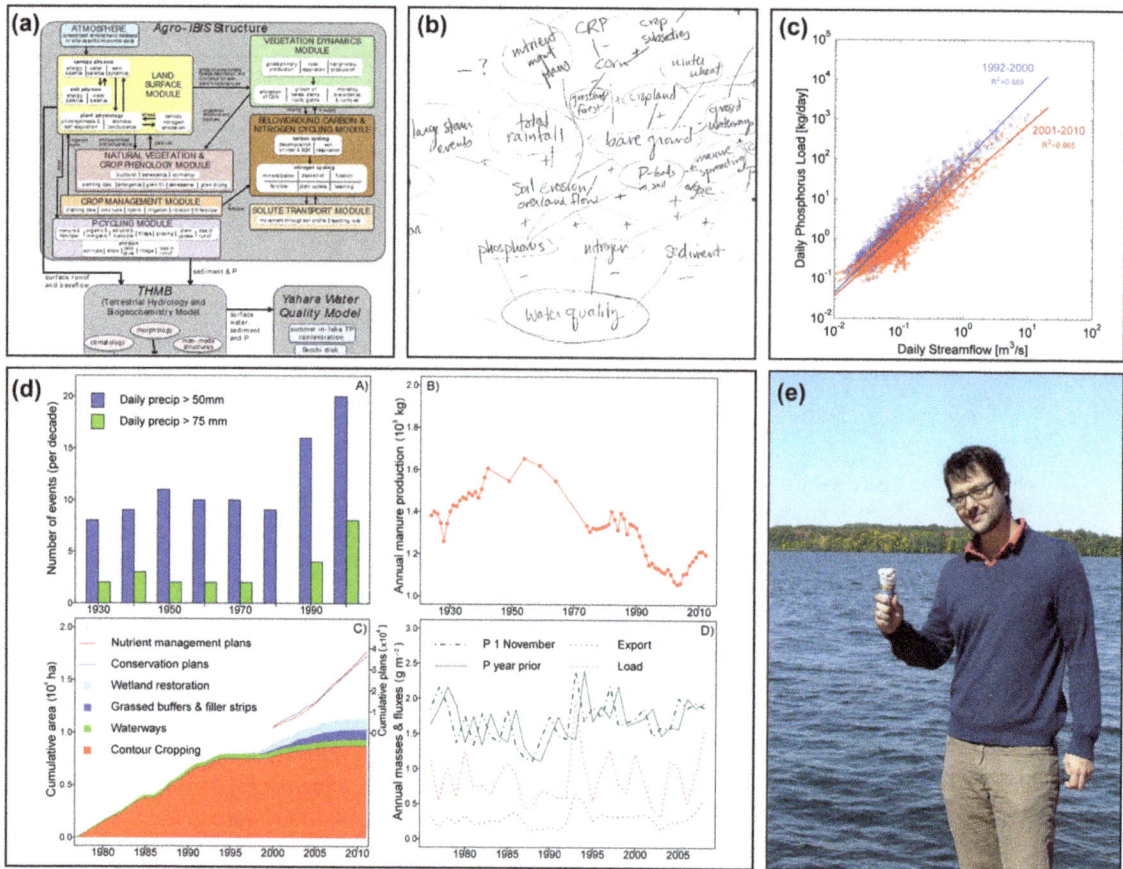

Figure 2 Methods for linking social and ecological research with examples from the Yahara Watershed, Wisconsin including (a) **models** of land use, plant and soil dynamics, and water, (b) **causal loop diagrams** such as this drawing from a participatory workshop, (c) **quantitative correlations** such as this phosphorus load-streamflow correlation showing improvement after new practice implementation, (d) **separate quantitative measures** presented together of precipitation, manure, conservation practices, and phosphorus loads, and (e) **rich description** that probes underlying causation – can we have ice cream from cows upstream and clean lakes too? (Carpenter *et al.* 2015; Gillon *et al.* 2015).

Causal loop diagrams

One-fifth (18%) of articles presented system dynamics through causal loop diagrams (called influence diagrams, cognitive mapping, or mental maps). Causal loop diagrams typically translate qualitative information into simple quantitative relationships with negative and positive feedback loops. For instance, Fazey *et al.* (2011) developed a conceptual model of SES change in the Solomon Islands based on 76 focus groups, which revealed that population growth and a desire for monetary prosperity "act synergistically to generate stress in communities" and result in "maladaptive trajectories of change" including declining provision of ecosystem services from fisheries, forests, and subsistence gardens. An advantage of causal loop diagrams is their relatively accessible communication of complex feedback dynamics involving many variables; a disadvantage is that the simple positive or negative polarity of loops may inadequately describe complex and variable relationships and obscure threshold effects (Richardson 1986).

Quantitative correlations

Only 8% of articles relied on statistical analysis to correlate social and ecological variables, with or without causal inferences. For example, in a study of over 50 marine reserves, multiple linear regression identified relationships between fish biomass (from visual census) and indices of socioeconomic development, compliance with reserve rules, human population, and democratic participation (largely from surveys) (Pollnac *et al.* 2010). Statistical correlations provide a quantifiable measure of association among variables but often exclude difficult-to-measure system components, ignore cross-scale dynamics and threshold effects, and may not allow for causal inference.

Where are Ecology and Biodiversity in Social–Ecological Systems Research? A Review of Research...

107

Separate quantitative measures

In 13% of articles, quantitative data on different social and ecological variables were linked through narrative or graphic visualization. For example, an interdisciplinary project in the Siberian Arctic narratively tied together three types of results that together told a nuanced story of social and environmental change: anthropological observation of migratory herders; remote sensing of land cover change; and field sampling of vegetation biomass (Forbes *et al.* 2009). Separate quantitative measures can accommodate differences in measurement, timing, and scale among disciplines but do not provide estimates of the mean, variance, or causal relationships among system components.

Indicators of the link

In 14% of articles, an index or composite indicator linked social and ecological variables through a mathematical calculation. Vulnerability indices, sustainability indices, resource metabolism rates, and per capita resource abundance levels quantify the link between humans and environments. For example, human appropriation of net primary production indicates the extent to which humans have appropriated natural resources (Krausmann *et al.* 2012). Indicators summarize complex information, attract public attention, and communicate a simple "big picture" to policymakers. However, they can encourage simplistic policy conclusions, obscure complex dynamics and subindicator variability, and ignore local context (European Commission Composite Indicators Research Group 2015).

Social–ecological rich description

One-third (37%) of articles provided a rich narrative, or what could be termed a social–ecological thick description, of the complex intertwining of social and environmental phenomena with little to no quantitative data. Rich or thick description involves detailed information that ascribes intentionality to behavior with specific information about the context of a situation (Creswell & Miller 2000). For example, through ethno-biological interviews and analysis of historical documents and maps, researchers traced a century of decline and renewal of managed forests of chestnuts and holm-oak-associated truffles, shaped by cross-scale socio-political legacies, values, knowledge, technology, economies, and institutions (Aumeeruddy-Thomas *et al.* 2012). Rich description allows for causal analysis and the interplay of complex system components, but does not quantify the importance of factors and may not systematically define system components to facilitate comparisons.

Making recommendations

Most articles (71%) included recommendations for policymakers, managers, or community members. Articles with social and ecological variables were significantly more likely than social-only articles to make recommendations that addressed interlinked social and ecological systems (Table 2). Articles tended to recommend big-picture shifts in governance, calling for institutional change or capacity increases rather than specific changes in policy or practice. Biodiversity conservation was uncommon as a focus of recommendations. We found only one statistically significant association between the six linking method and type of recommendation: quantitative indicator articles were more likely than SES articles with other linking methods to make recommendations for integrative social–ecological change ($n = 80$, $P = 0.018$), although small sample sizes suggest caution in interpretation (see Table S3 for additional information).

Framing human–environment relationships

Social–ecological articles were more likely than social-only articles to consider environmental benefits to humans and how humans negatively impact the environment (Table 3). Social–ecological and social-only articles were equally likely to consider environmental risks to humans. In contrast, social-only articles were more likely to consider how humans affect other humans, suggesting a trade-off between examining social equity and social–ecological dynamics.

Discussion: insights, challenges, and opportunities for ecology in SES

Research methodologies matter for the relevance of research findings. Articles with ecological and social variables were over twice as likely to make recommendations addressing integrated SESs, compared to social-only articles. One-third of articles did not link social and ecological variables, and only 12% included biodiversity measures. This confirms the concern that many SES publications do not couple social and ecological aspects of systems (Epstein *et al.* 2013). By focusing on SES research, we likely included more social science than in "coupled human and natural systems" or natural resource management literatures, which could be compared in the future. In order to make strong recommendations about improving SES sustainability, we need to understand how ecological dynamics relate to social conditions.

Approaches for linking social and ecological variables are widely divergent. We identified six primary linking methods that could help interdisciplinary teams

Table 2 Articles with social and ecological variables were more likely to have recommendations for social–ecological systems; statistical tests compare social–ecological with social-only articles (record numbers refer to articles in Supplement 2)

	Overall (n = 120)	Percent of social–ecological articles (n = 80)	Percent of social-only articles (n = 39)	χ^2	P-value
Any recommendation	71%	73%	69%	0.137	0.711
Social–ecological system recommendation	29%	38%	13%	7.692	**0.006**
Social system recommendation	55%	54%	59%	0.290	0.590

Social–ecological system recommendation examples:

"Long-term solutions to **scale mismatch problems** will depend on **social learning** and the development of **flexible institutions** that can **adjust and reorganize in response to changes in ecosystems**" (Record 24).

"Unless one **ensures the livelihoods** of those living around or within a forest, a major **investment in monitoring** alone is **not a sufficient, long-run management strategy** and may even be counterproductive" (Record 76).

"While we do not expect farmers or agricultural communities to 'turn back the clock' on time and technology, **initiatives that foster diversification** within and among agricultural landscapes, rather than their further homogenization, may be more likely to achieve the common goal of enhancing agricultural sustainability" (Record 13).

Social system recommendation examples:

"Addressing issues such as **corruption, transparency, and stability** of national governments will be key to building **effective social organization and adaptive capacity** at all scales" (Record 21).

"The results suggest **development policy** . . . needs to . . . increase **emphasis on well-being** aspects of development rather than **income generation** per se" (Record 109).

"We urge other researchers and practitioners to focus more strongly on **human relationships** and **capacity** and the **flexibility** they can create in other conservation initiatives in which local governance may be an option" (Record 9).

Table 3 Framing human–environment relationships; statistical tests compare social–ecological with social-only articles

	Percent of all articles (n = 120)	Percent of social–ecological articles (n = 80)	Percent of social-only articles (n = 39)	χ^2	P-value
How humans manage the environment	82%	86%	72%	3.635	**0.057**
Benefits or services the environment provides to humans	57%	65%	41%	6.153	**0.013**
How humans negatively impact the environment	54%	63%	36%	7.465	**0.006**
How humans affect other humans	48%	43%	62%	3.804	**0.051**
Environmental or ecosystem change and its consequences for ecosystem service provision to humans	45%	56%	23%	11.640	**0.001**
Environmental harms, threats, or risks to humans	43%	44%	41%	0.079	0.778
The authors conceptualize human and environmental systems as separate	42%	36%	54%	3.332	**0.068**
The authors suggest that humans and environment cannot be analyzed independently	32%	39%	18%	5.219	**0.022**

conceptualize their collaborative articles and help policymakers decide what research to prioritize. Further research on these linking methodologies could examine the trade-offs that individuals and research teams face when selecting methodologies and the contributions of analyses for diverse decision contexts.

Elements of ecosystems such as land cover, resource production, and abiotic conditions were most commonly measured, followed distantly by biodiversity. When biodiversity is measured directly, studies can reveal promising social pathways for biodiversity protection (López-

Angarita *et al.* 2014). The spatial and temporal scale of variables measured is also important for coupling social and ecological data (Gillon *et al.* 2015). Few analyses address the feedbacks among social dynamics, conservation initiatives, and linked social–ecological outcomes, although such analyses may effectively inform future policy (Miller *et al.* 2012).

Methodological pluralism is likely an advantage to the field since different methods answer questions in different situations better than others. Narrative, visual, and mathematical links between social and ecological can produce fascinating insights, and research scope influences the resulting recommendations. Quantitative studies rely on modeling and statistics to link social and ecological variables, indicating the necessary or likely characteristics that predict sustainable resource or livelihood outcomes. For instance, models have substantial utility for testing the effects of policy options (e.g., Guzy et al. 2008) and often underpin environmental policies (Rissman & Carpenter 2015). Beyond direct utility as perceived by decision makers, SES research often critiques existing environmental or economic policies as overly simplified, unsustainable, insufficiently protective of people or environments, and oblivious to local context and culture (e.g., Zimmerer 2011). Rich descriptions, for example, may explore policy outcomes and explain how and why change occurs. Our findings might provide better understanding of the implications of diverse research approaches for generating information and recommendations. Clear analytical pathways would provide structure for researchers and science funders seeking best practices, comparative analysis, and usable research.

The framing of human–environment relationships varied widely in the SES literature. Humans were equally likely to be framed as threats to ecosystems or as beneficiaries of ecosystem services, while fewer articles examined environmental threats to humans. Social–ecological articles were less likely to consider relations among humans. This finding illustrates a potential trade-off in focusing on social–ecological linkages or on complex dynamics among social groups including issues of equity and power.

Social–ecological research faces obstacles, including disciplinary incentives, cultures, epistemologies, funding, and transaction costs (Metzger & Zare 1999; Pooley et al. 2014). Analysis that connects changes in environments, technology, knowledge, organization, and values faces an internal tension between mind-opening integration and the analytical categorization that supports empirical scholarship (Kallis & Norgaard 2010). We experienced this tension in categorizing articles, some of which aimed to break down the very categories of social and ecological.

Many opportunities exist for social scientists and ecologists to increasingly engage in cross-domain research. Ecologists are undertaking innovative SES research, and the sophistication and growth of SES research is increasing (Binder et al. 2013; McGinnis & Ostrom 2014). Greater systemization of methodologies is allowing for comparative work (Cox 2014), and new syntheses of research approaches are being developed to train scholars (Wiek et al. 2011). Transdisciplinary teams would benefit from increased resources and from reflection on how methodological routes impact knowledge destinations (Mattor et al. 2014).

In conclusion, better integration of ecology is needed in SES research. We have demonstrated that variable selection and methodology influence SES recommendations. Further development of quantitative and qualitative linking methodologies is needed, as is increased effort in research coordination and training. Biodiversity measures remain uncommon in the SES literature, which suggests a potential challenge and opportunity for integration. Clear links between ecological and social dynamics, feedbacks, and outcomes are needed to effectively inform policy and management and improve the social–ecological fit of conservation strategies (Bodin et al. 2014). Achieving this will require continued attention to methodological choices, shifts in incentives and training, and engagement with policymakers and practitioners.

Acknowledgments

We thank the National Science Foundation Water Sustainability and Climate project DEB-1038759. We also thank E.G. Booth for design assistance; Ostrom Workshop participants, Stephen R. Carpenter, Sedra Shapiro, and Chloe Wardropper for their input on an earlier draft; and E. Geisler, B. Laursen, A. L'Roe, C. Locke, S. Shapiro, C. Wardropper, and S. Wilkins for research assistance.

Supporting Information

Supplement 1. Sampling Methodology, Variables, Definitions, and Additional Analysis
Supplement 2. Articles Reviewed

References

Anderies, J.M., Janssen, M.A. & Ostrom, E. (2004) A framework to analyze the robustness of social-ecological systems from an institutional perspective. Ecol. Soc., **9**, 18.

Aumeeruddy-Thomas, Y., Therville, C., Lemarchand, C., Lauriac, A. & Richard, F. (2012) Resilience of sweet chestnut and truffle holm-oak rural forests in Languedoc-Roussillon, France: roles of social-ecological legacies, domestication, and innovations. Ecol. Soc., **17**, 12.

Berkes, F., Colding, J. & Folke, C. (2003) Navigating social-ecological systems: building resilience for complexity and change. Cambridge University Press, Cambridge, UK.

Binder, C.R., Hinkel, J., Bots, P.W.G. & Pahl-Wostl, C. (2013) Comparison of frameworks for analyzing social-ecological systems. Ecol. Soc., **18**, 26.

Bodin, Ö., Crona, B., Thyresson, M., Golz, A.-L. & Tengö M., (2014) Conservation success as a function of good alignment of social and ecological structures and processes. Conserv. Biol., **28**, 1371-1379.

Carpenter, S.R., Booth, E.G., Gillon, S. *et al.* (2015) Plausible futures of a social-ecological system: Yahara watershed, Wisconsin, USA. *Ecol. Soc.*, **20**, 10.

Cox, M. (2014) Understanding large social-ecological systems: introducing the SESMAD project. *Intl. J. Commons.*, **8**, 265-276.

Creswell, J.W. & Miller, D.L. (2000) Determining validity in qualitative inquiry. *Theor. Pract.*, **39**, 124-131.

Epstein, G., Vogt, J.M., Mincey, S.K., Cox, M. & Fischer, B. (2013) Missing ecology: integrating ecological perspectives with the social-ecological system framework. *Intl. J. Commons.*, **7**, 432-453.

European Commission Composite Indicators Research Group. (2015) *What are the pros and cons of composite indicators?* European Commission Joint Research Centre, Ispra, Italy.

Fazey, I., Pettorelli, N., Kenter, J., Wagatora, D. & Schuett, D. (2011) Maladaptive trajectories of change in Makira, Solomon Islands. *Global Environ. Chang.*, **21**, 1275-1289.

Folke, C. (2006) Resilience: the emergence of a perspective for social–ecological systems analyses. *Global Environ. Chang.*, **16**, 253-267.

Folke, C., Hahn, T., Olsson, P. & Norberg, J. (2005) Adaptive governance of social-ecological systems. *Annu. Rev. Environ. Res.*, **30**, 441-473.

Forbes, B.C., Stammler, F., Kumpula, T., Meschtyb, N., Pajunen, A. & Kaarlejärvi, E. (2009) High resilience in the Yamal-Nenets social–ecological system, West Siberian Arctic, Russia. *Proc. Natl. Acad. Sci. USA*, **106**, 22041-22048.

Gillon, S., Booth, E.G. & Rissman, A.R. (2015) Shifting drivers and static baselines in environmental governance: challenges for improving and proving water quality outcomes. *Region Environ. Chang.*, **16**, 759-775.

Guzy, M.R., Smith, C.L., Bolte, J.P., Hulse, D.W. & Gregory, S.V. (2008) Policy research using agent-based modeling to assess future impacts of urban expansion into farmlands and forests. *Ecol. Soc.*, **13**, 37.

Kallis, G. & Norgaard, R.B. (2010) Coevolutionary ecological economics. *Ecol. Econ.*, **69**, 690-699.

Krausmann, F., Gingrich, S., Haberl, H. *et al.* (2012) Long-term trajectories of the human appropriation of net primary production: lessons from six national case studies. *Ecol. Econ.*, **77**, 129-138.

Liu, J., Dietz, T., Carpenter, S.R. *et al.* (2007) Complexity of coupled human and natural systems. *Science*, **317**, 1513-1516.

López-Angarita, J., Moreno-Sánchez, R., Maldonado, J.H. & Sánchez, J.A. (2014) Evaluating linked social–ecological systems in marine protected areas. *Conserv. Lett.*, **7**, 241-252.

MA (Millennium Ecosystem Assessment). (2005) *Ecosystems and human well-being: synthesis*. Island Press, Washington, DC.

Mattor, K., Betsill, M., Huber-Stearns, H. *et al.* (2014) Transdisciplinary research on environmental governance: a view from the inside. *Environ. Sci. Policy*, **42**, 90-100.

McClanahan, T., Cinner, J., Maina, J. *et al.* (2008) Conservation action in a changing climate. *Conserv. Lett.*, **1**, 53-59.

McGinnis, M.D. & Ostrom, E. (2014) Social-ecological system framework: initial changes and continuing challenges. *Ecol. Soc.*, **19**, 30.

Metzger, N. & Zare, R.N. (1999) Interdisciplinary research: from belief to reality. *Science*, **283**, 642-643.

Miller, B.W., Caplow, S.C. & Leslie, P.W. (2012) Feedbacks between conservation and social-ecological systems. *Conserv. Biol.*, **26**, 218-227.

Miller, T.R., Baird, T.D., Littlefield, C.M., Kofinas, G., Chapin, III F.S. & Redman, C.L. (2008) Epistemological pluralism: reorganizing interdisciplinary research. *Ecol. Soc.*, **13**, 46.

Ostrom, E. (2007) A diagnostic approach for going beyond panaceas. *Proc. Natl. Acad. Sci. USA*, **104**, 15181-15187.

Ostrom, E. (2009) A general framework for analyzing sustainability of social-ecological systems. *Science*, **325**, 419-422.

Palmer, M.A., Bernhardt, E.S., Chornesky, E.A. *et al.* (2005) Ecological science and sustainability for the 21st century. *Front Ecol. Environ.*, **3**, 4-11.

Pollnac, R., Christie, P., Cinner, J.E. *et al.* (2010) Marine reserves as linked social–ecological systems. *Proc. Natl. Acad. Sci. USA*, **107**, 18262-18265.

Pooley, S.P., Mendelsohn, J.A. & Milner-Gulland, E. (2014) Hunting down the chimera of multiple disciplinarity in conservation science. *Conserv. Biol.*, **28**, 22-32.

Richardson, G.P. (1986) Problems with causal-loop diagrams. *Syst. Dynam. Rev.*, **2**, 158-170.

Rissman, A.R. & Carpenter, S.R. (2015) Progress on nonpoint pollution: barriers and opportunities. *Daedalus*, **144**, 35-47.

Schlüter, M., McAllister, R., Arlinghaus, R. *et al.* (2012) New horizons for managing the environment: a review of coupled social-ecological systems modeling. *Nat. Resour. Model*, **25**, 219-272.

Turner, B.L., Kasperson, R.E., Matson, P.A. *et al.* (2003) A framework for vulnerability analysis in sustainability science. *Proc. Natl. Acad. Sci. USA*, **100**, 8074-8079.

Vogt, J.M., Epstein, G.B., Mincey, S.K., Fischer, B.C. & McCord, P. (2015) Putting the" E" in SES: unpacking the ecology in the Ostrom social-ecological system framework. *Ecol. Soc.*, **20**, 55.

Wiek, A., Withycombe, L. & Redman, C.L. (2011) Key competencies in sustainability: a reference framework for academic program development. *Sustain. Sci.*, **6**, 203-218.

Wilson, J., Yan, L. & Wilson, C. (2007) The precursors of governance in the Maine lobster fishery. *Proc. Natl. Acad. Sci. USA*, **104**, 15212-15217.

Zimmerer, K.S. (2011) The landscape technology of spate irrigation amid development changes: assembling the links to resources, livelihoods, and agrobiodiversity-food in the Bolivian Andes. *Global Environ. Chang.*, **21**, 917-934.

13

Costs and Opportunities for Preserving Coastal Wetlands under Sea Level Rise

Rebecca K. Runting[1,2], Catherine E. Lovelock[3], Hawthorne L. Beyer[2,3], & Jonathan R. Rhodes[1,2]

[1] School of Geography, Planning and Environmental Management, The University of Queensland, Brisbane 4072, Australia
[2] ARC Centre of Excellence for Environmental Decisions, The University of Queensland, Brisbane 4072, Australia
[3] School of Biological Sciences, The University of Queensland, Brisbane 4072, Australia

Keywords
Payments for ecosystem services (PES); climate change adaptation; carbon sequestration; SLAMM; nursery habitat; co-benefits; spatial conservation planning; opportunity cost; mangroves; Moreton Bay.

Correspondence
Rebecca K. Runting, School of Geography, Planning and Environmental Management, University of Queensland, Brisbane, Queensland, Australia.
E-mail: r.runting@uq.edu.au

Editor
Rudolf de Groot

Abstract

Rises in sea level can alter the distribution of coastal wetlands through migration landward and loss due to inundation. The expansion of coastal developments can prevent potential wetland migration, exacerbating loss as sea levels rise. Pre-emptive planning to set aside key coastal areas for wetland migration is therefore critical for the long-term preservation of species habitat and ecosystem services, yet we have little understanding of the economic costs and benefits of doing so. Using data and simulations from Queensland, Australia, we show that the opportunity cost of preserving wetlands is likely to be much higher under sea level rise than under current sea levels. However, we find that payments for ecosystem services can alleviate these costs, and in many cases may make expanding the reserve network profitable in the long run. This highlights the need to develop markets and payment mechanisms for ecosystem services to support climate change adaptation policies for coastal wetlands.

Introduction

Coastal ecosystems have important biodiversity values, with ~2,700 threatened species globally using these habitats for at least part of their life cycle (IUCN 2013). Additionally, coastal wetlands provide substantial benefits to humans through the provision of ecosystem services, such as the maintenance of fisheries, coastal protection, and carbon sequestration (Barbier et al. 2011). However, under sea level rise, coastal wetlands can be lost through inundation (Lovelock et al. 2015), but they can also migrate landward in the absence of steep gradients in topography or anthropogenic barriers, such as built structures (Kirwan & Megonigal 2013). The establishment of anthropogenic barriers to wetland migration could be prevented by pre-emptively expanding the coastal reserve network (i.e., adding to the set of protected areas) to accommodate wetland response to sea level rise. However, we know little about the likely costs and benefits of such an approach.

Global sea level rise is one impact of climate change that has seen recent upward revisions as further information becomes available (IPCC 2007; Church et al. 2013). These revisions, combined with the accelerated subsidence of deltas from anthropogenic activity (such as fossil fuel and water extraction and the trapping of sediment in reservoirs) (Syvitski et al. 2009), warrants urgent attention and the development of sound pre-emptive adaptation strategies. Despite this imperative, current spending on climate change adaptation remains low relative to the anticipated future costs (Parry et al. 2009). However, emerging markets for ecosystem services, such as

| **1. Sea level rise scenarios** (by 2100) | **2. Coastal impact model** (SLAMM - wetland distributions) | **4. Expand protected area** |

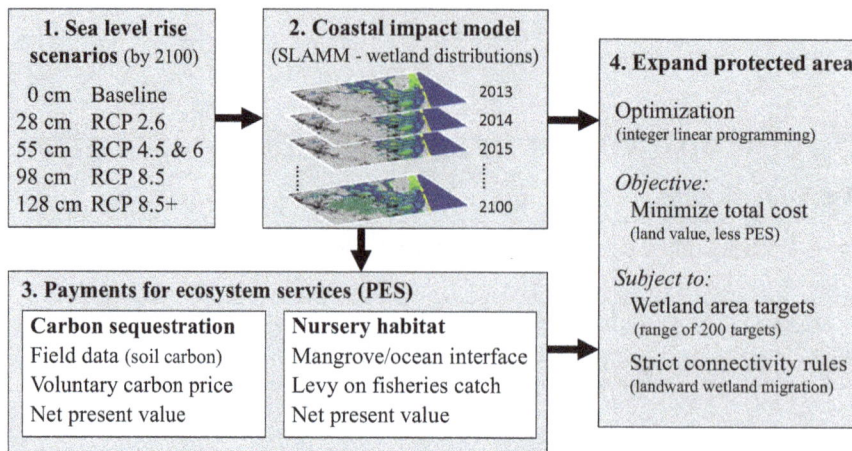

Figure 1 Diagram of the methodology used to expand the reserve network under a range of sea level rise scenarios and potential payments for ecosystem services. The sea level affecting marshes model (SLAMM) was used to simulate coastal wetland change under a range of sea level rise projections. This produced a map of coastal wetlands for each year to 2100 for as section of Moreton Bay, Queensland, Australia. Based on these wetland distributions, we modeled the provision of ecosystem services (carbon sequestration and nursery habitat for commercially important species) at each time step, and calculated the net present value of potential payments for these services (Supporting Information). Using integer linear programming, we then optimized the selection of additional wetland sites under the range of sea level rise projections and compared the resulting opportunity cost under different combinations of payments for ecosystem services. This allowed us to determine the potential of payments for ecosystem services to compensate the cost of reserve expansion under sea level rise.

the carbon market (voluntary or otherwise), may have the potential to relieve the financial burden of preserving coastal wetlands under sea level rise.

Previous studies have estimated the impact of sea level rise on coastal ecosystems (FitzGerald & Fenster 2008; Craft *et al.* 2009) and the species that depend on them (Traill *et al.* 2011; Iwamura *et al.* 2013), but none have quantified the costs of preserving wetlands under increasing rates of sea level rise and the potential of payments for ecosystem services to mitigate this cost. There has been a focus on the costs arising from human displacement or damage to private property and infrastructure (Dasgupta *et al.* 2009; Bin *et al.* 2011; Arkema *et al.* 2013; Hinkel *et al.* 2014), but there has been little consideration of the costs of preserving wetlands to facilitate their migration. Setting aside land for wetland migration has an opportunity cost, as this land might have otherwise been developed (e.g., for urban use; Mills *et al.* 2014). While the human element is undoubtedly important, it is vital that strategies to preserve wetlands under climate change are considered alongside anthropocentric impacts in order to conserve species and ecosystem services.

The aims of this research were to (i) determine if the opportunity costs of preserving coastal wetlands is higher under sea level rise compared to current sea levels, and (ii) determine the extent to which potential payments for ecosystem services can alleviate these costs. Here we show that, because coastal land value increases with elevation, coastal wetlands are likely to migrate into more

expensive land with sea level rise, thus increasing the costs of pre-emptively preserving those wetlands. We also demonstrate that, even when the area of coastal wetlands is projected to expand under sea level rise, the cost of preserving these wetlands is still likely to be greater with sea level rise than without it. Despite the higher costs of preserving wetlands under sea level rise, we show that payments for ecosystem services have the potential to offset the opportunity cost of the reserve network.

Methods

To establish why preserving coastal wetlands might cost more under sea level rise we quantified the relationship between coastal land values and elevation for the state of Queensland, Australia. We then undertook a local scale case study to compare the cost of expanding the reserve system with and without sea level rise and payments for ecosystem services, to determine the change in costs and potential of ecosystem services (Figure 1).

Coastal land value and elevation

To understand how land values vary with elevation we quantified the relationship between coastal land values and elevation for the entire 6,973 km coastline of Queensland. This coastline traverses five global ecoregions (WWF 2000) and four climatic zones (equatorial, tropical, subtropical, and grasslands) (Stern *et al.* 2000),

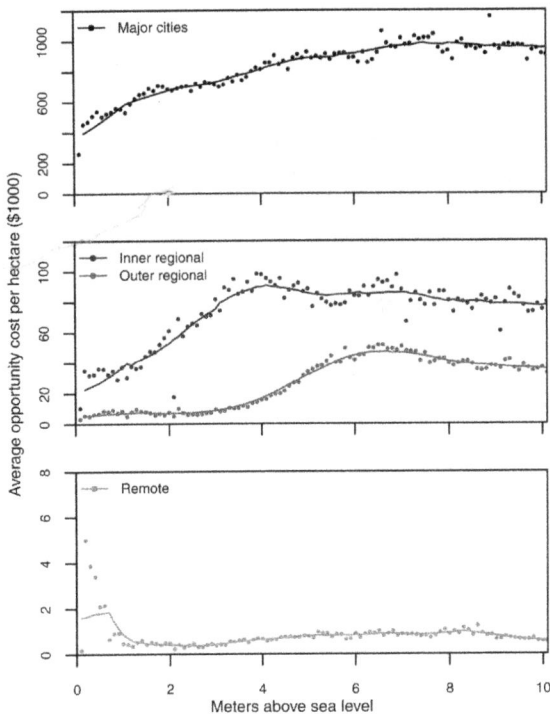

Figure 2 The average (mean) value of coastal land at increasing elevation in Queensland, Australia, separated by remoteness class. The remoteness classes are categorized based on the level of accessibility to remoteness to various service centers via the road network (Pink 2011). Trend lines indicate the moving average.

with human settlement patterns varying from urban to remote (Pink 2011). As extensive elevation data were required, we used a 1 second (~30 m) Digital Elevation Model (DEM; Gallant 2010). We obtained unimproved land values for 2012 from the Queensland Valuation and Sales database (DERM 2013) and converted these into a value per hectare at a resolution of ~30 m (to match the elevation data). We then categorized the DEM into 100 classes based on 10 cm elevation increments up to 10 m above sea level. These categories were used to derive the mean land value for each 10 cm interval of elevation. To determine the effect of urban, regional or remote areas on this pattern, we separated the results based on the remoteness classes from the Australian Statistical Geography Standard Remoteness Structure (Pink 2011).

Wetland transition model

The Sea Level Affecting Marshes Model (SLAMM; Clough et al. 2012) was used to predict wetland transitions under sea level rise for a 600 km² section of Moreton Bay, Australia (Figure 2). SLAMM simulates the main processes driving coastal wetland conversions and shoreline modifications under sea level rise, including salt water intrusion,

erosion and sedimentation, wetland transition dynamics, and anthropogenic barriers to these dynamics (Craft et al. 2009; Clough et al. 2012). When executed, SLAMM calculates the relative change in elevation and associated wetland transitions for each cell in each year through to 2100. The inclusion of these processes at a fine spatial and temporal resolution enables SLAMM to give an accurate assessment of sea level rise, particularly when combined with LiDAR-derived elevation data (McLeod et al. 2010; Geselbracht et al. 2011). We parameterized SLAMM for Moreton Bay with a combination of field based and remotely sensed data for the area (Supporting Information). This site was chosen because it is located near two urban centers (Brisbane to the north and the Gold Coast to the south) and contains a variety of ecosystem types, along with agricultural land. As the future rise in sea level is uncertain, we used a range of projections to 2100 (28, 55, 98, and 128 cm) from the IPCC's fifth assessment report (Church et al. 2013) to account for this variation (Supporting Information). This produced fine resolution (~5 m) simulations of changes in the distributions of wetlands for each year (2013–2100) for each sea level rise scenario.

Ecosystem services

While there are a range of ecosystem services provided by coastal wetlands, we focused on quantifying and valuing soil carbon sequestration and nursery habitat value for commercially important species. To quantify soil carbon sequestration, we used local field measurements for the different wetland types, and applied a range of carbon prices from the voluntary carbon market (mean \$6.1 AUD MgC^{-1}) and estimates of the social value of carbon (from \$10.94 to \$96.94 AUD MgC^{-1}, Supporting Information). To determine the value of nursery habitat, we linked a potential levy on the gross value of production of three mangrove-dependent and commercially important species (*Penaeus merguiensis*, *Scylla serrata*, and *Lates calcarifer*) to the area of mangroves that interface with the ocean. When combined with the simulations of wetland change, this produced an economic value in each year to 2100 for both services for all properties within the study site. These values were discounted to form a net present value that was appropriate to compare with land values (Supporting Information).

Finding the optimal reserve network

We used integer linear programming to find the optimal reserve network for a range of wetland area targets for the least cost under sea level rise (Supporting Information). Each property parcel was either set aside for

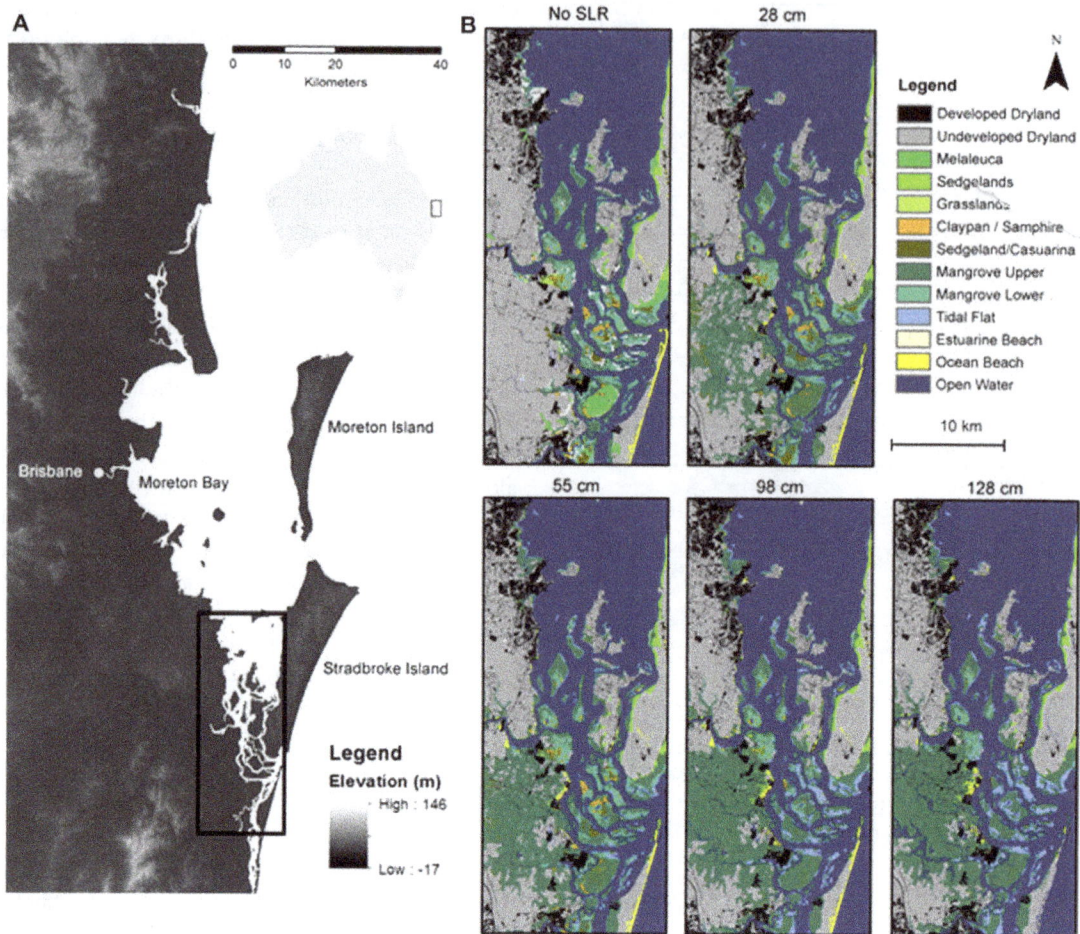

Figure 3 The distribution of coastal vegetation in the south of Moreton Bay, Australia. Panel (A) shows the location of the case study (specifically latitude 27.3°S to 27.5°S and longitude 153.15°E to 153.25°E), and panel (B) shows the distribution of coastal vegetation in 2100 based on no sea level rise (SLR), a rise of 28, 55, 98, and 128 cm.

wetlands (i.e., protected), or assumed to be lost to future development. The opportunity cost was initially based on unimproved land values (plus a transaction cost), but in subsequent scenarios the capitalized value of payments for ecosystem services was subtracted from the opportunity cost for each property. Spatial dependencies among planning units were enforced, to allow for the process of wetland migration. The resulting reserve networks were compared across sea level rise projections and ecosystem service payment schemes, based on the total cost of the solution and the area of wetlands preserved within the reserve network.

Results

Land value and elevation

Our analysis of coastal land values and elevation for the coastline of Queensland, Australia showed a generally positive association between land value and elevation in the narrow coastal strip (up to 10 m above sea level, Figure 2). The positive relationship was most apparent in major cities and regional settlements, but values were consistently low in remote areas (Figure 2). This rise in land values for cities and regional settlements is likely due to the declining flood risk with elevation. The shapes of the curves differ as the confounding drivers of land value (such as slope, accessibly, and amenity) are regionally variable.

Cost of reserve network

We predicted a substantial change in the distribution and extent of wetlands under sea level rise for our case study in Moreton Bay, Australia (Figure 3). Under the current reserve network, the landward movement of wetlands resulted in fewer wetlands protected under sea level rise. We estimated a loss of 4%–31% of the current area of

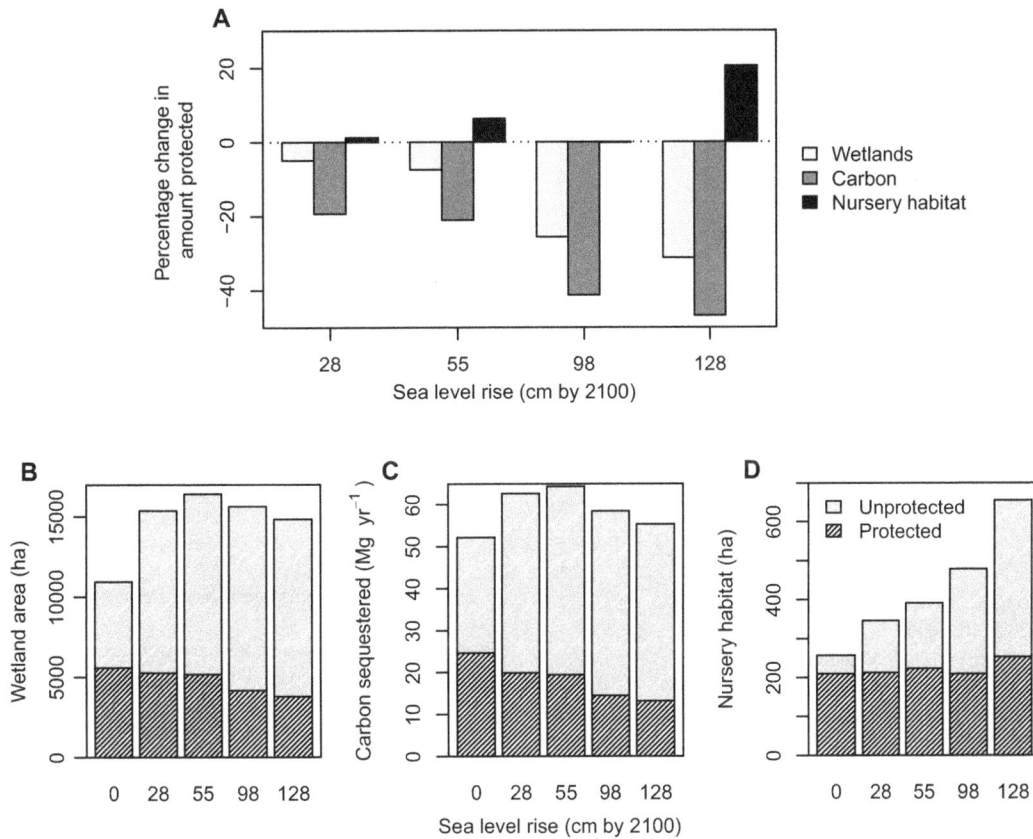

Figure 4 The change in the provision of wetlands and ecosystem services under sea level rise. Panel (A) shows the percentage change in the area of wetlands (wetlands), amount of carbon sequestration (carbon), and area of nursery habitat for commercially important species (nursery habitat) under sea level rise based on the current reserve network. The remaining panels show the area of wetlands (B), amount of carbon sequestration (C), and area of nursery habitat for commercially important species (D) that would be protected and unprotected in 2100 based on the current reserve network in Moreton Bay. "Protected" refers to areas that are currently contained within the reserve network, and "unprotected" refers to all other areas.

protected wetlands, with higher sea level rise scenarios resulting in lower levels of protection, despite an overall increase in wetland extent (Figure 4).

Therefore, to maintain the area of wetlands protected under future sea level rise, additional resources are required to expand the reserve network to allow for wetland migration. Under the lower rates of sea level rise (28 and 55 cm), matching the current level of protection would only require a modest additional investment (up to $40,000 AUD), yet a much larger investment is required under the higher rates of sea level rise (98 and 128 cm, a 377% [$151,000 AUD] and 677% [$271,000 AUD] increase, respectively, over lower rates of sea level rise) (Figure 5 and Figure S1). Further, increasing the level of protection beyond current levels exacerbates the increase in cost even further. For example, under current sea levels, a 20% increase in the area of wetlands protected would cost $105,000 AUD, with much of this target being met on public lands. However, as coastal wetlands move landward onto private land under the higher sea level rise scenarios, the required investment to match this target could be up to $1.3 million AUD (a 1,138% increase over current sea levels, Figure 5).

Payments for ecosystem services

Payments for ecosystem services have the potential to attenuate the opportunity costs of protection. We found that a carbon payment alone (at $6.11 MgC^{-1} AUD) completely compensated for the cost of protecting an additional 32–33 km^2 of wetlands (a ~60% increase over the current reserve network) under the baseline (0 cm) and lower sea level rise scenarios (28 and 55cm, Figure 5). However, under higher rates of sea level rise (98 and 128 cm), including a carbon payment only compensated for the cost of protecting an additional 20 and 15 km^2 (a 37% and 27% increase from the current reserve network), respectively (Figure 6). Stacking carbon

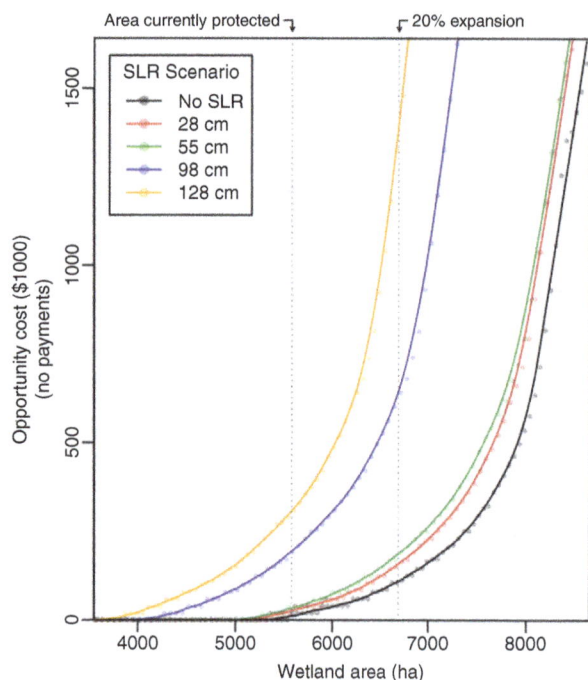

Figure 5 The total cost of preserving increasing wetlands under different rates of sea level rise (SLR) in the absence of payments for ecosystem services. Dotted lines indicate the area of wetlands that are currently contained within the reserve network (5,577 ha), and a 20% expansion of the area of wetlands protected (6,692 ha).

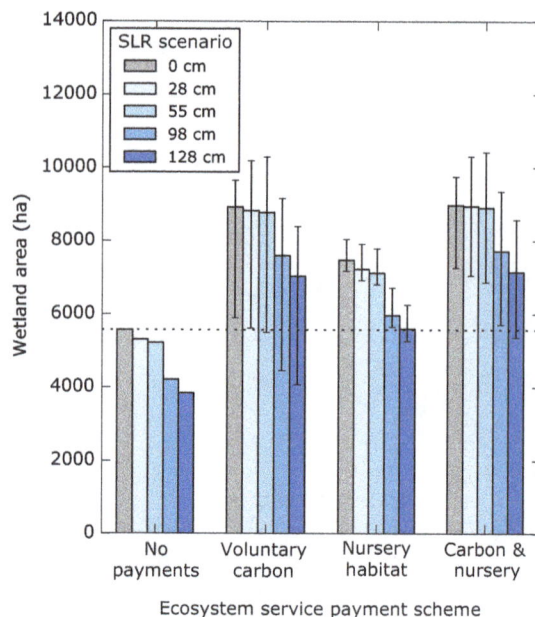

Figure 6 The maximum area of wetlands that can be preserved and still "break-even" ($0 cost) under different sea level rise (SLR) scenarios and payments for ecosystem services. The "break even" point is where the capitalized revenue from ecosystem service payments exceeds the opportunity cost of expanding the reserve network. "No payments" refers to the baseline case where there are no payments for any ecosystem services. "Voluntary carbon" is the result of an active voluntary carbon market with recent (2012) carbon prices. "Nursery habitat" refers to payments that could flow from a levy on the gross value of production for commercially important and mangrove dependent species. "Carbon & nursery" is the result from stacking payments for carbon and nursery habitat. Error bars represent the minimum and maximum wetland area based on variations in discount rates, voluntary carbon payments, and the method used to calculate the amount of nursery habitat. The dotted line indicates the wetland area that is currently contained within the reserve system (5,577 ha).

payments with a potential nursery habitat payment provided only a modest additional expansion over carbon payments alone (up to an additional 1.3 km² [~2% increase]), as the most cost-efficient areas for nursery habitat were already selected by a payment for carbon (Figure 6). Protecting a smaller area of wetlands (than given by the above values) would be more than compensated for by ecosystem service payments, as the capitalized value of the ecosystem services exceeded the opportunity cost of the reserve network (Figure S2).

Discussion

We have shown that substantial changes in the distribution of coastal wetlands under seal level rise are likely to lead to increases the costs of protecting them. Consistent with other studies, we predicted a landward movement of wetlands (particularly mangroves) under sea level rise (Traill *et al.* 2011; Di Nitto *et al.* 2014; Saintilan *et al.* 2014; Figure 3B). This landward movement, combined with the positive association between land values and elevation (Figure 2) drives the increase in cost of pre-emptively protecting wetlands to facilitate landward wetland migration under sea level rise. In fact we show that the higher the sea level rise projection, the higher the opportunity

cost of expanding the protected area network (Figure 5). This higher cost of preserving coastal wetlands is likely to be a general consequence of sea level rise, particularly in regions where the potential for urban development places further upward pressure on coastal land values.

Despite these higher costs, payments for ecosystem services have the potential to substantially reduce the net cost of expanding the reserve network under sea level rise. It is possible that the benefits from payments for ecosystem services could be further increased under different market conditions. For example, even more wetlands could be preserved if the carbon price reflected the social value of carbon (i.e., the total economic damages from emitting an additional 1 MgC^{-1}), or if these higher carbon payments were combined with those for the total value of nursery habitat. In both of these cases, the capitalized values of the services exceed the opportunity cost for all modeled wetland targets (up to

80% of the total wetland area in each scenario) (Table S2). Furthermore, including payments for additional ecosystem services not quantified here, such as storm protection or nutrient retention, would likely increase the economic benefits of coastal wetland protection.

While receiving payments for ecosystem services reduces the costs of coastal wetland protection for local planning authorities, this cost is shifted to the beneficiaries of the services. Carbon sequestration has potential buyers in both the public and private sectors, and transactions can be facilitated through the relatively well-established voluntary carbon market (Hamrick *et al.* 2015). In this case, shifting the cost burden to the buyer is unlikely to be problematic, as the buyers' participation is voluntary (such as individuals who purchase voluntary carbon offsets for air travel (Mair 2011)). In contrast, a nursery habitat payment shifts the costs to local fisheries via a compulsory levy. This may face opposition from commercial fishers if the additional cost is perceived to threaten the economic viability of their enterprise (Marshall 2007). Given that stacking nursery habitat payments with carbon payments facilitated only a modest additional expansion of the reserve network over carbon payments alone (~2%, Figure 6), the additional administrative burden and potential controversy of a nursery habitat levy might not be justified in this case.

It is imperative that local planning authorities pre-emptively limit development in dryland areas that are likely to transition to wetlands under climate change. The primary difficulty in implementing this strategy is that the opportunity costs of purchasing properties or re-zoning land are borne immediately, whereas the benefits take much longer to materialize and often flow to beneficiaries external to the local area (Friess *et al.* 2015). Even when the capitalized value of payments for ecosystem services exceed the opportunity cost of expanding the reserve network, the revenue from ecosystem service markets would not start flowing until the wetlands had migrated sufficiently landward. This delay in receiving benefit could explain why this strategy is not adopted in many vulnerable areas, despite the long term advantages. For example, local and state governments along the USA Atlantic coast plan to develop 60% of land below 1 m elevation (Titus *et al.* 2009), and Australian state governments across the eastern sea board have removed sea level rise from state planning policies (Bell & Baker-Jones 2014). However, climate change adaptation policies are emerging in other areas, such as the Thames Estuary 2100 plan (for London and the tidal reaches of the Thames river) which incorporates a projected sea level rise of up to 1.9 m and includes provisions for intertidal habitat creation (Environment Agency 2012).

Given the dynamic nature of land markets under sea level rise, coastal land may be cheaper in the future as flood risk increases (Bin *et al.* 2011). However, this does not necessarily justify local planning authorities delaying the purchase or re-zoning these areas. If new dwellings or other hard structures are permitted in the potential future locations of wetlands or their migration pathways, this will not only impact biodiversity through arresting wetland migration, but will also have socio-economic impacts. For example, the costs may be shifted to the coastal property owner who may face reduced property prices, periodic flooding, or relocation in a worst-case scenario. Furthermore, it may not always be the case that the cost of coastal land will decline. Despite increasing risks, coastal populations are large and growing (Martínez *et al.* 2007), which is likely to create upward pressure on land prices in future (Glaeser *et al.* 2005). Furthermore, future risks may not be given appropriate consideration (Newell *et al.* 2015), particularly if insurance companies are able to compensate damages (Bagstad *et al.* 2007) or the impacts of sea level rise are predicted to occur outside of the investors' outlook.

We have shown here that payments for ecosystem services can alleviate some of the costs of expanding the coastal reserve network under climate change, and in many cases may result in a profit in the long run. These cost reductions are possible because the costs are shifted from planning authorities to the beneficiaries of the services, which may not always be well received. Higher rates of sea level rise can reduce the effect of payments for ecosystem services, which highlights the importance of ambitious climate change mitigation efforts alongside adaptation plans. Although profits are possible in the long run, planning authorities may be strained in the short term, as some of the revenue from ecosystem service payments would not be received until wetland migration occurred. Alternatively, delaying the implementation of climate change adaptation policy may risk losing key areas of coastal wetlands, the species they support, and services they provide.

Acknowledgments

We would like to thank Yann Dujardin for assistance with formulating the integer linear programming problem, Lochran Traill and Karin Perhans for providing a parameterized sea level affecting marshes model, and Kerrie Wilson for comments on the manuscript. This research was funded by the Australian Research Council Centre of Excellence for Environmental Decisions and Australian Research Council Discovery Project DP130100218. RKR is supported by the University of Queensland – Commonwealth Scientific and Industrial Research Organisation

(CSIRO) Integrated Natural Resource Management Postgraduate Fellowship. CEL is supported by the Australian Research Council Superscience project FS100100024. HLB is supported by the Australian Research Council Discovery Early Career Researcher Award.

Supporting Information

Figure S1: The change in cost of preserving wetlands under increasing rates of sea level rise (SLR) and different market conditions when compared to the baseline (no sea level rise).

Figure S2: The variation in the potential for ecosystem services to attenuate the costs of preserving wetlands under sea level rise. The shaded areas for carbon and nursery habitat payments represent the uncertainty from varying the discount rate, the method for calculating nursery habitat, and the carbon price. Negative costs indicate a net gain (profit).

Table S1: The change in the provision of wetlands and ecosystem services under sea level rise.

Table S2: The variation in the potential for payments that reflect the social value of carbon and the total value of nursery habitat to attenuate the costs of preserving wetlands under sea level rise.

Table S3: The additional cost from using the strict connectivity requirement when compared to the more flexible connectivity requirement (in $1,000s 2012 AUD).

Table S4: The mean nursery habitat value and total site value based on the linear feature, 5 m strip and 10 m strip.

Table S5: The variation in, and combinations of, ecosystem value estimates and discount rates when capitalizing the value of ecosystem services to 2100.

References

Arkema, K.K., Guannel, G., Verutes, G. *et al.* (2013). Coastal habitats shield people and property from sea-level rise and storms. *Nat. Clim. Chang.*, **3**, 913-918.

Bagstad, K.J., Stapleton, K. & D'Agostino, J.R. (2007). Taxes, subsidies, and insurance as drivers of United States coastal development. *Ecol. Econ.*, **63**, 285-298.

Barbier, E.B., Hacker, S.D., Kennedy, C., Koch, E.W., Stier, A.C. & Silliman, B.R. (2011). The value of estuarine and coastal ecosystem services. *Ecol. Monogr.*, **81**, 169-193.

Bell, J. & Baker-Jones, M. (2014). Retreat from retreat—the backward evolution of sea-level rise policy in Australia, and the implications for local government. *Local Gov. Law J.*, **19**, 23-35.

Bin, O., Poulter, B., Dumas, C.F. & Whitehead, J.C. (2011). Measuring the impact of sea-level rise of coastal real estate: A hedonic property model approach. *J. Reg. Sci.*, **51**, 751-767.

Church, J.A., Clark, P.U., Cazenave, A. *et al.* (2013). *Sea level change*. In T.F. Stocker, D. Qin, G.-K. Plattner, M. Tignor, S.K. Allen, J. Boschung, A. Nauels, Y. Xia, V. Bex & P.M. Midgley, editors. *Clim. Chang. 2013 Phys. Sci. Basis. Contrib. Work. Gr. I to Fifth Assess. Rep. Intergov. Panel Clim. Chang.* Cambridge University Press, Cambridge, and New York, NY.

Clough, J.S., Park, R.A., Polaczyk, A. & Fuller, R. (2012). *SLAMM 6.2 Technical documentation*. Warren Pinnacle Consulting, Waitsfield, Vermont.

Craft, C., Clough, J., Ehman, J. *et al.* (2009). Forecasting the effects of accelerated sea-level rise on tidal marsh ecosystem services. *Front. Ecol. Environ.*, **7**, 73-78.

Dasgupta, S., Laplante, B., Meisner, C., Wheeler, D. & Yan, J. (2009). The impact of sea level rise on developing countries: a comparative analysis. *Clim. Change*, **93**, 379-388.

Department of Environment and Resource Management (DERM) & Department of Environment and Resource Mangement (DERM). (2013). *Queensland valuation and sales (QVAS)*. Queensland Government, Brisbane, Australia.

Environment Agency. (2012). *Thames Estuary 2100 (TE2100 Plan) November 2012*. Environment Agency, London, UK.

FitzGerald, D. & Fenster, M. (2008). Coastal impacts due to sea-level rise. *Annu. Rev. Earth Planet. Sci.*, **36**, 601-647.

Friess, D.A., Phelps, J., Garmendia, E. & Gómez-Baggethun, E. (2015). Payments for Ecosystem Services (PES) in the face of external biophysical stressors. *Glob. Environ. Chang.*, **30**, 31-42.

Gallant, J. (2010). *1 second SRTM Level 2 Derived Digital Elevation Model v1.0*. Geoscience Australia, Canberra.

Geselbracht, L., Freeman, K., Kelly, E., Gordon, D. & Putz, F. (2011). Retrospective and prospective model simulations of sea level rise impacts on Gulf of Mexico coastal marshes and forests in Waccasassa Bay, Florida. *Clim. Change*, **107**, 35-57.

Glaeser, E.L., Gyourko, J. & Saks, R.E. (2005). Urban growth and housing supply. *J. Econ. Geogr.*, **6**, 71-89.

Hamrick, K., Goldstein, A., Peters-Stanley, M. & Gonzolez, G. (2015). *Ahead of the curve: state of the voluntary carbon markets 2015*. Forest Trends' Ecosystem Marketplace, Washington, D.C.

Hinkel, J., Lincke, D., Vafeidis, A.T. *et al.* (2014). Coastal flood damage and adaptation costs under 21st century sea-level rise. *Proc. Natl. Acad. Sci. U.S.A.*, **111**, 3292-3297.

IPCC. (2007). *Climate change 2007: the physical science basis, contribution of working group i to the fourth assessment report of the intergovernmental panel on climate change*. Cambridge University Press, Cambridge and New York, NY.

IUCN. (2013). The IUCN Red List of Threatened Species. Version 2013.1. [WWW Document]. http://www.iucn redlist.org, Accessed, October 25, 2013.

Iwamura, T., Possingham, H.P., Chadès, I. et al. (2013). Migratory connectivity magnifies the consequences of habitat loss from sea-level rise for shorebird populations. Proc. R. Soc. B Biol. Sci., **280**, 20130325.

Kirwan, M.L. & Megonigal, J.P. (2013). Tidal wetland stability in the face of human impacts and sea-level rise. Nature, **504**, 53-60.

Lovelock, C.E., Cahoon, D.R., Friess, D.A. et al. (2015). The vulnerability of Indo-Pacific mangrove forests to sea-level rise. Nature, **526**, 559-563.

Mair, J. (2011). Exploring air travellers' voluntary carbon-offsetting behaviour. J. Sustain. Tour., **19**, 215-230.

Marshall, N.A. (2007). Can policy perception influence social resilience to policy change? Fish. Res., **86**, 216-227.

Martínez, M.L., Intralawan, A., Vázquez, G. et al. (2007). The coasts of our world: ecological, economic and social importance. Ecol. Econ., **63**, 254-272.

McLeod, E., Poulter, B., Hinkel, J., Reyes, E. & Salm, R. (2010). Sea-level rise impact models and environmental conservation: a review of models and their applications. Ocean Coast. Manag., **53**, 507-517.

Mills, M., Nicol, S., Wells, J.A. et al. (2014). Minimizing the cost of keeping options open for conservation in a changing climate. Conserv. Biol., **28**, 646-653.

Newell, B.R., Rakow, T., Yechiam, E. & Sambur, M. (2015). Rare disaster information can increase risk-taking. Nat. Clim. Chang., **6**, 158-161.

Di Nitto, D., Neukermans, G., Koedam, N. et al. (2014). Mangroves facing climate change: landward migration potential in response to projected scenarios of sea level rise. Biogeosciences, **11**, 857-871.

Parry, M., Lowe, J. & Hanson, C. (2009). Overshoot, adapt and recover. Nature, **458**, 1102-1103.

Pink, B. (2011). Australian statistical geography standard (ASGS): Volume 5: Remoteness structure. Aust. Bur. Stat. Australian Bureau of Statistics, Canberra, Australia.

Saintilan, N., Wilson, N.C., Rogers, K., Rajkaran, A. & Krauss, K.W. (2014). Mangrove expansion and salt marsh decline at mangrove poleward limits. Glob. Chang. Biol., **20**, 147-157.

Stern, H., Hoedt, G. de & Ernst, J. (2000). Objective classification of Australian climates. Aust. Meteorol. Mag., **49**, 87-91.

Syvitski, J.P.M., Kettner, A.J., Overeem, I. et al. (2009). Sinking deltas due to human activities. Nat. Geosci., **2**, 681-686.

Titus, J.G., Hudgens, D.E., Trescott, D.L. et al. (2009). State and local governments plan for development of most land vulnerable to rising sea level along the US Atlantic coast. Environ. Res. Lett., **4**, 044008.

Traill, L.W., Perhans, K., Lovelock, C.E. et al. (2011). Managing for change: wetland transition under sea level rise and outcomes for threatened species. Divers. Distrib., **17**, 1225-1233.

WWF. (2000). G200 Maps (1999-2000) [WWW Document]. WWF Glob. http://wwf.panda.org/about_our_earth/ecor egions/maps/, Accessed September 17, 2015.

14

Navigating the Space between Research and Implementation in Conservation

Anne H. Toomey[1,2], Andrew T. Knight[3,4,5,6], & Jos Barlow[7,8]

[1] Department of Environmental Studies and Science, Pace University, New York, New York, USA
[2] Center for Biodiversity and Conservation, American Museum of Natural History, New York, New York, USA
[3] Department of Life Sciences, Imperial College London, Ascot, Berkshire, United Kingdom
[4] ARC Centre of Excellence in Environmental Decisions, The University of Queensland, Brisbane, Queensland, Australia
[5] Department of Botany, Nelson Mandela Metropolitan University, Port Elizabeth, South Africa
[6] The Silwood Group, London, United Kingdom
[7] Lancaster Environment Centre, Lancaster University, Lancaster, UK
[8] Museu Paraense Emilio Goeldi, Belém, Brazil

Keywords
Knowing-doing; evidence-based conservation; society–science divide; research-implementation gap; social learning; alternative knowledges; research-implementation spaces.

Correspondence:
Andrew T. Knight, Department of Life Sciences, Imperial College London, Room W1.2, Kennedy Building, Silwood Park Campus, Buckhurst Road, Ascot, Berkshire SL5 7PY, UK.
E-mail: andrew.knight1@imperial.ac.uk

Abstract

Recent scholarship in conservation biology has pointed to the existence of a "research-implementation" gap and has proposed various solutions for overcoming it. Some of these solutions, such as evidence-based conservation, are based on the assumption that the gap exists primarily because of a communication problem in getting reliable and needed technical information to decision makers. First, we identify conceptual weaknesses with this framing, supporting our arguments with decades of research in other fields of study. We then reconceptualize the gap as a series of crucial, productive spaces in which shared interests, value conflicts, and complex relations between scientists and publics can interact. Whereas synonyms for "gap" include words such as "chasm," "rift," or "breach," the word "space" is connected with words such as "arena," "capacity," and "place" and points to who and what already exists in a specific context. Finally, we offer ways forward for applying this new understanding in practice.

Introduction

In the mid-1980s, the Society for Conservation Biology (SCB) was established, promoting a new kind of science whose success would be measured by the degree to which it could help to sustain the health and diversity of the natural world (Meine *et al.* 2006). However, such potential came with a warning, and Soule (1986, p. 4) cautioned that the new "mission-driven" discipline could remain in the "mental world of academia" if its followers did not actively engage with "real-world" problems, circumstances, and experiences. Three decades later, this warning has developed into a "vigorous debate" about the "gap" between research and implementation in conservation, leading some to question whether the

field has lost sight of its mission (Knight *et al.* 2008; Arlettaz *et al.* 2010). Much has been discussed and written about the gap, which has been described, rather generally, as a process by which "scientific information accumulates, but is not incorporated into management actions" (Matzek *et al.* 2014, p. 208).

The conceptualization of an issue greatly influences the ways in which it is perceived, framed, and bounded and hence the types of responses and solutions people create to address it (Nisbet & Scheulefe 2009; Newell *et al.* 2014). As such, conceptualizations reflect both how we "know" and the future knowledge that can be produced. While the literature on the research-implementation gap in the conservation sciences has focused on developing and promoting a litany of solutions for its "bridging" (e.g., evidence-based conservation, conservation evaluation, and science communication), very little effort has focused on whether the "gap" is an accurate description of the challenges we face. We argue that the way in which the research-implementation gap is conceptualized is a central but overlooked dimension within conservation science. Here, we: (1) offer a critique of current conceptualizations of the gap; (2) present an alternative framing that enables the identification of more carefully focused questions useful for improving our collective effectiveness; and (3) offer ways for applying this new understanding in practice.

The research-implementation gap as a linear model

From the early days of the SCB, the process for ensuring the persistence of nature was clearly framed: conservation scientists could, and should, be motivated by ethical concerns, but their work must be rooted in an objective "firm scientific basis" (Meine *et al.* 2006, p. 636). This requirement for problems to be quantitatively defined and their solutions founded upon science set the tone and trajectory for the evolution of conservation as a discipline. Currently, this often unstated perspective manifests as a belief that the world faces environmental problems best addressed by a self-selected group of experts (i.e., conservation scientists) providing evidence-based solutions to decision makers. Soule (1986, p.3) likened the operations of this process to "...a shuttle bus going back and forth, with a cargo of ideas, guidelines, and empirical results in one direction, and a cargo of issues, problems, criticism, constraints, and changed conditions in the other."

This conceptualization frames the relationship between research and practice as linear (Figure 1A–C): the influences of conservation problems proceed in one direction and the envisaged solutions—largely technical—proceed in another, emphasizing the role of conservation scientists as providing answers delivered as empirical information for "translation" into applied solutions by practitioners. This deficit-model of communication is supported by stronger calls for, and increasingly rapid advancement of, approaches promoting the primacy of quantitative scientific information, epitomized, for example, by the growth of evidence-based conservation as an approach to decision-making (Pullin *et al.* 2004; Adams & Sandbrook 2014; Walsh *et al.* 2014). Recent conservation science textbooks communicate and perpetuate this framing to students by emphasizing the importance that critical evidence syntheses, evaluations, and scientific consensus play in bridging the gap between scientists and practitioners (e.g., Macdonald & Willis 2013).

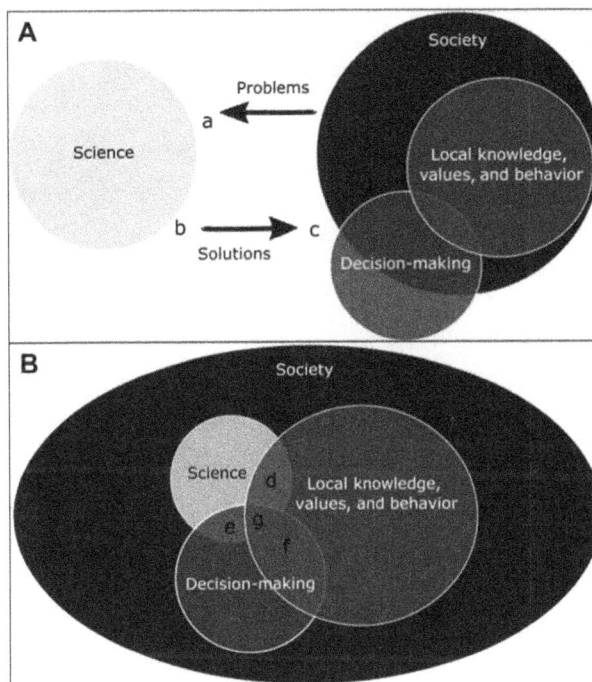

Figure 1 (A) The conventional conceptualization of research-implementation gaps in conservation: (a) in communicating societal concerns to science; (b) in translating scientific information to applicable recommendations; and (c) in disseminating scientific recommendations for policy-makers. (B) Research-implementation spaces: (d) public engagement in science, e.g., citizen science, (e) boundary work and organizations, (f) environmental activism, community-based conservation, (g) participatory action research, sustainability science.

However, the notion that larger quantities of evermore precise information inevitably lead to more effective outcomes in the "real-world" is refuted by well-established theories of human decision-making, such as value-action gap theory in social psychology and reflective practice theory in organizational management, along with contemporary understandings of the science–policy interface in communications and science and technology studies (Table 1). Scholars have convincingly shown that empirical "evidence" is only one factor (and often a minor one) influencing decision-making and change (Pielke 2007; Owens 2012). While the linear model depends upon the idea that science is a tool that enables the trained expert to uncover the "truth" about the universe (i.e., positivism), decades of science studies scholarship has demonstrated that "facts" are not perceived in the same way by different publics, but rather are filtered through existing beliefs, mental models, experiences, and concerns (Nisbet & Scheulefe 2009; Newell *et al.* 2014). Because of this, scientific information is unlikely to change the positions of stakeholders and decision makers

Table 1 Conceptualizations from various fields of study of the ways in which knowledge and action interact

Fields of study	Relationship between knowledge and action
Behavioral psychology	There is a "value-action gap" between the proenvironmental attitudes that people hold and the behaviors that they are willing to enact in order to address environmental issues, which cannot be overcome simply by using an "information deficit" model of individual participation. This is because there are many practical social or institutional constraints that may prevent people from adopting proenvironmental action, regardless of their attitudes or intentions. Thus, information accuracy is neither a necessary nor sufficient condition for producing desired behavioral outcomes and may in fact be irrelevant to decision-making (Blake 1999).
Organizational learning	In situations of uncertainty, instability, uniqueness, and value conflict, reflection-in-action is crucial. Reflection-in-action is at the core of "professional artistry" and the best professionals rely more on "tacit knowledge" than "technical-rationality," which erroneously maintains that problems are solvable through the rigorous application of science (Schön 1983).
Programme evaluation	Evaluation is an action-oriented process for identifying stakeholder's values and goals, understanding programmes and policies, clarifying options, identifying improvements, and facilitating judgment and decision-making. It applies highly diverse systematic techniques, qualitative and quantitative tools, and sources of knowledge through collaborative processes to create opportunities for dialogue "spaces" that promote utility for intended users (Patton 1997).
Policy sciences	Diverse groups of stakeholders should be involved in problematic policy situations. In this context, science is viewed as a narrowly focused, value-laden, explicitly subjective (not objective) process, and in its traditional form, undemocratic given its typically top-down approach. Instead, multiple realities of situations exist that are partially socially constructed and operationalized through collaborative, explicitly negotiated, social processes (Clark 2002).
Science and technology studies	Western science is viewed as a local form of knowledge that privileges certain ways of viewing the world over others. Effective research is always embedded in a normative understanding of what are the correct questions. As such, it is not accurate or useful to see science as existing separately from social, cultural, and political processes of decision-making and action (Harding 2006).
Science communication	Research shows that, in the highly political environments in which many health and environmental issues are nested, scientific information will not be successful in swaying individuals holding factually incorrect beliefs, and may even reinforce those beliefs. As such, translation and dissemination of activities designed to supply more accurate "facts" is a relatively ineffective way to influence public judgments and decisions. Instead, there is a need to enhance civic capacities for discussing, debating, and participating in collective decision-making (Nisbet & Scheulefe 2009).
"Soft" systems thinking	The realization that "hard" reductionist thinking applied through highly quantitative tools is of limited utility for understanding and solving real-world problems led to the development of "soft" systems thinking. This field recognizes the necessity of incorporating multiple knowledge types, perspectives, and realities in decision-making, and hence the importance of structured debate as a "space" for identifying and implementing desirable and feasible change (Checkland 1984).

Note: Knowledge is typically seen to comprise subjectively held information (beliefs, values, experiences, and rules), not simply reductionist science. Linear "information-deficit" models of knowledge transfer from science to decision makers apply only in very specific contexts.

in complex value laden situations and may even reinforce competing interests and alternate interpretations (Sarewitz 2004; Pielke 2007). While exceptions exist in which the linear model of scientific communication seems to apply, in fact such cases demonstrate how scientific knowledge can synergize with existing societal concerns and political readiness. For example, Arlettaz *et al.* (2010) describe how the practical involvement of

researchers in the implementation stage of a scientific study of the endangered hoopoe (*Upupa epops*) population in the Swiss Alps led it to its rapid recovery. While the knowledge used to determine appropriate conservation actions was evidence-based, the authors attribute social and political factors as being key to the project's success, citing "the tremendous support from regional authorities, and second, an incredible enthusiasm—after some initial skepticism—of the local farmers" (p. 839).

Unfortunately, such synergetic spaces of accord are uncommon for most environmental challenges. "Wicked problems" such as climate change and biodiversity loss are mired in debates fuelled by conflicting values regarding economics, social justice, and natural resources use. As a result, far from resolving discord, scientific information can further polarize debates around these issues (Pielke 2007; Nisbet & Scheulefe 2009).

In contrast to the linear model, research from various fields of study (Table 1) suggests that effective decision-making is based upon clear understandings of values, knowledge, rules, behaviors and actions, and the complex interactions between them. Such insights can assist in reconceptualizing and reframing our current understanding of the research-implementation gap by helping us envision impact as an ongoing emergent property of human-managed systems, one that manifests not only upon the completion of research, but also throughout the social and policy processes within which useful research is necessarily engaged and embedded (Clark 2002; Ascher *et al.* 2010; Jasanoff 2010).

From a "gap" to be bridged to "spaces" of interaction

Describing the challenge of research informing action as a "gap" suggests that there is something missing or lacking, and that there is a void that needs to be filled or a divide to be bridged (Van Kerkhoff & Lebel 2015). The concept of "space" more accurately describes that which exists between research and implementation, because it points to whom and what already exist in a specific context, conjures the excitement of discovery, and highlights the importance of values, ethics, institutions, and time. While synonyms for "gap" include words such as "chasm," "discontinuity," "rift," or "breach," the word "space" is connected with words such as "arena," "capacity," "leeway," and "place" and implies multiple dimensions. This conceptualization of the ways in which research and action coexist and interact in spaces also highlights the importance of entities and phenomena that emerge throughout the process of producing scientific knowledge—not just that of the end phase of disseminating or mainstreaming it. These include improvements in the number and utility of social interactions, attitudes, and institutional knowledge (Bottrill *et al.* 2012). Scientific research is a socioeconomic activity laden with power relations, cultural understandings (or misunderstandings), social interactions, and political consequences (Harding 2006; Jasanoff 2010). The extent to which local

values, knowledge, and behaviors are acknowledged, understood, and given due recognition *during the process* of conducting research can have great bearing on whether or not the results of such research are accepted or rejected in a given political or cultural context (Toomey 2016).

This directly points to the importance of thinking about *who* is involved in the production of knowledge. As Roux *et al.* (2006) argue, knowledge is not a "thing" to be transferred, but rather a "process of relating that involves negotiation of meaning among partners" (p. 11). This necessitates a reconceptualization of the "gap" as series of interactive spaces (Figure 1d–g), emphasizing who operates within it, including those who live on the land where fieldwork takes place, the history of outsiders in that place, as well as consideration of present-day sociocultural relations and political context (Toomey 2016). In this sense, the idea of "space" requires a reconsideration of who has traditionally been included or excluded from decision-making, how and by whom research is conducted, and for what purposes (Harding 2006). Much existing research on the "gap" typically focusses on the knowledge available to, and applied by, "conservation practitioners" or "resource managers" (e.g., Pullin *et al.* 2004; Matzek *et al.* 2014; Walsh *et al.* 2014). The structure of such research implicitly applies a conceptualization of the interaction between research and action framed as a relationship between two self-defined stakeholder groups (scientists and conservation managers) that are connected through a unidirectional flow of scientific (and purportedly reliable) information. The voices, ideas, knowledges, and concerns of other types of stakeholders (e.g., farmers, indigenous and traditional communities, park guards, community activists, and teachers) are typically missing in such investigations (Smith *et al.* 2009).

In contrast, scholars in the fields of anthropology, ethnobiology, and political ecology have long examined the diverse ways people come to know and adapt to their environments, and more recently, the conservation social sciences have focused inquiry on the varied perceptions of the policies and practices of conservation itself (Bennett *et al.* 2016). Van Kerkhoff & Lebel's (2015) guest editorial in *Ecology and Society* presents case studies from across the world in which researchers engaged with other stakeholder groups with the aim to better connect the links between scientific knowledge, sociopolitical conditions, and environmental governance at multiple scales. They demonstrate the importance of understanding the diverse ways in which different groups negotiate the spaces between science and governance, and explain why prior experiences, preconceptions, and expectations of stakeholders can have important implications for the extent to which people are willing to participate in new collaborative research processes. Thus, promoting a notion of interconnectedness at the heart of the science–society relationship can help to promote the effective positioning of all stakeholders within policy and social processes aimed at ensuring the persistence of nature.

From a new way of knowing to a new way of doing

The conceptualization of the research-implementation gap currently dominant within conservation science provides a paradox that hinders both our individual and collective effectiveness. On the one hand, rhetoric expressed in peer-reviewed journals argues strongly for stakeholder collaboration; breaking down disciplinary barriers; integration of local, traditional, indigenous, and scientific knowledge systems; and "extra-academic" activities such as outreach work (Balmford & Cowling 2006). On the other hand, the prevailing dominance of the linear conceptualization used for translating research into action undermines these objectives and prescribes, a priori, how conservation professionals design and implement their research; how collaboration is framed, operationalized, and who is involved; which studies are published; and how impact is evaluated. The reconceptualization of the research-implementation gap as a space encourages conservation professionals and their partners to engage and collaborate and to more effectively identify and understand *for whom* and *what* knowledge is produced, and the diversity of ways that this can be achieved. Thus, we propose replacing the terminology of the "research-implementation *gap*" with that of "*research-implementation spaces*" (alternately, research-practice spaces or knowledge-action spaces) as a starting point for reconceptualizing the diversity of ways of knowing and doing.

Instead of a linear, knowledge deficit-based model of scientific impact, we envisage the embedding of conservation science into the collaborative social and decision processes comprising the spaces where policy scenarios and grassroots action play-out (*sensu* Clark 2002). This reframing recognizes that conservation is a social process that engages science, not a scientific process that engages society (Balmford & Cowling 2006; Adams & Sandbrook 2014). Urging a reconceptualization of the research-implementation gap as a series of crucial, productive spaces in which shared interests, value conflicts, and complex relations between conservation biologists and the public can interact is not merely a conceptual shift, but also a practical one. In moving forward, we point to two interconnected areas in which the conservation community can begin to make changes in how we operate as professionals: in the field and within our educational institutions.

In the field, conservation scientists can broaden their roles by creating spaces in which interested stakeholders (farmers, ranchers, communities, and hunters) can engage with research in a diversity of ways. This most fundamentally begins by identifying conservation challenges with stakeholders and collaboratively developing research plans. Collaborative research approaches, such as participatory action research (PAR)—a longstanding research approach in the social sciences that is characterized by a theory of change—can provide theories and methods for adapting research topics to those of direct relevance to people who live and work in a given region. In the conservation sciences, PAR philosophy and methods are often incorporated into biocultural approaches to conservation, which recognize and support the interplay between biological and sociocultural systems through locally grounded environmental research and action. For example, the Global Diversity Foundation (http://www.global-diversity.org) conducts community-led, environmental justice-oriented research in order to promote biological, agricultural, and cultural diversity around the world. One of their projects, "An integrated approach to plant conservation in the Moroccan High Atlas," works with partner communities to strengthen cultural practices of conservation and restore traditional water management systems for the protection of plant biodiversity and medicinal livelihoods in the Mediterranean.

To effectively create, facilitate, and participate in research-implementation spaces, conservation scientists require education and training that goes beyond scientific positivism for a reemphasis on the building of a wide spectrum of *conservation capacities* (see also "coproductive capacities," Van Kerkhoff & Lebel 2015). This requires a renewed focus on the types of thinking, skills, and resourcing needed to more effectively navigate such spaces, rather than a continuous insistence on producing the "best available evidence" to fill a "gap" (Pullin et al. 2004). For example, students studying conservation science should be trained in professional problem-solving skills (e.g., creative and critical thinking, active listening, programme evaluation, participatory planning methods, and systems thinking) and to be presented with important perspectives from disciplines that have traditionally been marginalized in conservation science, such as psychology, sociology, anthropology, and development studies (Bennett *et al.* 2016). In order to achieve this, environmental studies departments need to move beyond multidisciplinary (where faculty brings expertise from multiple fields but tends to stay within the boundaries of its disciplines) toward inter- and transdisciplinary, where shared goals and values at the departmental level (and a carefully constructed curriculum) can train students to seek out, understand, integrate, and apply different types of knowledge gained in their courses to problems in the real world (Clark *et al.* 2011). More determined efforts from university departments to partner with community-based organizations, environmental nonprofits, and other stakeholder groups for undergraduate and graduate research collaborations will do much to ensure that students are already inhabiting research-implementation spaces as part of their formal training.

Importantly, while the contexts in which *research-implementation spaces* will emerge are innumerable, conservation scientists can seek to better understand the scales, boundaries, interrelationships, perspectives, and ethical parameters through which they can be navigated. To ensure that they are transformative, there is a need for deep self-reflection by conservation professionals, both within our organizations and through our broader collaborations with other stakeholders. In so doing, we can begin to understand how our own limitations (e.g., our worldviews, cognitive biases, and fears) hinder our effectiveness. Having the courage to act with humility, question our own assumptions and worldviews, and trial and learn from what we believe to be more effective approaches will provide the prerequisite for moving beyond the safe notion of a "gap" toward the uncertainty and complexity inherent in the inhabiting of new spaces.

Acknowledgments

The manuscript benefited from thoughtful comments by Chris Sandbrook, Daniel Tregidgo, and three anonymous reviewers, as well as conversations with Beth Brockett and Emily Adams. We also thank Rebecca Ellis and Saskia Vermeylen for their insights on earlier versions of this work.

References

Adams, W.M. & Sandbrook, C. (2014). Conservation, evidence and policy. *Oryx*, **47**, 329-335.

Arlettaz, R., Schaub, M., Fournier, J. *et al.* (2010). From publications to public actions: when conservation biologists bridge the gap between research and implementation. *BioScience*, **60**, 835-842.

Ascher, W., Steelman, T. & Healy, R. (2010). *Knowledge and environmental policy: re-imagining the boundaries of science and politics*. The MIT Press, Massachusetts, USA.

Balmford, A. & Cowling, R.M. (2006). Fusion or failure? The future of conservation biology. *Conserv. Biol.*, **20**, 692-695.

Bennett, N.J., Roth, R., Klain, S.C. *et al.* (2016). Mainstreaming the social sciences in conservation. *Conserv. Biol.*, **31**, 56-66.

Blake, J. (1999). Overcoming the 'value-action gap' in environmental policy: tensions between national policy and local experience. *Local Env.*, **4**, 257-278.

Bottrill, M.C., Mills, M., Pressey, R.L., Game, E.T. & Groves, C. (2012). Evaluating perceived benefits of ecoregional assessments. *Conserv. Biol.*, **26**, 851-861.

Checkland, P. (1984). *Systems thinking, systems practice*. John Wiley & Sons, Chichester, UK.

Clark, S.G., Rutherford, M.B., Auer, M.R. *et al.* (2011). College and university environmental programs as a policy problem (part 2): strategies for improvement. *Environ. Manage.*, **47**, 716-726.

Clark, T.W. (2002). *The policy process: a practical guide for natural resource professionals*. Yale University Press, New Haven.

Harding, S. (2006). *Science and social inequality: feminist and postcolonial issues*. University of Illinois Press, Chicago.

Jasanoff, S. (2010). A new climate for society. *Theor. Cult. Soc.*, **27**, 233-253.

Knight, A.T., Cowling, R.M., Rouget, M., Balmford, A., Lombard, A.T. & Campbell, B.M. (2008). Knowing but not doing: selecting priority conservation areas and the research–implementation gap. *Conserv. Biol.*, **22**, 610-617.

Macdonald, D.W. & Willis, K.J. (2013). *Key topics in conservation biology 2*. John Wiley & Sons, Chichester, UK.

Matzek, V., Covino, J., Funk, J.L. & Saunders, M. (2014). Closing the knowing–doing gap in invasive plant management: accessibility and interdisciplinarity of scientific research. *Conserv. Lett.*, **7**, 208-215.

Meine, C., Soule, M. & Noss, R.F. (2006). "A mission-driven discipline": the growth of conservation biology. *Conserv. Biol.*, **20**, 631-651.

Newell, B.R., McDonald, R.I., Brewer, M. & Hayes, B.K. (2014). The psychology of environmental decisions. *Annu. Rev. Environ. Resour.*, **39**, 443-467.

Nisbet, M.C. & Scheufele, D.A. (2009). What's next for science communication? Promising directions and lingering distractions. *Am. J. Bot.*, **96**, 1767-1778.

Owens S. (2012). Experts and the environment: the UK Royal Commission on Environmental Pollution 1970–2011. *J. Environ. Law*, **24**, 1-22.

Patton, M.Q. (1997). *Utilization-focused evaluation: the new century text*. 3rd edn. Sage Publications, Los Angeles, California.

Pielke, R.A.J. (2007). *The honest broker: making sense of science in policy and politics*. Cambridge University Press, Cambridge.

Pullin, A.S., Knight, T.M., Stone, D.A. & Charman, K. (2004). Do conservation managers use scientific evidence to support their decision-making? *Biol. Conserv.*, **119**, 245-252.

Roux, D.J., Rogers, K.H., Biggs, H.C., Ashton, P.J. & Sergeant, A. (2006). Bridging the science—management divide: moving from unidirectional knowledge transfer to knowledge interfacing and sharing. *Ecol. Soc.*, **11**(1): 4. URL http://www.ecologyandsociety.org/vol11/iss1/art4/

Sarewitz, D. (2004). How science makes environmental controversies worse. *Environ. Sci. Policy*, **7**, 385-403.

Schön, D.A. (1983). *The reflective practitioner: how professionals think in action*. Temple Smith, London.

Smith, R.J., Veríssimo, D., Leader-Williams, N., Cowling, R.M. & Knight, A.T. (2009). Let the locals lead. *Nature*, **462**, 280-281.

Soule, M.E. (1986). Conservation biology and the 'Real World'. In M.E. Soule, editor. *Conservation biology. The science of scarcity and diversity*. Sinauer Associates, Sunderland, 5-12.

Toomey, A.H. (2016). What happens at the gap between knowledge and practice? Spaces of encounter and misencounter between environmental scientists and local people. *Eco. Soc.*, **21**(2): 28. http://dx.doi.org/10.5751/ES-08409-210228.

Van Kerkhoff, L.E. & Lebel, L. (2015). Coproductive capacities: rethinking science–governance relations in a diverse world. *Ecol. Soc.*, **20**(1): 14. http://dx.doi.org/10.5751/ES-07188-200114.

Walsh, J.C., Dicks, L.V. & Sutherland, W.J. (2014). The effect of scientific evidence on conservation practitioners' management decisions. *Conserv. Biol.*, **29**, 88-98.

15

Social Outcomes of Community-based Rangeland Management in Mongolian Steppe Ecosystems

Tungalag Ulambayar[1], María E. Fernández-Giménez[2], Batkhishig Baival[3], & Batbuyan Batjav[1]

[1] Mongolian Institute of Geography and Geoecology, Ulaanbaatar, Mongolia
[2] Department of Forest and Rangeland Stewardship, Colorado State University, Campus Mail 1472, Fort Collins, CO, 80523–1472, USA
[3] Nutag Partners, Usnii St, Ulaanbaatar, Mongolia

Keywords
Community-based natural resource management; community-based conservation; pastoralism; integrated conservation and development; common pool resources; common property; environmental governance.

Correspondence
Maria E. Fernandez-Gimenez, Department of Forest and Rangeland Stewardship, Colorado State University, Campus Mail 1472, Fort Collins, CO 80523–1472, USA.
E-mail: maria.fernandez-gimenez@colostate.edu

Editor
Derek Armitage

Abstract

Community-based rangeland management (CBRM) has been promoted as a promising option for achieving both rangeland conservation and community well-being. However, research on its effectiveness is limited, and the reported outcomes are mixed, especially with regard to socioeconomic outcomes. We measured social outcomes of CBRM in Mongolia by comparing 77 formally organized pastoral groups with 65 traditional herder neighborhoods across four ecological zones. We used household surveys, focus groups, and interviews to measure livelihoods, social capital, and management behavior. Members of CBRM groups were significantly more proactive in addressing resource management issues and used more traditional and innovative rangeland management practices than non-CBRM herders. However, the group types did not differ in social capital or on most livelihood measures. Our results demonstrate that formal CBRM is strongly associated with herder behavior, but calls for consideration of how to reach livelihood outcomes, a key incentive for community-based conservation.

Introduction

Temperate grasslands are among the world's most imperiled ecosystems and Mongolia has one of the largest intact expanses of temperate grasslands on Earth. Occupying ~80% of Mongolia's territory, Mongolian rangelands are vulnerable to climate change (Batima 2006), extreme weather (Dagvadorj et al. 2009), degradation (Khishigbayar et al. 2015), and increasing rural poverty gaps (Janes 2010). Following the transition from Socialism to a market economy, Mongolia's livestock population boomed and use of traditional pastoral practices such as mobility and grazing reserves declined (Fernandez-Gimenez 2002; Fernandez-Gimenez & Batbuyan 2004). When droughts and harsh winters ensued in 1999–2003 and 2009–

2010, many herders suffered devastating losses (Sternberg 2010; Fernandez-Gimenez et al. 2012). In response, over 2,000 donor- and NGO-supported community-based rangeland management (CBRM) groups have organized in Mongolia since 1999, with the goals of improving rangeland conditions and herder well-being and livelihoods. This national-scale social experiment creates an unprecedented opportunity to assess the effectiveness of CBRM in meeting social development, livelihood and rangeland conservation goals.

Community-based conservation has been advanced as a win–win solution to biodiversity loss and poverty in communities that reside in and depend upon high-biodiversity ecosystems (Western & Wright 1994; Fabricius & Koch 2004). However, evidence about its effectiveness for rangeland conservation is scant. Most existing studies are limited by small sample sizes and few compare CBRM outcomes to alternative management regimes. Globally, studies of the social and economic effects of community-based natural resource management generally, and CBRM specifically, have shown mixed results (Dressler *et al.* 2010; Saito-Jensen *et al.* 2010; Suich 2010; Bowler *et al.* 2014). Even when measurable conservation benefits accrue, economic and social development benefits do not always occur or do not reach the wider community (Collomb *et al.* 2010; Nkhata & Breen 2010; Silva & Mosimane 2013; Bowler *et al.* 2014). Lack of or weak social and livelihood benefits of CBRM are concerning, because without direct benefits to households and communities, there is little incentive for local people to engage in or support conservation.

As elsewhere, research on Mongolian CBRM has been dominated by case studies (Upton 2008; Batkhishig *et al.* 2011; Dorligsuren *et al.* 2011; Baival & Fernandez-Gimenez 2012), most within a single ecological zone and lacking comparison sites without CBRM. Some studies have shown social benefits (Batkhishig *et al.* 2011; Baival & Fernandez-Gimenez 2012; Upton 2012; Fernandez-Gimenez *et al.* 2015) or livelihood improvements (Dorligsuren *et al.* 2011; Leisher *et al.* 2012), but others show no benefits (Addison *et al.* 2013) or negative CBRM effects (Upton 2008; Murphy 2011). Taking advantage of the social experiment underway in Mongolia, we report on the first large-*N*, case-control study of CBRM social outcomes that accounts for variability across different ecological contexts. Rigorous, broad-scale studies are essential to crafting conservation policies and providing donors with information to guide future investments (Bowler *et al.* 2014). As the first large-scale study of CBRM globally, it informs rangeland conservation strategy worldwide.

Our approach to assessing CBRM outcomes is informed by common pool resource (CPR) governance theory (Ostrom 1990; Agrawal 2002), and its application to CBNRM. CPR theory predicts that, given certain institutional attributes and external conditions, groups of resource users who share the same resources are capable of self-regulating resource use. We hypothesized that in the post-Socialist Mongolian context of ineffective national rangeland governance institutions and weakened customary use norms (Fernandez-Gimenez & Batbuyan 2004), formal CBRM groups aid herders to organize and self-govern pasture use, resulting in superior social and livelihood outcomes compared to traditional herder neighborhoods that lack strong or formal organization (Figure 1). Specifically, we expected that CBRM herders would have stronger and wider social networks, use more traditional and innovative management practices,

earn higher incomes, and hold more assets than herders from non-CBRM communities. Based on CPR theory, we also hypothesized that outcomes would differ by ecological zone. We expected that collective action would be more readily achieved, and social outcomes greater, where forage supplies are more abundant and predictable, CPRs are smaller, easier to delineate and monitor, and herders are less mobile (Ostrom 1990).

Methods

Sampling

Across four ecological zones—mountain and forest steppe, steppe, eastern steppe and desert steppe—we selected pairs of adjacent soums (districts) with ($n = 18$) and without ($n = 18$) formally organized CBRM (Figure 2). Within each soum, we randomly selected an average of five community groups sharing common grazing areas and water sources, and interviewed five households representing each group. We surveyed 706 herder households belonging to 142 groups; 65 traditional herder neighborhoods (hereon non-CBRM) and 77 formally organized CBRM groups. CBRM groups averaged 5 years since formation (range <1 to 14).

Because sampling before and after CBRM was established was impossible, we used community-level poverty and leadership indicators and group-level demographic indicators to assess whether CBRM communities had other characteristics that predisposed them to higher social outcomes, and found none (Supplementary Table 1). Selecting matched sites within ecological zones and provinces further reduced potential for pre-existing environmental and governance differences. Although we controlled for potential confounding differences between CBRM and non-CBRM sites in this manner, our single-point-in-time observational measurements preclude causal inferences about CBRM effects.

Data collection

We collected data using household surveys, focus groups and interviews. Individual households were surveyed using quantitative questionnaires measuring household demographics, income and expenditures, management practices and behaviors, and social norms and networks. Questionnaires were designed to investigate whether formal organization influenced household-level practices and socioeconomic conditions. At the group level, we interviewed group leaders and held focus groups with members. Qualitative data from interviews and focus groups were synthesized into an organizational database for each group. Instrument design was

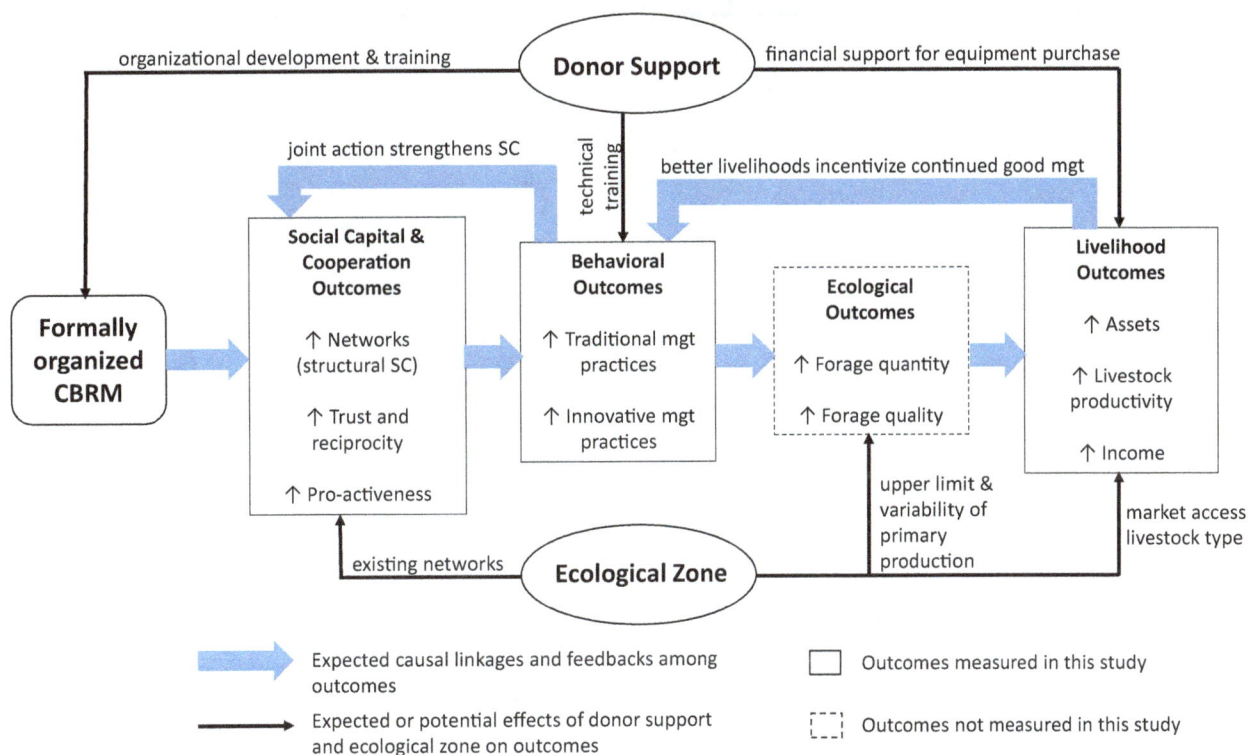

Figure 1 Theorized relationships between donor support, formal community-based rangeland management (CBRM), ecological zone, and social and ecological outcomes of CBRM.

based on prior studies in Mongolia (Fernandez-Gimenez 2002; Fernandez-Gimenez & Batbuyan 2004; Fernandez-Gimenez et al. 2015) and guided by the International Forestry Resources and Institutions protocols for data collection at the community level (IFRI 2013).

Variables

Organization status (CBRM vs. non-CBRM) and ecological zone were the independent variables. Dependent social outcomes included livelihoods, social capital, and rangeland practices and behaviors. Livelihoods were measured using three variables: household assets, annual per capita net cash income, and livestock number per household member in sheep forage units (SFU), a standardized livestock unit used in Mongolia. We measured two types of social capital (SC). Cognitive SC measured levels of trust and norms of reciprocity among group members. Structural SC indicated the number of bonding and linking social ties (Grootaert & Van Bastelaer 2002). Bonding ties are those with individuals of similar social position such as family, friends and neighbors, and linking SC refers to vertical ties with government, banks and NGOs, etc. (Woolcock 2001). Herder practices

and behavior were measured with three variables: (1) traditional rangeland and herd management practices in place during collectivization or earlier, (2) recently introduced innovative management practices, and (3) proactive actions and engagement in local rangeland-related initiatives. Except for per capita income and livestock number, all variables are indices calculated from multiple survey items. All outcome variables were calculated from the household survey and aggregated to the organization level by taking the mean value for the sampled households within each group. Group-level structural and demographic characteristics from the organizational database were used to assess whether systematic differences existed between CBRM and non-CBRM groups that could contribute to any observed differences in social outcomes (Supplementary Table 1).

Analyses

All data were inspected, and two variables (SFU per capita and net income per capita) log transformed and winsorized following Vaske (2008) to meet normality assumptions. We used t tests, chi-square tests and two-factorial ANOVA to compare groups by organization type

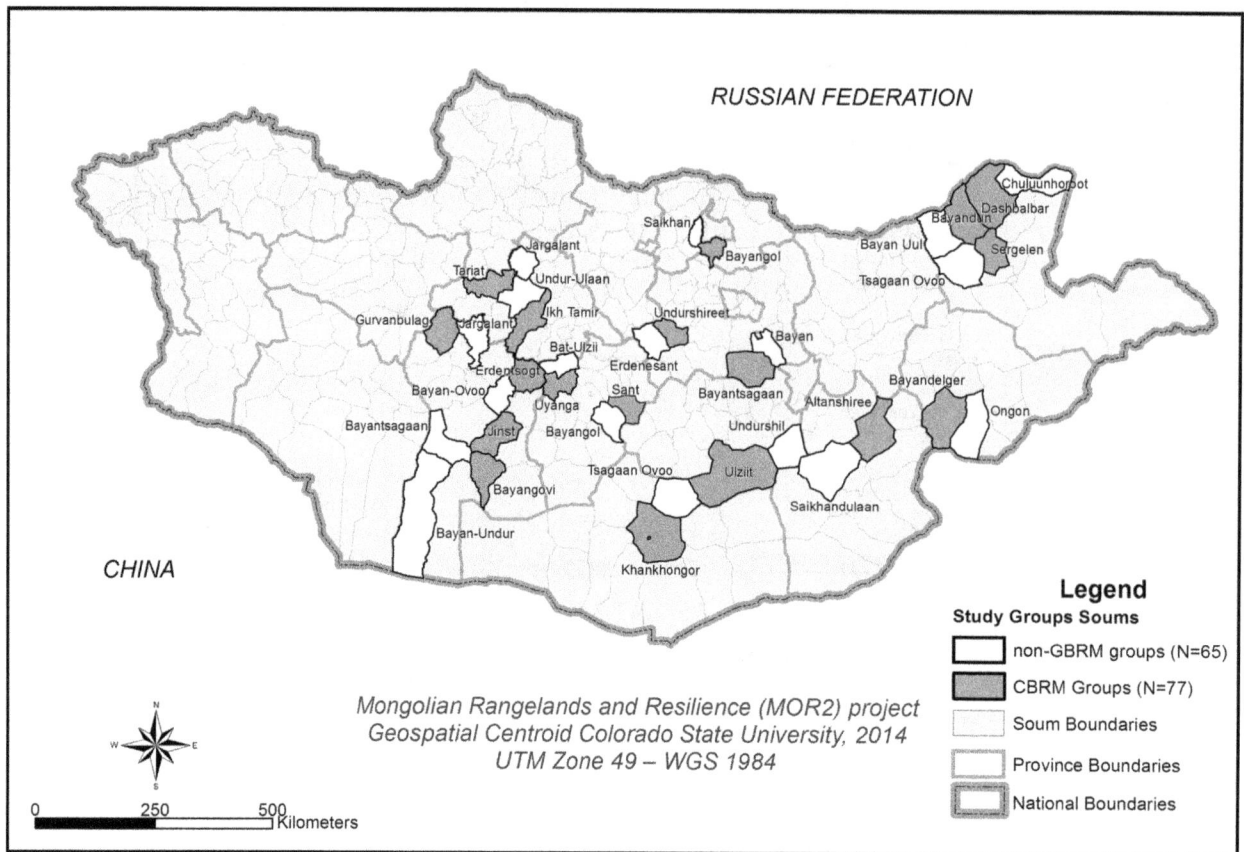

Figure 2 The location of the paired study soums (districts) with community-based rangeland management (CBRM) groups ($n = 18$) and without (non-CBRM) ($n = 18$).

(CBRM vs. non-CBRM) and ecological zone. All analyses were conducted using SPSS 22.0 (IBM Corp. 2013). We used the Bonferroni post hoc test for multiple comparisons. Statistical effect sizes were assessed using eta-squared for ANOVA and phi for chi-square and cross-tab tests (Vaske 2008). When outcomes differed significantly ($P < 0.05$) or nearly so ($0.05 < P < 0.10$) between group types or zones, we examined the constituent items in the index to identify the specific attributes or behaviors responsible for the observed differences.

Results

Social outcomes by group type

Members of formal CBRM groups demonstrated greater proactive behavior, used more innovative and traditional rangeland management practices, and possessed more household assets compared to non-CBRM herders (Figure 3). Effect sizes for proactiveness, traditional and innovative practices were moderate to substantial. For-

mal CBRM and non-CBRM groups did not differ in cognitive ($P = 0.06$) or structural ($P = 0.07$) social capital, cash income or livestock number. There were no interactions between ecological zone and organization for any of the outcome variables, indicating that CBRM is consistently associated with similar social outcomes across differing ecological contexts.

CBRM members were more proactive than traditional group herders on three measures—talking with experts about rangeland issues, joining local initiatives to improve resource use, and acting to address local problems (Table 1). Among traditional practices, more CBRM herders reported reserving winter and spring pastures, culling unproductive animals before winter, making hay and hand fodder, and digging new wells. More CBRM members used 11 out of 19 types of innovative practices. Among these, fencing critical grazing areas and hay fields, growing fodder plants, vegetable gardening, and monitoring rangeland resources had higher effect sizes.

Although the group types did not differ significantly on either SC index, they varied on some individual items.

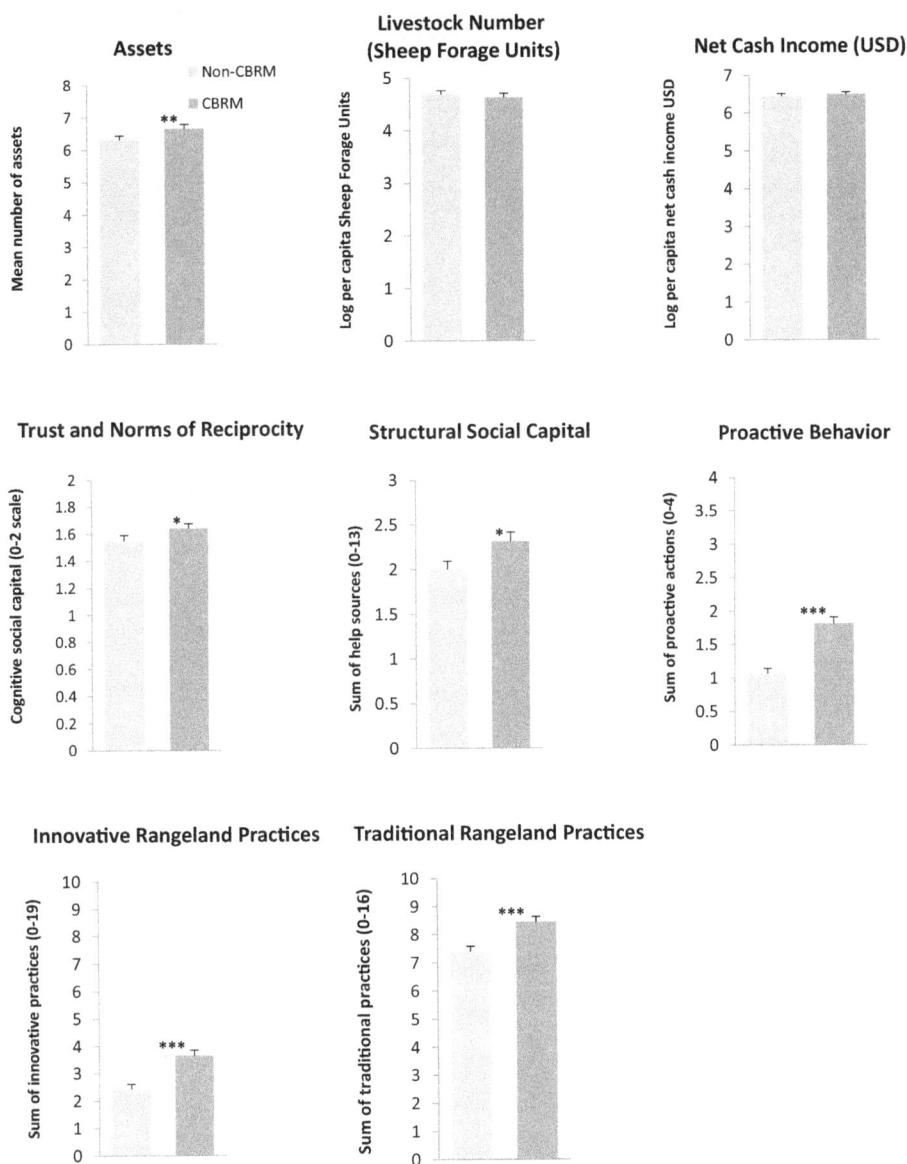

Figure 3 Comparisons of social outcomes between two types of pastoral organizations: traditional informal (non-CBRM) groups ($n = 65$) and formal CBRM groups ($n = 77$). Error bars show one standard error. *, **, and *** indicate the statistical significance at $P < 0.10$, 0.05, and 0.01, respectively.

More CBRM members receive help from CBRM organizations and their connections with religious leaders were modestly higher than herders from traditional neighborhoods (Table 2). More CBRM herders believed that people in their area always try to help each other and disagreed that people in their area are generally selfish (Table 3).

Social outcomes by ecological zone

Ecological zone significantly affected five of eight social outcome variables including assets, cash income, social capital, and traditional management practices (Table 4). In all instances where there were differences, groups from the mountain and forest steppe had lower outcomes, and there were no differences among the other zones. Households from the eastern steppe had more assets than those from the mountain and forest steppe. Herders from the steppe had significantly higher cash income than mountain and forest steppe herders. Pastoralists from the steppe and the desert steppe had greater levels of trust and reciprocity than mountain and forest steppe herders. Desert steppe herders also had more bonding and linking networks. Herders from the mountain and forest steppe were

Table 1 Comparison of proactive behavior and management practice items for non-CBRM and CBRM member households

Items	% non-CBRM ($n = 314$)	% CBRM ($n = 392$)	χ^2	P	φ
Proactive behavior					
Talked to experts about rangeland issues	19	33	16.11	<0.01	0.15
Talked to local authority about problems	47	53	2.31	2.31	0.06
Joined in rangeland improvement initiatives	19	52	81.52	<0.01	0.34
Joined community to address other problems	21	42	33.14	<0.01	0.22
Traditional rangeland management practices					
Reserve winter pasture	47	57	7.43	0.02	0.10
Reserve spring pasture	38	49	10.96	<0.01	0.13
Reserve dzud pasture	24	32	5.72	0.06	0.09
Do fall (or summer) otor	37	39	2.47	0.29	0.06
Do winter otor	17	22	4.29	0.12	0.08
Cull (sell/slaughter) unproductive animals	55	68	14.76	<0.01	0.15
Cut hay	63	79	24.02	<0.01	0.18
Prepare hand fodder	34	48	13.75	<0.01	0.14
Purchase and store grain	76	77	0.50	0.78	0.03
Purchase and store concentrate	35	40	2.41	0.30	0.06
Purchase other feed	19	24	3.06	0.22	0.07
Vaccinate livestock	88	90	3.19	0.20	0.07
Deworm livestock	87	89	0.33	0.85	0.02
Treat livestock for external parasites	57	61	0.83	0.66	0.03
Dig a new well	13	21	7.60	<0.01	0.10
Repair existing well	48	51	2.84	0.24	0.06
Innovative rangeland management practices					
Improve camel breed	2	2	0.10	0.76	−0.01
Improve horse breed	15	15	0	0.95	−0.00
Improve cattle breed	11	13	0.87	0.35	0.04
Improve sheep breed	34	45	4.74	0.03	0.08
Improve goat breed	36	41	1.80	0.18	0.05
Intentionally change species proportion	25	27	2.23	0.33	0.06
Sell animals to reduce herd size	21	28	5.88	0.02	0.09
Intentionally not breed animals due to dzud	19	18	0.30	0.58	−0.02
Fence critical pasture area	6	22	40.87	<0.01	0.23
Fence hay area	3	16	36.55	<0.01	0.22
Fence or improve natural water sources	15	22	5.29	0.07	0.09
Plant fodder or grass	4	13	17.98	<0.01	0.16
Use fertilizer	7	17	17.67	<0.01	0.13
Use irrigation	5	12	12.12	<0.01	0.13
Plant garden for food	16	30	19.82	<0.01	0.17
Take other action to protect key resources	10	19	10.95	<0.01	0.13
Take action to reduce soil erosion	4	8	6.06	0.05	0.09
Take action to restore damaged lands	5	4	0.22	0.64	−0.02
Take part in monitoring of resources	5	13	17.21	<0.01	0.16

Note: Dzud refers to a winter weather disaster. Otor are long-distance movements to fatten animals or avoid bad weather or drought. The term φ (phi) indicates effect size with 0.10 interpreted as minimal, 0.30 as typical, and 0.50 or greater as substantial (Vaske 2008).

less likely to use seven traditional practices compared to those in other ecological zones. We found no difference in per capita livestock holdings, use of innovative rangeland practices, or proactiveness by ecological zone. More eastern steppe pastoralists had animal carts, butter churns, books and cell phones. Fewer mountain and forest steppe households reported doing fall *otor* (long-distance moves to fatten livestock), reserving grain and other feed, and digging and repairing wells than the herders in other ecological zones. Herders in the desert steppe had more bonding and linking network ties than those in any other zone (Supplementary Tables 2–4).

Table 2 Comparison of structural social capital and household asset items among members of non-CBRM and CBRM pastoral groups

Items	% non-CBRM (n = 314)	% CBRM (n = 392)	χ^2	P	φ
Structural social capital					
Bonding social capital					
Help from neighbors	67	64	0.32	0.57	−0.03
Help from family in district (soum)	60	63	0.47	0.49	0.03
Help from family in province (aimag) center or capital city	39	42	0.73	0.39	0.04
Help from distant relatives	16	21	2.55	0.11	0.07
Help from friends	52	56	0.56	0.45	0.03
Linking social capital					
Help from local/national government	63	64	0.01	0.92	0.01
Help from politicians	38	31	2.45	0.12	−0.07
Help from religious leaders	4	8	3.80	0.05	0.09
Help from CBRM organization	5	44	95.43	<0.01	0.43
Help from development or aid organization	33	36	0.49	0.48	0.03
Help from nongovernmental organization	10	15	3.61	0.06	0.08
Help from banks	19	22	0.62	0.43	0.04
Help from insurance companies	5	8	1.28	0.26	0.05
Household assets					
Mobile phone	92	94	1.81	0.28	0.04
Radio	59	62	0.72	0.04	0.03
Television	84	89	2.93	0.09	0.06
Motorcycle	82	77	2.41	0.12	−0.06
Car	20	28	6.48	0.01	0.10
Truck or tractor	27	28	0.22	0.64	0.02
Cart: cattle, horse, or camel	32	31	0.05	0.82	−0.01
Refrigerator	12	20	9.25	<0.01	0.12
Butter churn	5	10	5.63	0.02	0.09
Electricity generator (portable)	17	23	3.6	0.06	0.07
Windmill	5	9	4.46	0.04	0.08
Solar panel	91	87	2.61	0.11	−0.06
Electric lights	69	67	0.25	0.62	−0.02
Books	26	38	11.53	<0.01	0.13
Computer	5	9	2.51	0.11	0.06

Note: The term φ (phi) indicates effect size with 0.10 interpreted as minimal, 0.30 as typical, and 0.50 or greater as substantial (Vaske 2008).

Discussion

Mongolian herders belonging to formally-organized CBRM groups demonstrate significantly greater social outcomes than herders in traditional neighborhoods on all behavioral measures, and one livelihood measure. CBRM herders had slightly higher outcomes on some social capital metrics, but not on overall social capital indices. This pattern holds across four ecological zones. CBRM was most strongly associated with proactive behavior and both traditional and recently adopted resource management practices. Although most practices measured are associated with adaptive capacity or livelihood diversification, and therefore can be considered social benefits of CBRM, the conservation effects of some practices, such as irrigation, fertilization and fencing,

are debatable or potentially negative. These results confirm earlier studies in Mongolia that documented positive social outcomes of CBRM (Batkhishig et al. 2011; Dorligsuren et al. 2011; Baival & Fernandez-Gimenez 2012; Upton 2012; Fernandez-Gimenez et al. 2015), and provide evidence that community-based approaches to rangeland conservation may effectively influence resource users' behavior. Lack of difference in social capital between group types may be due to strong traditions of reciprocity across all types of herder communities in Mongolia (Batkhishig et al. 2011). Our focus groups and interviews also suggest that while generalized trust and reciprocity are relatively high, as our survey data show, specific trust related to economic cooperation and financial risk is not well developed. We expect that if CBRM is successful, levels of specific trust and economic

Table 3 Comparison of trust and norms of reciprocity items among members of non-CBRM versus CBRM pastoral groups using ANOVA

Items	Non-CBRM ($n = 314$)	CBRM ($n = 392$)	F	P	η^2
People always try to help each other	1.65	1.78	7.60	<0.01	0.01
People help each other in times of need	1.64	1.74	5.19	0.02	0.01
People mainly look out for themselves*	1.21	1.42	11.17	0.01	0.02
Most people are trustworthy	1.73	1.78	1.61	0.21	0
People will take advantage of others*	1.57	1.63	1.31	0.25	0
Our community is getting less friendly*	1.44	1.49	0.77	0.38	0

Note: All items are on a 0–2 scale; *indicates items were reverse coded: 0 = agree, 1 = neutral, and 2 = disagree. The term η^2 is interpreted as the percentage of variation in the dependent variable explained by the independent variable (CBRM).

Table 4 Results of two-way ANOVA showing effects of ecological zone on social outcome variables

Ecological zone Variable name	Desert steppe ($n = 47$)	Steppe ($n = 31$)	Eastern steppe ($n = 11$)	Mountain and forest steppe ($n = 53$)	F	P value	η^2
Livelihood							
Assets (sum of 15 items)	6.62	6.58	7.24*	6.20*	3.43	0.02	0.07
Log net cash income per capita in USD	6.40*	6.79*†	6.62	6.33†	5.50	<0.01	0.11
Log livestock per capita	4.50	4.72	4.78	4.71	1.36	0.26	0.03
Social capital							
Cognitive (0–2 scale)	1.66*	1.68†	1.59	1.49*†	4.57	<0.01	0.09
Structural (sum of 13 items)	2.45*	2.19	1.90	1.95*	4.35	<0.01	0.09
Behavior							
Traditional practices (sum of 16 items)	8.24*	8.52†	9.14‡	7.1*†‡	9.92	<0.01	0.16
Innovative practices (sum of 19 items)	3.12	2.71	3.71	3.15	1.15	0.33	0.02
Proactiveness (sum of four items)	1.65	1.34	1.41	1.39	1.80	0.15	0.03

Note: Pairs of means in the same row that share superscripts (*, †, and ‡) differ at $P < 0.05$ using the Bonferroni multiple comparison test. The term η^2 is interpreted as the percentage of variation in the dependent variable explained by the independent variable (CBRM).

cooperation will increase over time. The limited progress of Mongolian CBRM groups toward improving herder livelihoods contradicts prior studies (Dorligsuren et al. 2011; Leisher et al. 2012) and deserves further consideration.

CBRM member households possessed more assets than non-CBRM households, but did not differ in per capita income or livestock holdings. Donor contributions toward buying and maintaining capital assets such as wells and tractors, which freed herders to use income to buy assets like automobiles and refrigerators, may explain greater assets in CBRM households. CBRM training also may have influenced herders' investments in technology to improve production and marketing. Lack of differences between group types in income or livestock holdings have three possible explanations. First, the primary mechanism through which CBRM is expected to influence income and livestock holdings is via improved grassland quality and production, which result from more sustainable grazing practices such as increased mobility and grazing reserves (Figure 1). Increased forage quality

and quantity, in turn, should lead to greater livestock productivity in terms of number of animals (reproductive success) and individual animal productivity and quality. Because many CBRM groups are relatively young, and changes in management recent, the ecological and livestock productivity responses may lag behind behavioral changes. Alternatively, the influence of CBRM on ecologically beneficial management practices may not be sufficiently strong to result in forage quality and quantity increases that affect livestock production, regardless of time since establishment. A companion study of ecological outcomes (Chantsallkham 2015) shows slight ecological benefits of CBRM, consistent with these explanations. Finally, if CBRM and non-CBRM herders have similar herd sizes, but CBRM livestock produce more or higher quality meat, milk or fiber, we would expect CBRM incomes to be higher despite similar herd sizes. However, livestock markets in Mongolia are poorly developed and as yet do not differentiate on product quality. Herders do not receive more money for high quality products than for the same quantity of low-quality product. This is a serious obstacle to the long-term effectiveness of CBRM in Mongolia. A differential economic return is likely an essential incentive to sustain herder commitment to CBRM.

Social outcomes of CBRM in Mongolia varied considerably by ecological zone, suggesting that resource system characteristics and geography influence group outcomes. Contrary to theoretical predictions, mountain and forest steppe communities had lower outcomes than other regions on many indicators. Mountain and forest steppes have more productive and predictable forage supplies, and are associated with smaller and more easily bounded CPRs, which we theorized would enhance collective action due to lower transaction costs and easier monitoring. However, some of these areas are also the most socially isolated and remote from markets, which may limit social networks and constrain their ability to implement some practices. Alternatively, the more productive and predictable forage supply may lead to greater competition for and conflict over pastures, as some of our focus groups suggest. Steppe herders, located closest to major markets in the capital, had the highest incomes. Steppe and desert steppe herders had higher levels of trust and reciprocity, and desert steppe herders had more bonding and linking network ties. The higher levels of social capital in less predictable and highly variable semiarid and arid zones align with findings from other pastoral regions, including Australia, where herders in variable environments form more extensive social networks with strong reciprocal arrangements to access pasture during periodic droughts (McAllister et al. 2011). Earlier case studies in Mongolia found a similar pattern in social capital, with the highest levels in the desert steppe (Fernandez-Gimenez et al. 2012; Fernandez-Gimenez et al. 2015).

Our findings have several important implications for conservation policy, practice and research in Mongolia and beyond. First, our results show that CBRM is strongly associated with herders' management practices and proactive behavior, which are critical steps toward improving ecological and social conditions on Mongolian rangelands. As such they demonstrate the potential social return on donors' initial conservation investment in CBRM. These findings have implications beyond Mongolia, since CBRM has been widely advocated in other dryland regions (Turner 2011), and has recently increased rapidly in southern and eastern Africa (Fabricius & Koch 2004; Bennett 2013).

Second, the lack of clear social capital and livelihood benefits from CBRM in Mongolia indicates that CBRM outcomes may take time to achieve, especially when they depend on a series of linked feedbacks, each of which is also affected by exogenous factors such as climate, weather, and markets (Oldekop et al. 2010).

Additionally, current levels of management may be insufficient to strongly affect pasture and livestock conditions. This suggests that ongoing technical support for CBRMs is needed, with a focus on promoting practices that have clear conservation as well as livelihood benefits.

Third, the possibility that lack of strong livelihood outcomes is due to insufficient market price differentiation for quality livestock products deserves further investigation. Livestock markets have long been recognized as a critical constraint to sustainable management of Mongolia's rangelands. If CBRM is to succeed in Mongolia over the long term, herders must be able to earn more with fewer animals through a premium price for higher quality or sustainably produced livestock products, value-added processing, or alternative rural livelihoods.

Fourth, Mongolia's CBRM movement was largely catalyzed by external actors–donors. Although the movement is widespread, it has reached only a fraction of Mongolia's herders. Whether this externally initiated movement will lead to scaling-out of endogenous CBRM—self-organized groups following the lead of earlier externally initiated groups—remains unknown, although self-organized groups have been observed (Undargaa 2006). The intensive donor-provided support to the initial CBRM groups is likely cost-prohibitive to scale out to the entire country. Therefore, both government policy-makers and conservation-minded donors and NGOs should consider how to cooperate to support grassroots CBRM initiatives through lower cost peer-to-peer training and education programs coupled with appropriate policy incentives and legislative reforms. Finally, if new CBRM groups are established with external facilitation, a randomized trial approach incorporating control sites and pre-implementation measurement of baseline ecological and social conditions will permit more rigorous assessment of CBRM effectiveness than was possible in our post-hoc study.

Further research is needed to (1) establish clear causal relationships between CBRM and social outcomes, (2) examine the process through which CBRMs achieve these outcomes and the factors that predict success, (3) determine whether CBRM performance varies with institutional design, (4) assess CBRM ecological outcomes, and (5) monitor performance of CBRM institutions over time.

Acknowledgments

This research was sponsored by National Science Foundation Award No. BCS-1011801, with additional support from The World Bank, US AID, American Association of University Women, Open Society Institute, Center for Collaborative Conservation, Colorado State

University, and Reed Funk Account, Utah State University. We sincerely thank the following MOR2 team members for data collection and entry assistance: Amanguli, Ariuntuya, Azjargal, Battuul, Enkhmunkh, Erdenechimeg, Ganjargal, Gankhuyag, Gantsetseg, Gantsogt, Khishigdorj, Khishigjargal, Khishigsuren, Narantuya, Nomin-Erdene, Odgarav, Pagmajav, Solongo, Tamir, Tsengelmaa, Unurzul, Uuganbayar, Vandandorj.

References

Addison, J., Davies, J., Friedel, M. & Brown, C. (2013). Do pasture user groups lead to improved rangeland condition in the Mongolian Gobi Desert? *J. Arid. Environ.*, **94**, 37-46.

Agrawal, A. (2002). Common resources and institutional sustainability. Pages 41-85 in E. Ostrom, editor. *The drama of the commons*. National Research Council, Washington, D.C.

Baival, B. & Fernandez-Gimenez, M.E. (2012). Meaningful learning for resilience-building among Mongolian pastoralists. *Nom. Peoples*, **16**, 53-77.

Batima, P. (2006). Climate change vulnerability and adaptation in the livestock sector of Mongolia: a final report submitted to Assessments of Impacts and Adaptations to Climate Change (AIACC), Projet No. AS 06. International START Secretariat, Washington, D.C.

Batkhishig, B., Oyuntulkhuur, B., Altanzul, T. & Fernández-Giménez, M.E. (2011). A case study of community-based rangeland management in Jinst Soum, Mongolia. Pages 113-135 in M.E. Fernandez-Gimenez, X. Wang, B. Batkhishig, J. Klein, R.S. Reid, editors. *Restoring community connections to the land: building resilience through community-based rangeland management in China and Mongolia*. CABI, Wallingford, UK.

Bennett, J.E. (2013). Institutions and governance of communal rangelands in South Africa. *Afr. J. Range Sci.*, **30**, 77-83.

Bowler, D.E., Buyung-Ali, L.M., Healy, J.R., Jones, J.P.G., Knight, T.M. & Pulling, A.S. (2014). Does community forest management provide global environmental benefits and improve local welfare? *Front. Ecol. Environ.*, **10**, 29-36.

Chantsallkham, J. (2015). Effects of grazing and community-based management on rangelands of Mongolia. *Dept. of Forest and Rangeland Stewardship*, Colorado State University, Fort Collins, CO.

Collomb, J.G.E., Mupeta, P., Barnes, G. & Child, B. (2010). Integrating governance and socioeconomic indicators to assess the performance of community-based natural resources management in Caprivi (Namibia). *Environ. Conserv.*, **37**, 303-309.

Dagvadorj, D., Natsagdorj, L., Dorjpurev, J. & Namkhainyam, B. (2009). Mongolia: assessment report on climate change. p. 228. Ministry of Environment,Nature and Tourism, Ulaanbaatar.

Dorligsuren, D., Batbuyan, B., Bulgamaa, D. & Fassnacht, S.R. (2011). Lessons from a territory-based community development approach in Mongolia: Ikhtamir Pasture User Groups. Pages 166-188 in M.E. Fernandez-Gimenez, X. Wang, B. Baival, J. Klein, R. Reid, editors. *Restoring community connections to the land: learning from community-based rangeland management in China and Mongolia*. CABI, Wallingford, UK.

Dressler, W., Buscher, B., Schoon, M.L. *et al.* (2010). From hope to crisis and back again? A critical history of the global CBNRM narrative. *Environ. Conserv.*, **37**, 5-15.

Fabricius, C. & Koch, E., editors. (2004). *Rights, resources and rural development: community-based natural resource management in southern Africa*. Earthscan, London.

Fernandez-Gimenez, M.E. (2002). Spatial and social boundaries and the paradox of pastoral land tenure: a case study from postsocialist Mongolia. *Hum. Ecol.*, **30**, 49-78.

Fernandez-Gimenez, M.E. & Batbuyan, B. (2004). Law and disorder: local implementation of Mongolia's land law. *Dev. Change*, **35**, 141-165.

Fernandez-Gimenez, M.E., Batkhishig, B. & Batbuyan, B. (2012). Cross-boundary and cross-level dynamics increase vulnerability to severe winter disasters (dzud) in Mongolia. *Global Environ. Change*, **22**, 836-851.

Fernandez-Gimenez, M.E., Batkhishig, B., Batbuyan, B. & Ulambayar, T. (2015). Lessons from the dzud: community-based rangeland management Increases the adaptive capacity of Mongolian herders to winter disasters. *World Dev.*, **68**, 48-65.

Grooteart, C. & van Bastelaer, T. (2002). *Understanding and measuring social capital: a multidisciplinary tool for practitioners*. The World Bank, Washington, D.C.

IBM Corp. (2013). *IBM SPSS Statistics for Windows*, Version 22.0. IBM Corp., Armonk, NY.

IFRI. (2013). International Forestry Resources and Institutions (IFRI) network: research methods. www.ifriresearch.net. Accessed January 9, 2016

Janes, C.R. (2010). Failed development and vulnerability to climate change in Central Asia: implications for food security and health. *Asia Pac. J. Publ. Health*, **22**, 2365-2455.

Khishigbayar, J., Fernandez-Gimenez, M.E., Angerer, J.P. *et al.* (2015). Mongolian rangelands at a tipping point? Biomass and cover are stable but composition shifts and richness declines after 20 years of grazing and increasing temperatures. *J. Arid Environ.*, **115**, 100-112.

Leisher, C., Hess, S., Boucher, T.M., van Beukering, P. & Sanjayan, M. (2012). Measuring the impacts of community-based grasslands management in Mongolia's Gobi. *PLoS ONE*, **7**, e30991. Doi: 30910.31371/journal.pone.0030991

McAllister, R.R.J., Tisdell, J.G., Reeson, A.F. & Gordon, I.J. (2011). Economic behavior in the face of resource variability and uncertainty. *Ecol. Soc.* **16**, 6 [online], http://dx.doi.org/10.5751/ES-04075-160306.

Murphy, D.J. (2011). Going on otor: disaster, mobility and the political ecology of vulnerability in Uguumur, Mongolia. p. 585. *Department of Anthropology*. University of Kentucky, Lexington, KY.

Nkhata, B.A. & Breen, C.M. (2010). Performance of community-based naturel resource governance for the Kafue Flats (Zambia). *Environ. Conserv.*, **37**, 296-302.

Oldekop, J., Bebbington, A., Brockington, D. & Prieziosi, R. (2010). Understanding the lessons and limitations of conservation and development. *Conserv. Lett.*, **24**, 461-469.

Ostrom, E. (1990). *Governing the commons: the evolution of institutions for collective action*. Cambridge University Press, Cambridge.

Saito-Jensen, M., Nathan, I. & Treue, T. (2010). Beyond elite capture? Community-based natural resource management and power in Mohammed Nagar village, Andhra Pradesh, India. *Environ. Conserv.*, **37**, 327-335.

Silva, J.A. & Mosimane, A.W. (2013). Conservation-based rural development in Namibia: a mixed-methods assessment of economic benefits. *J. Environ. Dev.*, **22**, 25-50.

Sternberg, T. (2010). Unravelling Mongolia's extreme winter disaster of 2010. *Nom. Peoples*, **14**, 72-86.

Suich, H. (2010). The livelihood impacts of the Namibian community-based natural resource management programme: a meta-synthesis. *Environ. Conserv.*, **37**, 45-53.

Turner, M.D. (2011). The new pastoral development paradigm: engaging the realities of property institutions and livestock mobility in dryland Africa. *Social. Natur. Resour.*, **24**, 469-484.

Undargaa, S. (2006). Community organization: a policy level study of "community organization" as a grass-roots institution that contributes to strengthen co-management of sustainable pastoralism and nature conservation. New Zealand Nature Institute, Initiative for People Centered Conservation, Ulaanbaatar.

Upton, C. (2008). Social capital, collective action and group formation: developmental trajectories in post-socialist Mongolia. *Hum. Ecol.*, **36**, 175-188.

Upton, C. (2012). Adaptive capacity and institutional evolution in contemporary pastoral societies. *Appl. Geogr.*, **33**, 135-141.

Vaske, J.J. (2008). *Survey research and analysis: applications in parks, recreation and human dimensions*. Venture Publishing, State College, PA.

Western, D. & Wright, R.M., editors. (1994). *Natural connections: perspectives in community-based conservation*. Island Press, Washington D.C.

Woolcock, M.M. (2001). The place of social capital in understanding social and economic outcomes. *Isuma: Can. J. Policy Res.*, **2**(1), 11-17.

16

Projecting Global Biodiversity Indicators under Future Development Scenarios

Piero Visconti[1,2], Michel Bakkenes[3], Daniele Baisero[2], Thomas Brooks[4,5,6], Stuart H. M. Butchart[7], Lucas Joppa[1], Rob Alkemade[3,8], Moreno Di Marco[2], Luca Santini[2], Michael Hoffmann[4,9], Luigi Maiorano[2], Robert L. Pressey[10], Anni Arponen[11], Luigi Boitani[2], April E. Reside[12], Detlef P. van Vuuren[3,13], & Carlo Rondinini[2]

[1] Microsoft Research Computational Science Laboratory, 21 Station Road, Cambridge, CB1 FB, UK
[2] Global Mammal Assessment Program, Department of Biology and Biotechnologies, Sapienza University of Rome, Viale dell'Università 32, Rome, 00185, Italy
[3] PBL, Netherlands Environmental Assessment Agency, PO Box 303, 3720, AH, Bilthoven, The Netherlands
[4] IUCN Species Survival Commission, International Union for Conservation of Nature, 28 rue Mauverney, CH-1196, Gland, Switzerland
[5] World Agroforestry Center (ICRAF), University of the Philippines Los Baños, Laguna, 4031, Philippines
[6] School of Geography and Environmental Studies, University of Tasmania, Hobart, TAS 7001, Australia
[7] BirdLife International, Wellbrook Court, Cambridge, CB3 0NA, UK
[8] Environmental Systems Analysis Group, Wageningen University, P. O. Box 47, 6700, AA, Wageningen, The Netherlands
[9] United Nations Environment Programme World Conservation Monitoring Centre, 219c Huntingdon Road, Cambridge, CB3 0DL, UK
[10] Australian Research Council Centre of Excellence for Coral Reef Studies, James Cook University, Townsville, QLD 4811, Australia
[11] Metapopulation Research Group, Department of Biosciences, University of Helsinki, P.O. Box 65, Helsinki, 00014, Finland
[12] Centre for Tropical Environmental & Sustainability Sciences, James Cook University, QLD, 4811, Australia
[13] Copernicus Institute of Sustainable Development, Department of Geosciences, Utrecht University, Heidelberglaan 2, 3584 CS, Utrecht, The Netherlands

Keywords

Biodiversity scenarios; biodiversity indicators; carnivores; climate change; extinction risk; land-use change; Geometric Mean Abundance; Red List Index; ungulates.

Correspondence

Piero Visconti, Microsoft Research Computational Science Laboratory, 21 Station Road, Cambridge CB1 FB, UK.
E-mail: pierovisconti@gmail.com

Editor

Edward Game

Abstract

To address the ongoing global biodiversity crisis, governments have set strategic objectives and have adopted indicators to monitor progress toward their achievement. Projecting the likely impacts on biodiversity of different policy decisions allows decision makers to understand if and how these targets can be met. We projected trends in two widely used indicators of population abundance Geometric Mean Abundance, equivalent to the Living Planet Index and extinction risk (the Red List Index) under different climate and land-use change scenarios. Testing these on terrestrial carnivore and ungulate species, we found that both indicators decline steadily, and by 2050, under a Business-as-usual (BAU) scenario, geometric mean population abundance declines by 18–35% while extinction risk increases for 8–23% of the species, depending on assumptions about species responses to climate change. BAU will therefore fail Convention on Biological Diversity target 12 of improving the conservation status of known threatened species. An alternative sustainable development scenario reduces both extinction risk and population losses compared with BAU and could lead to population increases. Our approach to model species responses to global changes brings the focus of scenarios directly to the species level, thus taking into account an additional dimension of biodiversity and paving the way for including stronger ecological foundations into future biodiversity scenario assessments.

Introduction

Growing concerns over the loss of biodiversity and the goods and services it provides to humankind have prompted the United Nations to establish the Intergovernmental Platform on Biodiversity and Ecosystem Services (IPBES), to inform global environmental decision making (Brooks *et al.* 2014). The main function of IPBES is to produce regional and global assessment on status, trends, and future scenarios of biodiversity and ecosystem services. These assessments will advise on the policies required to achieve sustainable development goals, including the Convention on Biological Diversity (CBD) Aichi targets for 2020 and the CBD vision for 2050. These targets have an associated set of biodiversity indicators to monitor progresses (Tittensor *et al.* 2014).

Both IPBES and the CBD require a framework for producing projections about future trends in biodiversity loss under alternative policy scenarios. Until now, such projections measured via biodiversity indicators adopted by the CBD have been limited to a single study of African-protected area scenarios (Nicholson *et al.* 2012).

In order for any biodiversity scenario projection to be relied upon, it is important that the ecological response models are known to provide accurate estimates of past trends; surprisingly, there are no studies hindcasting terrestrial ecological models from past to present for models calibration and validation. Moreover, most biodiversity scenario studies have used indicators such as total number of species derived from species-area curves (Van Vuuren *et al.* 2006) and naturalness via Mean Species Abundance (MSA; Alkemade *et al.* 2009) that do not use species-specific responses to anthropogenic pressures. Species-specific ecological models improve predictions of ecological responses to global change by accounting for life-history traits, and allow understanding which species are at higher risk and why (Pearson *et al.* 2014).

Here, we assess the ecological impact of different human development scenarios with species-specific ecological models and two established species-level indicators, the Red List Index (RLI; an aggregate measure of species' extinction risk) and species Geometric Mean Abundance, (GMA) an indicator equivalent to the Living Planet Index (which is based on observed trends of populations of vertebrates species) for 440 terrestrial mammalian carnivores and ungulates (89% of the species in these groups). These two complementary indicators have been adopted by the CBD to measure progress toward global biodiversity targets (Butchart *et al.* 2010). We validate our models through hindcasting species distributions and biodiversity indicators from 1970 to the present and we provide confidence intervals around past and future modeled trends. We conclude by highlighting the step-changes required for achieving conservation goals for large mammals based on our scenario projections.

Methods

Scenario storylines

The "Business-as-usual" (BAU) scenario explores the effects of economic growth, consumption patterns, and

energy mix in the absence of new policies (PBL 2012). Growing human population and economic development will increase the demand for food, energy, and other essential goods, such as clean water, fibers, and wood. These demands are satisfied by increasing agricultural productivity and expanding agricultural land and freshwater consumption; by expanding fisheries and aquaculture; by increasing the use of fossil fuels and wood products (PBL 2012). These trends largely satisfy human needs, reduce extreme poverty, and improve human health. However, they also result in ongoing decline of biodiversity measured as MSA (PBL 2012) and large increases of greenhouse gas emissions.

An alternative scenario, "Consumption Change" belongs to a family of scenarios designed to achieve a set of sustainable development goals on human well-being, climate change, and biodiversity simultaneously. Consumption Change, does so by limiting meat intake per capita, reducing waste in the agricultural production chain and adopting a less energy-intensive lifestyle (PBL 2012). The rapid adoption of these societal changes make this scenario possible but ambitious (PBL 2012). Scenario assumptions are in Table 1, trends in major land-uses projected for both scenarios are in Figures S1 and S2.

Climate change

Our baseline climate was an average of the observed bioclimatic variables between 1975 and 2005. We considered two Intergovernmental Panel on Climate Change - Assessment Report 4 climate scenarios: A1B, associated with BAU and B1 associated with Consumption Change (PBL 2012). Raw monthly temperature minimum and maximum were obtained from http://climascopewwwfus.org at a resolution of 0.5° (Price *et al.* 2012). Standard bioclimatic variables (Table S1) were generated using the "climates" package in R (VanDerWal *et al.* 2011).

Habitat loss

We used the outputs from Integrated Model to Assess the Global Environment (IMAGE) version 2.5 (Bouwman *et al.* 2006) as an estimate of the area converted to or from cropland, pasture, plantation or forestry, in 24 world macroregions at any time step. These estimates from the IMAGE agroeconomic model were used as input into the GLOBIO land-use change model to derive fractions of different land-cover and land-uses within 6' grid cells (approximately 10 by 10 km at the equator) for the years 1970–2050 at decadal interval (Alkemade *et al.* 2009; Visconti *et al.* 2011). The GLOBIO land-cover and land-use data (see, e.g., in Table S6) were used together with the relevant climate for projecting species responses to global

Table 1 Assumptions of Business-as-usual and Consumption Change scenarios for the year 2050 (PBL 2012)

Assumption	Business-as-usual	Consumption Change
Access to food	250 million people globally have insufficient access to food in 2050	Inequality in access to food due to income inequality converges to zero by 2050
Consumption	+65% energy consumption, +50% food consumption	Meat consumption per capita levels off at twice the consumption level suggested by a supposed healthy diet (Stehfest *et al.* 2009) which would imply reducing meat and egg consumption in all regions by 76–88%.
Waste	Stable 30% of total production	Waste is reduced by 50% with respect to BAU by 2030
Agricultural productivity	Yield increase by 0.06% annually (+27% by 2050)	In all regions, 15% increase in crop yields by 2050, compared with the BAU scenario
Protected areas	No further protected areas respect to 2010	17% of each of the 65 realm-biomes. Expansion allocated close to existing agriculture to protect areas currently most threatened by habitat loss
Forestry	+30% in clear-cut, +35% plantation, −12.5% selective logging. No reduced impact logging.	Forest plantations supply 50% of timber demand; almost all selective logging based on reduced impact logging by 2020.

changes from past to present, thereby validating our model results against known trends ("model hindcast" Supporting Information: S5), and to model the impacts of future global change on large terrestrial mammals.

Species' response to climate and land-use change

We followed a hierarchical approach to model species distribution (Pearson & Dawson 2003). Bioclimatic envelope models were used to estimate species past, present, and future extent of occurrence (EOO) reflecting the known relationship between climate and species geographic range, (Soberón & Nakamura 2009). Habitat suitability models were used to identify the areas potentially occupied by the species within the EOO (i.e., Extent of Suitable Habitat; ESH) based on habitat preferences coded in the IUCN Red List (RL) database (IUCN 2012b) and projected land-cover and land-use.

Species data

We focused on all extant terrestrial carnivore and ungulate species of the orders Carnivora, Cetartiodactyla, Perissodactyla, and Proboscidea for which the geographic range was known and available from the IUCN (IUCN 2012b), and sufficiently large to obtain an adequate sample of presence points for fitting bioclimatic envelope models. In total, we projected the responses to climate and land-use change impact of 440 of the 493 species in these orders for which range data were available.

Bioclimatic envelope models

We simulated climate change effects on species distribution by fitting bioclimatic envelopes at 30' resolution and by projecting spatial changes to this envelope at 10-year intervals. We used seven statistical models with the R package BIOMOD (Thuiller *et al.* 2009) to fit current bioclimatic envelopes and to project these envelopes into future climatic scenarios (see Supporting Information: S2.1). The variables selected (Table S1) are those usually considered most important for modeling species distributions at large scale (Guisan & Zimmermann 2000). We transformed the probabilistic output of these models to a binary (presence/absence) output selecting for each species and model the probability threshold that maximized True Skill Statistic TSS (Allouche *et al.* 2006). We obtained a single model output by calculating the mode of the seven binary values (presence/absence) derived from individual models.

We accounted for species ability to track climate using two dispersal assumptions. The first represents a pessimistic scenario where species were unable to disperse and adjust their EOO according to climate; hence species could only lose suitable climate space within their present EOO. In the second assumption, species were allowed to track climate at the speed of one dispersal distance per generation (see Supporting Information, S2.2). This equals to assuming an intergenerational relay race to track climate change. We also considered a climate adaptation scenario in which we assumed species to be able to adapt locally to climate change and persist in their present EOO wherever the habitat is suitable.

Habitat suitability models

We used the IUCN Global Mammal Assessment habitat suitability models (Rondinini *et al.* 2011; Visconti *et al.* 2011) to quantify the ESH for each species within a species' EOO. Each combination of land-cover, land-use, and elevation within a grid cell was scored as either suitable or not according to the land-cover and altitudinal preferences of species and their sensitivity to different land-uses reported by IUCN taxonomic experts (Rondinini *et al.* 2011). The land-use classification system was a modification of the 23 Global Land Cover 2000 adopted by the GLOBIO model which included grazing areas and subclasses related to the type and intensity of agriculture and forestry, yielding a total of 66 classes (Table S5). For each species, suitable habitat within the 6′ cells inside a species EOO was calculated as the proportion of suitable land-cover/land-use within the cell multiplied by the proportion of suitable altitude within the cell. The ESH was the sum of all suitable habitat within a species' EOO.

The suitable area does not reflect the actual occurrence of the species because parts might not be occupied due to other biophysical, ecological, or anthropogenic factors, including habitat fragmentation and isolation. We accounted for this by correcting the ESH with an occupancy factor ϕ to derive the Area of Occupancy (AOO = ESH*ϕ). To account for uncertainty in this parameter, we ran 1,000 Monte Carlo simulations in which ϕ was drawn from a distribution U (0.1, 1).

Red List Index

The RLI shows trends in aggregate extinction risk of species, as measured using the categories of the IUCN Red List of Threatened Species, and ranges from 0 if all species are Extinct to 1 if all are assigned the lowest possible extinction risk category ("Least Concern") (Butchart *et al.* 2007). RL categories are broad classes of extinction risk (IUCN 2012a) assigned on the basis of criteria relating to the size, structure, and trends in population and geographic range (IUCN 2012a). We estimated each species' RL category at each time-step by comparing the projected EOO, AOO, and estimated population size (number of mature individuals) against RL criteria A2 (trends in population size), B1 (EOO size), B2 (AOO size), C1 (small and declining population), and D1 (very small population). We followed IUCN guidelines (IUCN 2012a) to assign each species to the following RL categories: Least Concern, Near Threatened, Vulnerable, Endangered, Critically Endangered, and Extinct (including Extinct in the Wild), according to IUCN criteria and thresholds. To project RL categories according to criteria C1 and D1, we

first estimated the potential global population size of a species by multiplying the AOO by the average population density of the species. To account for the uncertainty in the realized density of the species and that mature individuals are a fraction of the whole population, we multiplied the observed and estimated mean density by a correction factor δ (see Supporting Information: S4). We drew 1,000 values of δ from a distribution U (0.1, 1) and applied these values to Monte Carlo simulations (together with ϕ that was sampled independently). For each combination of socioeconomic scenario and dispersal assumption, we thus obtained 1,000 time series of EOO, AOO, and population size which we used to calculate the RLI (see Supporting Information: S3.1).

After transforming RL categories into weights W from 0 (LC) to 5 (EX), we calculated the RLI following Butchart *et al.* (2007)

$$\mathrm{RLI}_t = 1 - \frac{\sum_{s}^{S} W_{c(s,t)}}{W_{\mathrm{EX}} S}, \qquad (1)$$

where $W_{c(s,t)}$ is the weight applied to category c of species s at time t, S is the total number of species modeled, and W_{EX} is the weight applied to extinct species.

We also created spatial maps of RLI for 2010 and 2050 and its difference for each scenario and dispersal assumption (see Supporting Information: S3.2).

Geometric Mean Abundance

The GMA at time t is the geometric mean across a group of species S of the ratio between their population size at time t and their population size in 1970. This is the equivalent at the species to the LPI which is instead based on trends of single populations (Collen *et al.* 2009).

$$\mathrm{GMA}_t = \sqrt[s]{\prod_{s=1}^{S} \frac{p_{s,t}}{p_{s,1970}}}, \qquad (2)$$

where $p_{s,t}$ is the total expected population size of species s at time t, across its whole EOO obtained by multiplying the AOO of species s for its expected population density.

Indicator trends validation

We hindcasted species' responses to global changes from 1970 to 2010 and compared the predicted and observed trends in GMA and RLI for model validation and calibration (see Supporting Information: S5).

Results

When assuming that species can adapt locally to climate change, for example, through phenotypic plasticity or

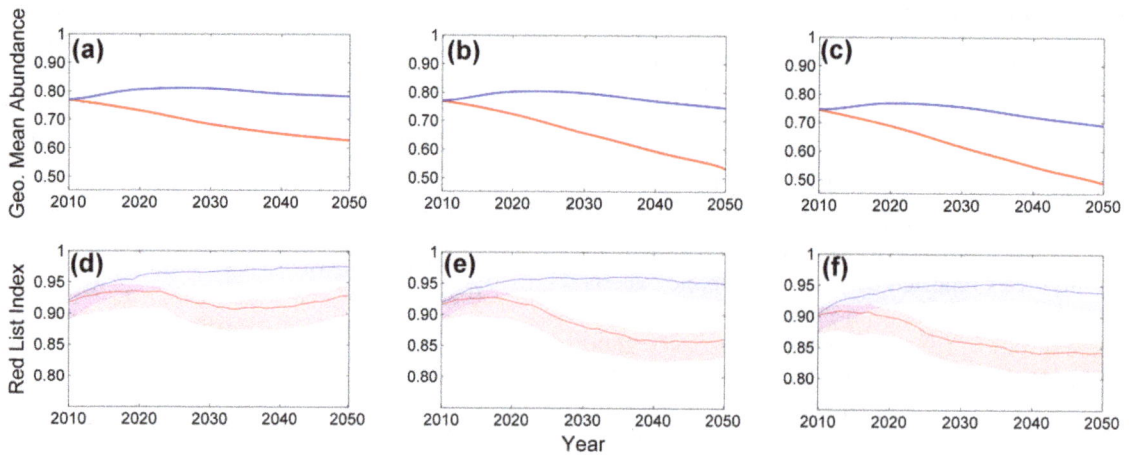

Figure 1 Projected GMA (a, b, and c) and RLI (d, e, and f) for terrestrial carnivores and ungulates under two global socioeconomic scenarios. Business-as-usual in red and Consumption Change in blue. (A and D) Species can adapt to climate change, (B and E) maximum dispersal under land-use and climate change, (C and F) and no dispersal under land-use and climate change. Shading indicates 95% confidence intervals in RLI, the dark lines within the shading represent the median RLI values. The GMA trends do not show confidence intervals because, contrary to RLI, the correction factors for population density and area occupied did not affect these indicators. This is because these factors were applied to obtain population estimates at both numerators and denominator of Equation (2), thereby cancelling each other and generating only one GMA value across all parameter tested in the Monte Carlo simulations. The GMA and RLI values in 2010 vary depending on the dispersal assumption; this affected species range dynamics during the period 1970–2010 used as "burn-in" phase thereby influencing GMA and RLI in 2010.

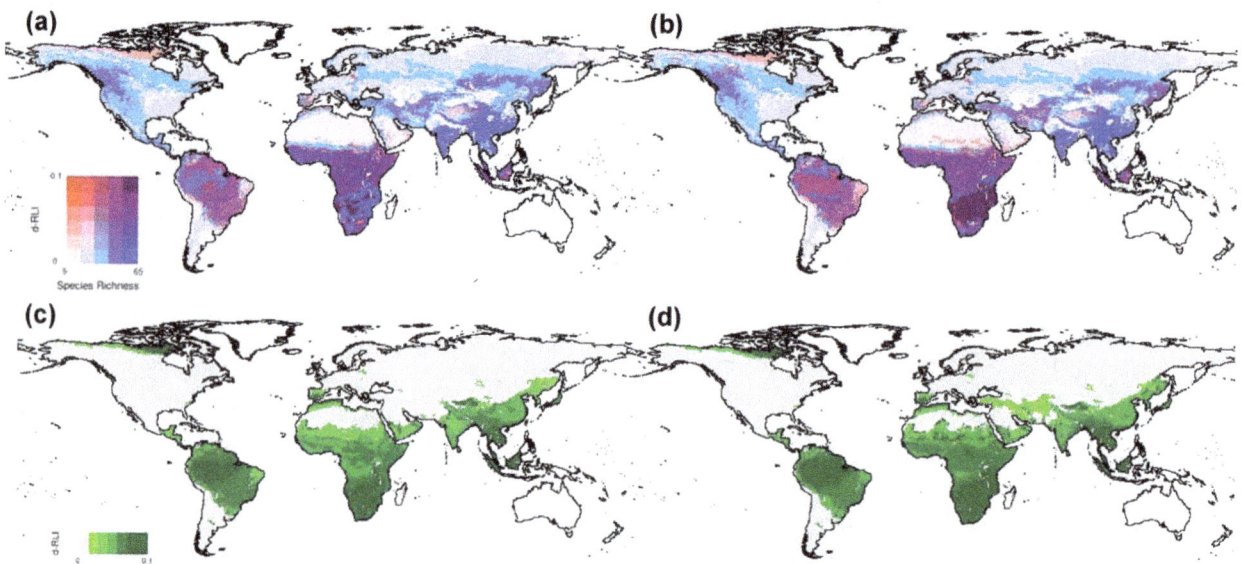

Figure 2 Spatial patterns of trends in Red List Index. (a and b) Bivariate plot showing spatial pattern in species richness and trends in the Red List Index (d-RLI) between 2010 and 2050 under the BAU scenario, with land-use and climate change and assuming maximum dispersal (a) and no dispersal (b). (c and d) Relative improvements in d-RLI for an alternative scenario, Consumption Change relative to Business-as-usual for year 2050 under maximum dispersal (c) and no dispersal (d). Areas in white contain fewer than five species per grid cell modeled in 2010.

microevolution (Boutin & Lane 2014), their EOOs remain stable, and only the area occupied within varies due to land-use and land-cover changes. Under a combination of this assumption and the BAU scenario, we found a steady global decline in mean population abundance of large mammals over the next 40 years (Figure 1a).

The 18% projected GMA decline until 2050, however is comparable to the rate observed in the last 40 years (Figure 3b), therefore it does not lead to changes in RL categories, which require an increase in the rate of decline (Figure 1d). When assuming that species cannot adapt locally to climate change, we found a decline in the GMA in the period 2010–2050 of 31–34%, assuming respectively that species disperse to their maximum physiological capacity, or cannot disperse at all (Figure 1b and c). Extinction risk increased by at least one category for 21–23% of the species. The RLI is projected to decline by 0.055–0.0582 points (Figure 1e and f) which equates to 27.5–29.1% of the species moving one RL category closer to extinction over the time period, a trend comparable to that of the last 40 years (Di Marco et al. 2014). In the BAU scenario, climate change is predicted to outpace the ability of many species to shift their distributions even under the maximum physiological dispersal assumption (Figure 1b and e, Supporting Information: S2.2).

The Consumption Change scenario, regardless of assumptions concerning climate change impact and dispersal, results in an initial improvement and then stabilization in the RLI and GMA until 2030 brought about by habitat regeneration and human-driven habitat restoration assumed for this scenario (PBL 2012). When accounting for species responses to climate change, this initial improvement is followed by a decline due to the later onset of EOO contractions caused by climate change (Figure 1b, c, e, and f); which poses at risk the achievement of long-term conservation goals.

The overall trends shown by the RLI do not follow the same monotonic decline as the GMA for the BAU scenario. As the rate of habitat loss slows down toward 2050 in the BAU scenario, these species decline more slowly, eventually qualifying for lower categories of extinction risk under RL criteria A and C (Figure 1d and e). This leads to an improvement in the RLI trend, which contrasts with the GMA trend, in which the magnitude of decline reflects the total reduction in population abundance of the set of species by 2050.

Under the BAU scenario, increases in species extinction risk (i.e., declining RLI trends) are predicted in all regions of high-current mammal richness (Figure 2, Figure S6). However, particularly steep declines are predicted in the Amazon, a region with very low spatial climatic gradients that is predicted to experience no-analog future climates (Williams et al. 2007) and with a high richness of mammal species whose dispersal abilities are insufficient to keep pace with projected climate change (Schloss et al. 2012). Large declines are also predicted in sub-Saharan Africa under all scenarios. This hotspot of carnivore and ungulate species diversity is predicted to double its human population size and experience a rapid increase in per-capita growth rate from 2030 leading to a tripling of per-capita calorie consumption, and rapidly reducing the extent of natural vegetation (PBL 2012). Insular Southeast Asia, which holds many currently threatened and restricted-range species (Schipper et al. 2008), due to the highest rates of deforestation globally (Hansen et al. 2013;

Abood et al. 2014) is also expected to face an increase in overall extinction risk under the BAU scenario due to continued deforestation (PBL 2012). Improvements in extinction risk are expected in continental South-East Asia due to the slowdown of deforestation with respect to the past 40 years (Figure S6). Compared with the BAU scenario, Consumption Change reduces aggregate extinction risks under all dispersal assumptions. Reductions are more pronounced in the tropics (Figure 2c and e), driven by measures to reduce deforestation, such as reduced meat consumption, reduced impact logging, and setting aside areas for protection (PBL 2012).

When comparing modeled and observed responses to recent (1970–2010) land-use and climate change, we found that the modeled GMA is within the large confidence intervals of the observed LPI for the subset of species for which population trends were available (Collen et al. 2009) (Figure 3a). Aggregate extinction risk for the period 1996–2008 (corresponding to the two published global mammal assessments), was estimated accurately and without bias after accounting for all uncertainties in parameters and models (Figure 3b).

Discussion

Our analyses show that a scenario with aggressive policies to eradicate hunger, ensure universal health, and access to modern energy can be compatible with short-term biodiversity goals. This is the first quantitative analysis in demonstrating that these potentially conflicting goals are not mutually exclusive. These ambitious goals will require rapid and widespread implementation of sustainable production practices, for example, adoption of reduced impact logging and sustainable agricultural intensification to increase crop yields. It will also require changes in consumption: low-energy lifestyle, reduce waste, and consumption of meat from industrially farmed animals. Finally, it will require progressive environmental legislation: carbon taxation (including emission from land-use change), and strategic placement of protected areas where habitat loss poses the highest threat to biodiversity. Our results also show that this might not be sufficient to stabilize long-term trends, due to the lasting effect of increased carbon emissions. BAU instead, will fail to meet both short- and long-term CBD goals.

We explored species responses to future global changes by projecting two indicators adopted by the CBD to monitor progress toward the Aichi targets. However, our approach lends itself to project any indicator based on species distribution and population abundance and is potentially applicable to any taxonomic group for which distribution data and habitat requirements are known.

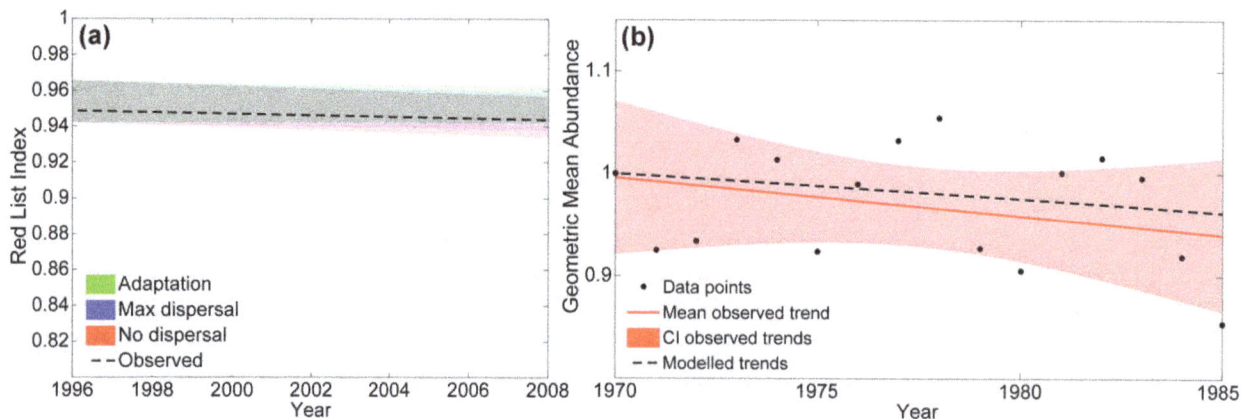

Figure 3 Validation of modeled trends in Red List Index (a) and Geometric Mean Abundance, (b). The colored bands in panel A represent the 95% confidence interval from the Montecarlo simulations of the modeled RLI (hindcast) under three assumptions of species responses to climate change: species can adapt locally to climate change (Adaptation, green band), species can colonize new suitable climatic areas that are within their maximum physiological dispersal abilities (Max. dispersal, blue band), species cannot colonize new suitable climatic areas (No dispersal, red band). The black dashed line represents the observed RLI from 1996 to 2008. The points in panel B represent observed annual GMA values for the set of carnivore and ungulate species for which population data were available. The solid line represents the trend in GMA (line of best fit across the points) and has a slope $\beta_o = -0.0037 \pm 0.061$. The modeled GMA (calculated in the same way as the LPI) has slope $\beta_m = -0.0021$ within the confidence intervals of the line of best fit of GMA values.

Our modeled responses to global changes may be overoptimistic for some species because we did not account for all threats to mammals, especially hunting which is a major threat to many of the species considered here (Hoffmann *et al.* 2011). However, we do not expect this to change the qualitative differences in projected trends between scenarios. Rather it might widen the difference between BAU and Consumption Change because low food security, poor access to food markets, a high proportion of people living in rural areas, and poorer environmental governance are more likely to exacerbate hunting in the former than the latter (PBL 2012).

Our results illustrate how detailed biodiversity indicators can be used in conjunction with coupled socioeconomic and environmental scenarios to inform the development of policies that achieve future sustainability goals. Our approach brings the focus of scenarios directly to the species level, thus taking into account an additional dimension of biodiversity and paving the way for incorporating stronger ecological foundations into future biodiversity assessments.

Acknowledgments

We thank Ben Collen, Robin Freeman and Georgina Mace for insightful discussions and comments to prior version of this manuscript. We acknowledge the CINECA Award Try 2013 for the availability of high performance computing resources and support. This work was possible because of the work and dedication of hundreds of experts of the IUCN Species Survival Commission's mammal specialist groups who provide their data and expert knowledge on species distribution and life-history traits.

Supporting Information

S1–S6 Methods.

Table S1. Bioclimatic variables used in the species distribution models.

Table S2. Eighteen general circulation models used in the analyses.

Table S3. Description of Red List criteria and parameters used for projecting species Red List categories.

Table S4. Mean annual change in forest cover from observed and modeled data.

Table S5. Description of GLOBIO land-use categories and how they relate to GLC2000

Table S6. An extract of the GLOBIO land-cover and land-use tables for year 2000.

Table S7. Summary of the Red List Index validation for the period 1996–2008, $n = 245$ species.

Table S8. Number of species in IUCN-threatened categories (VU, EN, CR) listed under each criterion in the Global Mammal Assessment of 2008, in the model hindcast and one model forecast.

Figure S1–S2. Global trends in area covered by the two socioeconomic scenarios as modeled by the IMAGE Integrated Assessment Model platform.

Figure S3. Boxplots of Red List Index in 2008 across 1000 Monte Carlo simulations with and without the calibration of deforestation data using satellite imagery

Figure S4. Predictive ability of bioclimatic envelope models measured by True Skill Statistic.

Figure S5. Comparison of projected Geometric Mean Abundance and Red List Index for the set of species included in the validation analyses, and for the full data set of 440 species (presented in Figure 1 in the main text).

Figure S6. Richness of species subject to downlisting and uplisting of IUCN Red List categories between 2010 and 2050 under the BAU scenario.

Supplementary Data File: Supplementary Table 1 Data Sufficient Species Validation.xlsx

References

Abood, S.A., Lee, J.S.H., Burivalova, Z., Garcia-Ulloa, J. & Koh, L.P. (2014). Relative contributions of the logging, fiber, oil palm, and mining industries to forest loss in Indonesia. *Conserv. Lett. this is actually still in press despite online since April 2014. doi: 10.1111/conl.12103.*

Alkemade, R., van Oorschot, M., Miles, L., Nellemann, C., Bakkenes, M. & ten Brink, B. (2009). GLOBIO3: a framework to investigate options for reducing global terrestrial biodiversity loss. *Ecosystems*, **12**, 374-390.

Allouche, O., Tsoar, A. & Kadmon, R. (2006). Assessing the accuracy of species distribution models: prevalence, kappa and the true skill statistic (TSS). *J. Appl. Ecol.*, **43**, 1223-1232.

Boutin, S. & Lane, J.E. (2014). Climate change and mammals: evolutionary versus plastic responses. *Evol. Applic.*, **7**, 29-41.

Bouwman, A., Kram, T. & Klein Goldewijk, K. (2006). *Integrated modelling of global environmental change: an overview of Image 2.4.* Netherlands Environmental Assessment Agency, Bilthoven.

Brooks, T.M., Lamoreux, J.F. & Soberón, J. (2014). IPBES ≠ IPCC. *Trends Ecol. Evol.*, **29**, 543-545.

Butchart, S.H.M., Akçakaya, H.R., Chanson, J. et al. (2007). Improvements to the red list index. *PLoS ONE*, **2**, e140.

Butchart, S.H.M., Walpole, M., Collen, B. et al. (2010). Global biodiversity: indicators of recent declines. *Science*, **328**, 1164-1168.

Collen, B., Loh, J., Whitmee, S., McRAE, L., Amin, R. & Baillie, J. (2009). Monitoring change in vertebrate abundance: the Living Planet Index. *Conserv. Biol.*, **23**, 317-327.

Di Marco, M., Boitani, L., Mallon, D. et al. (2014). A retrospective evaluation of the global decline of carnivores and ungulates. *Conserv. Biol.*, **28**, 1109-1118.

Guisan, A. & Zimmermann, N.E. (2000). Predictive habitat distribution models in ecology. *Ecol. Model.*, **135**, 147-186.

Hansen, M., Potapov, P., Moore, R. et al. (2013). High-resolution global maps of 21st-century forest cover change. *Science*, **342**, 850-853.

Hoffmann, M., Belant, J.L., Chanson, J. et al. (2011). The changing fates of the world's mammals. *Philos. Trans. Royal Soc. B*, **1578**, 2598-2610.

IUCN. (2012a). *IUCN Red List Categories and Criteria: Version 3.1.* Second edition. p. iv + 32 pp. IUCN Species Survival Commission, Gland, Switzerland and Cambridge, UK. This is the recommended citation by IUCN for this document http://www.iucnredlist.org/documents/redlist_cats_crit_en.pdf

IUCN. (2012b). IUCN Red List of Threatened Species Version 2012.2. http://www.iucnredlist.org. Accessed 2 November 2012.

Nicholson, E., Collen, B., Barausse, A. et al. (2012). Making robust policy decisions using global biodiversity indicators. *PLoS ONE*, **7**, e41128.

PBL. (2012). *Roads from Rio+20 Pathways to achieve global sustainability goals by 2050*, The Hague, The Netherlands.

Pearson, R.G. & Dawson, T.P. (2003). Predicting the impacts of climate change on the distribution of species: are bioclimate envelope models useful? *Global Ecol. Biogeogr.* **12**, 361-371.

Pearson, R.G., Stanton, J.C., Shoemaker, K.T. et al. (2014) Life history and spatial traits predict extinction risk due to climate change. *Nat. Clim. Change*, **4**, 217-221.

Price, J., Warren, R. & Goswami, S. (2012). Climascope. http://climascopewwfusorg. Accessed 22 May 2012.

Rondinini, C., Di Marco, M., Chiozza, F. et al. (2011) Global habitat suitability models of terrestrial mammals. *Philos. Trans. Royal Soc. B: Biol. Sci.* **366**, 2633-2641.

Schipper, J., Chanson, J.S., Chiozza, F. et al. (2008) The Status of the World's Land and Marine Mammals: Diversity, Threat, and Knowledge. *Science*, **322**, 225-230.

Schloss, C.A., Nuñez, T.A. & Lawler, J.J. (2012). Dispersal will limit ability of mammals to track climate change in the Western Hemisphere. *Proc. Nat. Acad. Sci.*, **109**, 8606-8611.

Soberón, J. & Nakamura, M. (2009). Niches and distributional areas: concepts, methods, and assumptions. *Proc. Nat. Acad. Sci.*, **106**, 19644-19650.

Stehfest, E., Bouwman, L., van Vuuren, D.P., den Elzen, M.G., Eickhout, B. & Kabat, P. (2009). Climate benefits of changing diet. *Clim. Change*, **95**, 83-102.

Thuiller, W., Lafourcade, B., Engler, R. & Araújo, M.B. (2009). BIOMOD–a platform for ensemble forecasting of species distributions. *Ecography*, **32**, 369-373.

Tittensor, D.P., Walpole, Matt, Hill Samantha, L.L. et al. (2014) A mid-term analysis of progress toward international biodiversity targets. *Science*, **346**, 241-244.

Van Vuuren, D., Sala, O. & Pereira, H. (2006). The future of vascular plant diversity under four global scenarios. *Ecol. Soc.*, **11**, 25. http://www.ecologyandsociety.org/vol11/iss2/art25/ES-2006-1818.pdf

VanDerWal, J.J., Beaumont, L.J. & Zimmermann, N.E. (2011). R package 'climates': methods for working with weather and climate. www.rforge.net/climates/.

Visconti, P., Pressey, R.L., Giorgini, D. *et al.* (2011). Future hotspots of terrestrial mammal loss. *Philos. Trans. Royal Soc. B: Biol. Sci.*, **366**, 2693-2702.

Williams, J.W., Jackson, S.T. & Kutzbach, J.E. (2007). Projected distributions of novel and disappearing climates by 2100 AD. *Proc. Nat. Acad. Sci.*, **104**, 5738-5742.

17

Dealing with Cumulative Biodiversity Impacts in Strategic Environmental Assessment: A New Frontier for Conservation Planning

Amy L. Whitehead*, Heini Kujala, & Brendan A. Wintle

School of BioSciences, The University of Melbourne, Parkville, VIC 3010, Australia

Keywords

Spatial conservation prioritization; conservation planning; biodiversity; development; strategic environmental assessment; cumulative impact assessment; species distribution model; irreplaceability; complementarity.

Correspondence

Amy Whitehead, School of BioSciences, The University of Melbourne, Parkville, VIC 3010, Australia. E-mail: amy.whitehead@niwa.co.nz

Editor

Amanda Lombard

Abstract

Biodiversity impact assessments under threatened species legislation often focus on individual development proposals at a single location, usually for a single species, leading to inadequate assessments of multiple impacts that accumulate over large spatial scales for multiple species. Regulations requiring ad-hoc assessments can lead to "death by a thousand cuts," where biodiversity is degraded by many small impacts that individually do not appear to threaten species' persistence. Spatial prioritization methods can improve the efficiency of decision-making by explicitly considering cumulative impacts of multiple proposed developments on multiple species over large spatial scales. We present an assessment approach and a unique case study in which spatial prioritization tools were used to support strategic assessment of a large development plan in Western Australia. The application of the approach helped identify relatively minor alterations to development plans that resulted in reductions in biodiversity impacts and informed expansion of the protected area network. Using these tools to assess trade-offs between conservation and development will help identify planning footprints that minimize biodiversity losses.

Introduction

Evaluating the environmental impact of projects is a critical prerequisite for sustainable development. Anticipating and acting on foreseeable development-conservation conflicts helps cost-effectively reduce biodiversity losses because the cost of conserving species and communities increases rapidly as they become less widespread and options for their conservation narrow (Mills *et al.* 2014). In many countries, existing legislation and regulation requires formal assessment of activities deemed likely to pose some risk to threatened species (Chaker *et al.* 2006; Connelly 2011). Threatened species and environmental protection regulations have traditionally been operationalized by focusing on project-by-project impact assessments, aiming to approve, reject, or condition development projects in order to minimize impacts on threatened species or ecological communities (e.g., Mörtberg *et al.* 2007; Atkinson & Canter 2011). These approaches have been criticized for failing to deliver adequate biodiversity protection due to the ad-hoc and local nature of assessments (Hawke 2009). Concern has been raised about the impact of multiple discrete and/or consecutive projects that, when individually assessed, may be approved but that cumulatively will lead to "death by a thousand cuts" (Hawke 2009).

Impacts on biodiversity accumulate in multiple ways: (i) at a site, impacts may accumulate due to the effect of multiple stressors (e.g., cumulative impacts of habitat fragmentation, noise, and pollution; Halpern *et al.* 2008), (ii) over larger, heterogeneous environments, individual or multiple impacts may accumulate spatially as an impact footprint expands under multiple independent developments, changing habitat extent and connectivity (Hawke 2009), and (iii) over time impacts may accumulate at a site or across a landscape (e.g., accumulation of heavy metals in an ecosystem). Impacts of multiple individual stressors or threats can be synergistic or antagonistic. Impacts may accumulate both linearly or nonlinearly over time and space (Halpern & Fujita 2013). The cumulative impact assessment literature primarily focus on evaluating the impacts of multiple interacting stressors at a site level, using coarse surrogates for species persistence such as ecosystem types, with little or no reference to the specific requirements, distributions, or persistence of individual species (e.g., Halpern et al. 2008; Halpern & Fujita 2013; Andersen et al. 2015). These approaches tend not to explicitly consider spatial accumulation of impacts, nor the impacts that multiple developments might have on the connectivity and viability of species populations, rendering them inadequate for assessing impacts under threatened species legislation. Here, we focus on the less examined spatial accumulation of multiple individual development impacts across large regions.

Many spatially-explicit environmental assessments use single species or biodiversity surrogates to assess impacts due to development (Mörtberg et al. 2007; Atkinson & Canter 2011). Single-species assessments are not useful for decision makers determining acceptability of impacts at a regional scale when multiple threatened species with differing initial rarity occur disparately across the landscape (Connelly 2011; Duinker et al. 2012). Spatial prioritization planning tools such as Zonation (Moilanen et al. 2005) or Marxan (Ball et al. 2009) can support regional-scale assessments by characterizing conservation outcomes under planning options for multiple species over large landscapes. Outputs can identify areas of high biodiversity value where development should be avoided or areas where development may impact biodiversity least (Bekessy et al. 2012; Kareksela et al. 2013). Spatial impact assessments during planning help planners identify development footprints that minimize cumulative impacts on biodiversity and adhere to threatened species regulations. There is an opportunity to embed these approaches within statutory strategic environmental assessment (SEA; Therivel & Paridario 2013) that sets out to reconcile environmental, social, and economic impacts of proposed developments.

We demonstrate the use of spatial prioritization tools in a region-wide land-use planning process in Western Australia (WA), where the Australian Government (AG) is assessing the impacts of a 30-year development plan for ~8,500 km[2] around the city of Perth. This statutory strategic assessment, under Australia's Environment Protection and Biodiversity Conservation (EPBC) Act (1999),

seeks to assess the cumulative spatial impact of multiple individual developments on multiple species and ecological communities. We employed a three-step approach to analyze and minimize cumulative impacts of multiple residential, industrial, infrastructure, and extractive development actions on 227 biodiversity features over approximately one million landscape elements (1 ha cells) (Figure 1). We discuss the importance of sound technical solutions combined with timely, consistent, and thorough engagement with planners and decision makers to ensure that best-practice analyses were understood, accepted, and integrated into the decision-making process. We show how strong engagement contributed to significantly reduced impacts on biodiversity and informed a significant expansion of the conservation estate.

Methods

Perth and Peel Strategic Assessment

Australia's EPBC Act allows for approval of policies, plans, or programs, under which future actions may be undertaken without the need for individual referral and approval (DSEWPAC 2012). The Perth and Peel Strategic Assessment[1,2] (PPSA) seeks to assess the biodiversity impacts of ~570 km[2] of proposed development that would add to ~1,390 km[2] of existing development approvals. The total increase in developed area of ~1,960 km[2] is to accommodate a doubling of the population to ~3.5 million people, within 8,500 km[2] of south-western Australia. Development options initially proposed under the WA Planning Commission's (WAPC) *Directions 2031 and Beyond* (WAPC 2010) and other key planning documents[3] were refined to reduce environmental impacts, leading to the 2015 release of the WAPC's *Perth and Peel at 3.5 million* suite of planning documents and the draft *Perth and Peel Green Growth Plan* (GGP).[4] The PPSA region is part of the Southwest Australia Ecoregion, one of the 35 global biodiversity hotspots (Myers et al. 2000). Approximately 70% of the region contains important native vegetation, wetlands, and other habitats supporting threatened and endemic species and ecological communities.

Characterizing regional biodiversity

We mapped the distributions of 189 threatened species and 38 threatened ecological communities (TECs) listed under national and state legislation. Presence-only point data for species were obtained from online public databases,[5] while TEC and habitat polygon data were provided by government agencies. To identify suitable

1) Species distribution modeling

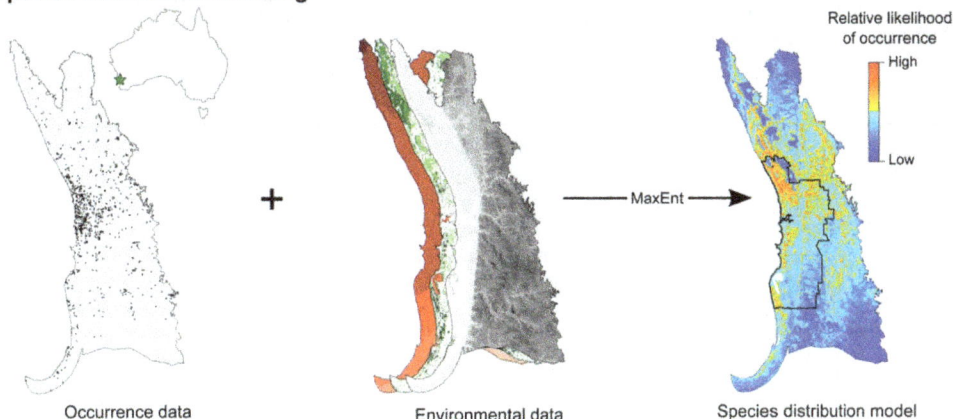

2) Spatial prioritization

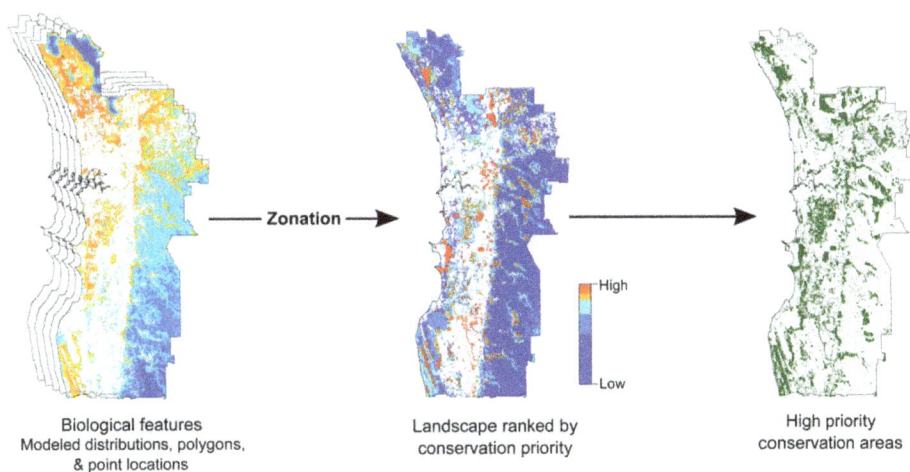

3) Strategic Impact Assessment

Figure 1 Schematic diagram representing the three-step modeling process used to undertake the spatial cumulative impact assessment in the Perth-Peel region of Western Australia. (A) Species occurrence data were obtained from online databases and combined with environmental data to produce species distribution models for 61 threatened species using MaxEnt at the scale of the surrounding bioregions. (B) These models were clipped to the Perth-Peel region and combined with additional biological features, including threatened ecological communities, in the spatial conservation prioritization software, Zonation, to rank the landscape for its conservation value for 227 biodiversity features. Priority conservation areas were identified as the best 30% of the landscape for conservation. (C) We used Zonation to assess the impacts of spatially explicit proposed development types, identifying areas of potential conflict with high priority areas for conservation. This was undertaken as part of an iterative process, where planning footprints were refined to avoid areas of high conflict and reassessed.

Table 1 The relative impact of development can be assessed by examining the area of habitat cleared, conflicts with existing protected areas and identified priority conservation areas and the estimated loss of biodiversity feature distributions under different development types within the final cumulative planning footprint based on the draft Green Growth Plan (DPC 2015). The mean and maximum distribution loss for each development type is based on the distributions of 227 biodiversity features within the PPSA region. Locally extinct features are those losing all of their known occurrences in the region due to development, with the number of features losing at least 50% of their distribution also shown.

Development type	Area developed (ha)	Habitat cleared (ha)	Development conflicts with (%)		Loss of feature distributions (%)		Locally extinct features	Features losing at least 50%
			Protected areas	Priority areas	Mean	Max		
1. Urban	121,737	34,016	0.72	7.90	7.24	100.00	1	5
2. Industrial	28,655	9,408	0.36	2.69	2.32	50.00	0	3
3. Rural Residential	11,007	5,176	0.02	1.51	0.63	33.33	0	0
4. Infrastructure	19,216	8,153	0.59	2.14	2.69	100.00	1	3
5. Forestry	11,527	11,036	2.82	0.02	0.16	3.41	0	0
6. Mining	20,284	16,282	3.71	0.99	0.68	50.00	0	1
7. Final cumulative	196,282	73,015	5.95	14.57	12.94	100.00	2	14

habitat for threatened species, we modeled distributions for 61 threatened species with at least 20 occurrence records within the PPSA region using MaxEnt (Appendix A; Phillips *et al.* 2006) (Figure 1A). Spatial data for 128 threatened species with less than 20 occurrence records and 38 TECs were included as binary maps. All data layers were clipped to areas of extant vegetation.

Conservation priority areas

We used the conservation prioritization software Zonation v.4.0 (Moilanen *et al.* 2005, 2014) to identify areas of conservation priority within the PPSA region for the 227 biodiversity features described above (Figure 1B). Zonation is a maximum-coverage (Camm *et al.* 1996) tool that identifies areas maximizing the representation of suitable habitat for multiple species over large landscapes (Appendix B; Moilanen *et al.* 2005, 2014). It creates a hierarchical ranking of all sites across the landscape according to conservation priority. A relatively small proportion of the top-ranked sites typically represent all species and core habitats.

In collaboration with agency staff, we developed a scheme to weight endemic and/or threatened biodiversity features more highly in the prioritization process (Appendix C). For reporting purposes, we focus on the top 30% ranked sites within the PPSA region, hereafter referred to as *priority conservation areas*. Spatial prioritization outputs were compared to existing protected areas within the PPSA region to assess how well the identified priority conservation areas are currently protected under differ-

ent IUCN protected area categories. We also assessed the proposed expansion of IUCN category I-IVa protected areas in the PPSA region, a planning outcome of the strategic assessment process (Appendix E).

Strategically assessing the spatial accumulation of biodiversity impacts

We assessed the potential biodiversity impacts of both approved developments that are currently undeveloped and proposed developments being assessed under the PPSA. We used four spatially explicit planning footprints obtained from government from 2013 to January 2016 (Figure 1C; Table 1), including six development types: (1) urban, (2) industrial, (3) rural residential, (4) infrastructure, (5) exotic plantation forestry, and (6) basic raw materials (sand, limestone, clay, and hard rock) mining. Each development type was assumed to result in the absolute loss of biodiversity value within the impact footprint. Partial loss of value after an impact could be incorporated in Zonation where appropriate but was unnecessary here. Over the 2010–2016 planning period, footprints were refined through a review process that assessed potential biodiversity impacts. We spatially overlaid the six development types for each planning footprint to assess the spatial accumulation of impacts and to understand the iterative changes in biodiversity impact from the initial (WAPC 2010, 2012) to the current GGP proposals (DPC 2015).

We identified areas in the landscape where development scenarios overlapped with priority conservation

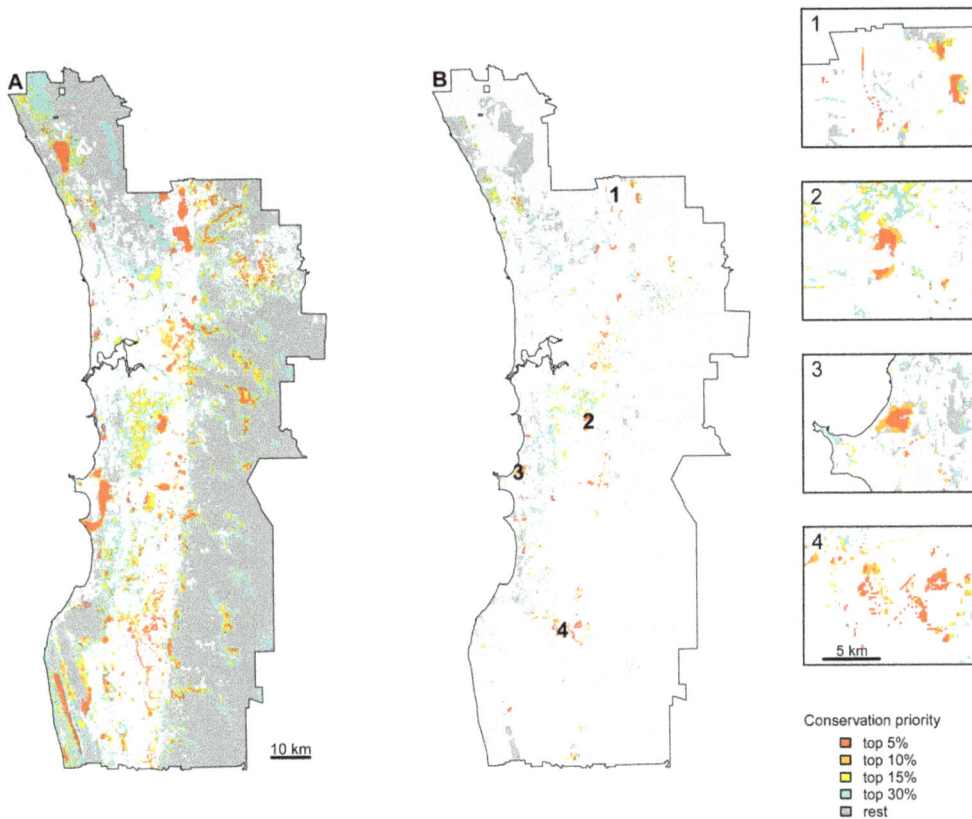

Figure 2 (A) Priority conservation areas within the PPSA region based on the best 30% of the landscape (red to blue) of a spatial prioritization for 227 biodiversity features, irrespective of current land tenure. (B) The conflicts between priority conservation areas and the final cumulative planning footprint based on the draft *Green Growth Plan* (DPC 2015) are highlighted by clipping the conservation priorities to the footprint. The inset boxes highlight areas of conservation priority that are likely to be developed for (1) mining, rural residential, and industrial; (2) urban and industrial; (3) urban, industrial, and infrastructure; and (4) urban and industrial.

areas. The potential biodiversity impact of each scenario was estimated by measuring proportional losses of each biodiversity feature (species habitat, point locations, or TECs), and by considering the overall proportion of biodiversity value lost across all species and threatened communities (Cabeza & Moilanen 2006). The latter analysis explicitly incorporates the spatial accumulation of impacts on individual biodiversity features through habitat loss and reduced connectivity between remaining habitats, and on the combined biodiversity value by considering the relative importance of each landscape cell for the overall representation of biodiversity in the landscape (Appendix B).

Results

The areas identified as priority conservation areas (Figure 2A) contain, on average, 87.3% of the mapped area of TECs, species habitat, and point locations within

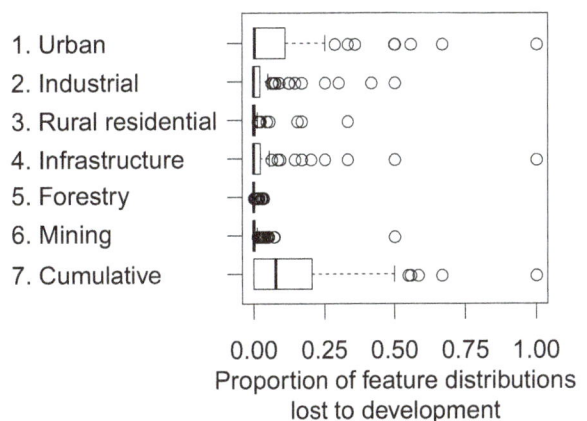

Figure 3 The proportion of feature distributions lost under each development type within the final cumulative planning footprint based on the draft *Green Growth Plan* (DPC 2015). Considering the average loss of habitat across the feature pool can provide insights into the relative impacts of proposed development but it is also important to consider how individual features are likely to be impacted.

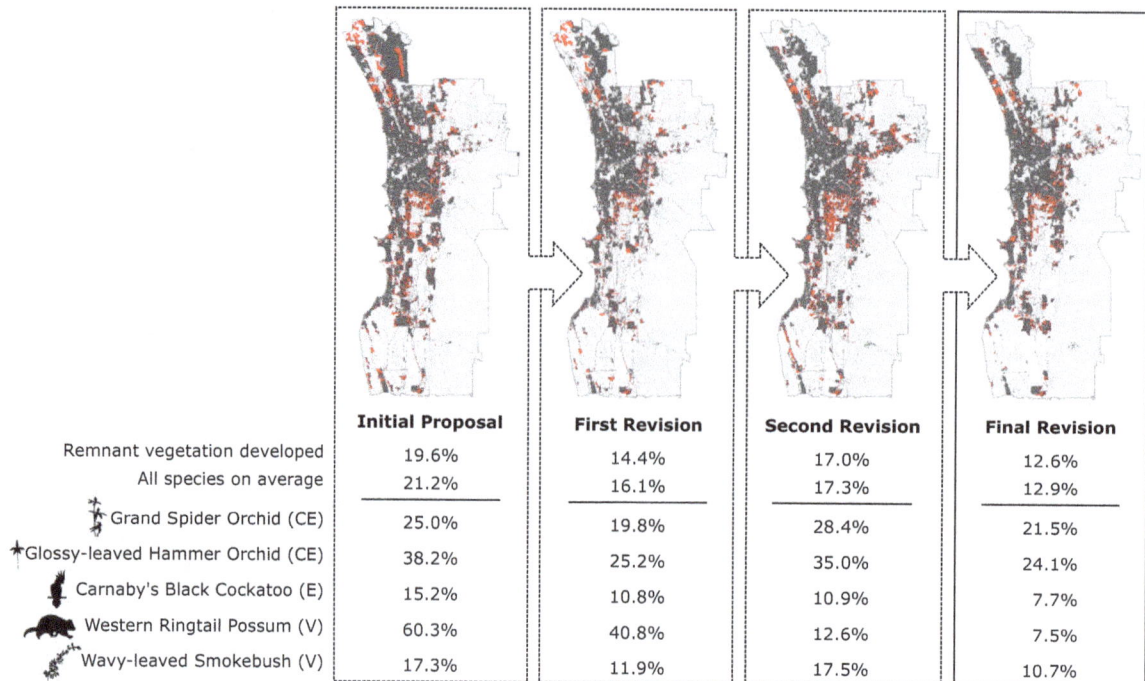

Figure 4 Using spatial conservation prioritization to identify priority conservation areas where development should not occur can help planners make biodiversity-sensitive revisions to planning footprints. The initial proposal is the *Directions 2031* plan, and first, second, and final revisions were part of the *GGP* planning process. Here, iterative changes to the planning footprint led to overall reductions in biodiversity loss, as well as reducing the impacts on key species of concern. Areas colored red represent priority conservation areas at risk of being developed under each proposal, while dark gray areas show development that will not result in the clearance of native vegetation or areas of significant values. Values describe the proportion of each feature's distribution within the PPSA region that will be lost to development.

the PPSA region. However, the occurrence of biodiversity features varied considerably within the priority conservation areas, with the ranges of some widespread and/or lower weighted features receiving proportionally lower representation (Appendix D). Only 26% of the identified priority conservation areas (<9% of the total region) are contained within the existing protected area network, with 18 biodiversity features not protected at all. The addition of proposed conservation reserves within the PPSA region under the GGP would increase IUCN category I-IVa protected areas from 99,812 to 177,497 ha (30% of the region's habitat), leaving fewer features without category I-IVa protection (29 vs. 63). However, only 11.7% of the priority conservation areas we identify would be protected under this proposal. The new proposal for protected areas improves the protected area network. However, further expansion of reserves in unprotected high priority areas is warranted (Appendix E).

The predicted impacts to biodiversity varied between development types and planning footprints with larger development areas leading to larger biodiversity impacts (Figure 2; Table 1). Under the final GGP cumulative plan-

ning footprint, urban development was predicted to have the largest impact, resulting in the loss of up to 34,000 ha of habitat and 7.9% of priority conservation areas. This is comprised of 13,500 ha of existing approvals and 20,500 ha proposed under the GGP. The plantation forestry and mining impacts were relatively minor (and overlapping) resulting in a mean habitat loss of 0.2–0.99%, though often occured in existing protected areas (2.8–3.7%, respectively). At the time of writing, the GGP cumulative planning footprint was estimated to result in the loss of 73,000 ha of habitat, including 14.8% of priority conservation areas (Figure 2B) and on average biodiversity features would lose 12.9% of their current distributions. However, the potential impact varies considerably among species and communities, and for the majority, individual development types pose moderate or low habitat impacts, emphasizing the importance of cumulative assessment across all development types (Figure 3). Carnaby's black cockatoo (*Calyptorhynchus latirostris*) is one species for which the cumulative impacts are higher (16.3% habitat loss) than anticipated by any individual development type (1.1–7.5%), though the

GGP introduces special measures to restore Carnaby's cockatoo habitat.

Two development types (urban, infrastructure) and the GGP cumulative planning footprint will likely result in the loss of all recorded locations of two plant species (*Thysanotus glaucus* and *Calothamnus macrocarpus*) within the PPSA region if mitigation measures are not implemented (Table 1, Figure 3, Appendix F). Fourteen biodiversity features may lose at least 50% of their local distributions under the GGP cumulative planning footprint.

There was an overall reduction in biodiversity impacts across the four cumulative planning footprints considered (Figure 4, Appendix F), although the area of habitat loss increased in the third revision when rural residential development was added to the planning process. Revisions decreased the proposed total area of habitat cleared from 114,065 ha (*Directions 2031*) to 73,015 ha (GGP), reducing the anticipated mean biodiversity losses from 21.2% to 12.9%, and more than halving the impact for some threatened species (Figure 4). Nearly 6,500 ha of avoided clearing in the proposed urban development zones resulted from the transition from *Directions 2031* to GGP, with an additional 163 ha set aside in *Retention Zones* that will not be developed (Table F4; Figure F2), further reducing the mean biodiversity loss from the newly proposed urban expansion from 1.11% to 0.21%.

Discussion

This study represents a rare case in which a systematic spatial prioritization was used to support a statutory land-use planning process; reducing losses through planning and informing protected area expansion (Figure 4 and Figures E1, E2, and F2). Assessment was based on a transparent and repeatable framework that provided planners and decision makers with comprehensive information for each species, TEC, and the combined biodiversity values. The value of repeatability was highlighted when analyses were repeated to incorporate new preferences, constraints, and features required by planners and policy officers.

Our approach efficiently visualized conflicts between proposed development and priority conservation areas, identifying the most impacted species under each scenario. By assessing the cumulative impacts, we were able to identify biodiversity losses that might not have been apparent through project-by-project assessments. The framework allowed decision makers to compare planning footprints and reduce biodiversity losses via plan revisions. Further revisions to the plan will occur after the date of publication, following public consultation.

Three key ingredients were essential for the successful application of our approach in this complex and contentious political process. First, rigorous repeatable, peer-reviewed methods that directly addressed the problem, but which were adaptable to the needs of planners and decision makers were the basic ingredient. Conservation planning tools based on irreplaceability and complementarity have 20 years of pedigree in peer-reviewed science literature and yet are intuitive enough that the relevance of these principles and tools can easily be described to planners and decision makers.

A second key ingredient was the joint willingness of planners and decision makers to engage with and explore technical, science-based inputs to the planning process. We were fortunate in this case study that the Australian and WA Government officers were eager to understand and utilize scientific approaches to assessment. Throughout the process, they maintained a high degree of engagement with our analyses, and facilitated data-sharing to ensure timely delivery of outputs.

Third, there was concerted and consistent engagement between researchers and planners throughout the project, including problem scoping, analyses and refinements, plan improvement, and communication of findings. This was achieved through over 40 face-to-face interactions, seminars to planners, decision makers, and stakeholders, and by being available to discuss ideas or concerns. While this interaction comes at a high time cost to researchers and planners, in the absence of relevant technical expertise within government agencies, it is the only way to ensure that the scientific methods are carried through the planning process.

Despite the success of this case study, we underline that our framework sets out a starting point for a comprehensive biodiversity impact assessment, adaptable to different circumstances. Other applications may require analysis of the cumulative impacts of a range of partially impacting stressors, which degrade habitats without totally destroying them, or where mitigation actions partially mitigate impacts. Partial losses and mitigation can already be handled within Zonation and related tools (Moilanen *et al.* 2011), and methods could be further developed to integrate site-level cumulative impacts (*sensu* Halpern *et al.* 2008). In this study, conservation priorities and impacts were estimated based on species occupancy records and species distribution modelss, ignoring many ecological processes. Therefore, we may have underestimated the impacts of development on species persistence. Refinements could consider gradual degradation of remaining habitats due to edge effects (Moilanen & Wintle 2007), nonlinear declines in species due to reduced habitat connectivity (Fordham *et al.* 2012), changes to resources that limit breeding potential or increase mortality (Bekessy *et al.* 2009), and spatial or temporal uncertainties about current and future distributions of species (Moilanen & Wintle 2006; Strimbu & Innes 2011; Kujala *et al.* 2013). By incorporating these factors in population viability analyses (PVAs), a more comprehensive evaluation of plan adequacy, which explicitly considered species persistence under development options, could be achieved (Wintle *et al.* 2005; Sebastián-González *et al.* 2011). Such analyses require species demography and movement information, rarely available for most species in a region, though implementing PVA for a subset of well-studied species could improve assessments. On the flip side, we have assumed that habitat within the impact footprint will be completely lost after development. This may not always be the case if appropriate in-development mitigation measures can be implemented, providing the possibility of better outcomes than our analyses predict.

A particular feature of this study was the focus on threatened species and TECs; a requirement of local threatened species legislation. Given the extent of planned development, species not currently listed as threatened could become so once all development occurs. Current studies provide little optimism about the degree to which individual taxon can be used as surrogates for others in spatial conservation analyses (Lentini & Wintle 2015). Expanding analyses to consider impacts on non-listed species is a logical extension of this work.

The analytical approach presented in this study is generalizable to any situation in which multiple biodiversity values (e.g., species and ecological communities) and anticipated impacts can be mapped. The most obvious application is in strategic assessments under threatened species legislations in other jurisdictions; however, the approach is not restricted to statutory assessments. There is significant potential in integrating spatially-explicit analyses with other types of cumulative assessment that consider additive and synergistic effects of multiple stressors on ecosystems and species (e.g., Halpern *et al.* 2008; Halpern & Fujita 2013).

Policy recommendations/conclusions

We recommend that our approach be adopted as a requirement of statutory strategic assessments of development impacts on biodiversity and equivalent planning processes in other jurisdictions. While our framework sets out a reasonable minimum standard for assessments, the approach does not deal explicitly with species persistence. Refining approaches to modeling persistence over large areas with multiple species is a pressing research priority. Implementation of our recommendations for large land-use planning and assessment exercises requires con-

sistent and dedicated engagement between researchers, planners, managers, and policy makers

Acknowledgments

NERP Decisions Hub supported this work. Wintle was supported by ARC Future Fellowship (FT100100819). Carolyn Cameron motivated and instigated the study. Simon Banks, Nicole Matthews, Hana McDonald, Jess Miller, Erin Pears (AG-DotE), Tahlia Rose, Simon Taylor, Sarah Woods (WA-DPC), Catherine Garlick (WA-OEPA), Bryce Bunny, Nicholas Dufty, Leo Peter, Aidan Power (WA-DoP), David Mitchell, and Christina Ramahlo (WA-DPaW) all provided data, knowledge and direction. Three anonymous reviewers provided valuable comments.

Supporting Information

Appendix A: Description of MaxEnt modeling.
Table A1. Abbreviated names and definitions of mapped environmental data used in species distribution models.
Table A2. AUC values and the relative contribution of each environmental variables for species distribution models summarized across taxonomic groups.
Figure A1. Boxplot of AUC values for 61 species distributions modeled using MaxEnt summarized across the five broad taxonomic groups.
Appendix B: Description of the Zonation prioritization algorithm.
Figure B1. Illustration of how the Zonation software undertakes spatial prioritization.
Appendix C: Development of a species weighting scheme.
Table C1. Criteria used to weight biodiversity features.
Appendix D: Assessing the outcomes of prioritization for individual species and TECs.
Figure D1. Proportion of each biodiversity feature's distribution captured by the priority conservation areas, plotted against distribution size.
Appendix E: Assessing the effectiveness of the existing protected area network.
Figure E1. Current and proposed protected areas within the PPSA region, and their overlap with priority conservation areas.
Figure E2. The proposed expansion of the conservation reserve system under the draft Green Growth Plan Conservation Program.

Table E1. The level of biodiversity protection within existing protected area network in the PPSA region.

Table E2. Biodiversity features (species or TECs) that do not occur within the existing protected area network in the PPSA region.

Table E3. The level of biodiversity protection achieved after proposed protected area expansion.

Table E4. Biodiversity features that would not occur within the expanded protected area network in the PPSA region.

Appendix F: Estimated biodiversity impacts of the four cumulative planning footprints.

Table F1. The relative impact of the four cumulative planning footprints.

Table F2. Biodiversity features (species or TECs) predicted to lose more than 50% of their habitat due to the different development types.

Table F3. Biodiversity features (species or TECs) predicted to lose more than 50% of their habitat under the cumulative planning footprints.

Figure F2. The proposed urban expansion zones within the initial Directions 2031 and Beyond proposal and Green Growth Plan.

Table F4. Changes in the proposed urban development impacts from initial to final revision of the footprint.

References

Andersen, J.H., Halpern, B.S., Korpinen, S., Murray, C. & Reker, J. (2015). Baltic Sea biodiversity status vs. cumulative human pressures. *Estuarine Coastal Shelf Sci.*, **161**, 88-92.

Atkinson, S.F. & Canter, L.W. (2011). Assessing the cumulative effects of projects using geographic information systems. *Environ. Impact Assess. Rev.*, **31**, 457-464.

Ball, I.R., Possingham, H.P. & Watts, M.E.J. (2009). Marxan and relatives: software for spatial conservation prioritisation. Pages 185–195 in A. Moilanen, K.A. Wilson, H.P. Possingham, editors. *Spatial conservation prioritisation: quantitative methods and computational tools.* Oxford University Press, Oxford, UK.

Bekessy, S.A., White, M., Gordon, A., Moilanen, A., Mccarthy, M.A. & Wintle, B.A. (2012). Transparent planning for biodiversity and development in the urban fringe. *Landscape Urban Plann.*, **108**, 140-149.

Bekessy, S.A., Wintle, B.A., Gordon, A., Fox, J.C., Chisholm, R., Brown, B. *et al.* (2009). Modelling human impacts on the Tasmanian wedge-tailed eagle (Aquila audax fleayi). *Biological conservation*, **142**, 2438-2448.

Cabeza, M. & Moilanen, A. (2006). Replacement cost: a practical measure of site value for cost-effective reserve planning. *Biol. Conserv.*, **132**, 336-342.

Camm, J.D., Polasky, S., Solow, A. & Csuti, B. (1996). A note on optimal algorithms for reserve site selection. *Biol. Conserv.*, **78**, 353-355.

Chaker, A., El-Fadl, K., Chamas, L. & Hatjian, B. (2006). A review of strategic environmental assessment in 12 selected countries. *Environ. Impact Assess. Rev.*, **26**, 15-56.

Connelly, R.B. (2011). Canadian and international EIA frameworks as they apply to cumulative effects. *Environ. Impact Assess. Rev.*, **31**, 453-456.

DPC. (2015). *Perth and peel green growth plan for 3.5 million.* Department of Premier and Cabinet, Perth, WA.

DSEWPAC. (2012). *A guide to undertaking strategic assessments.* Australian Government, Canberra.

Duinker, P.N., Burbidge, E.L., Boardley, S.R. & Greig, L.A. (2012). Scientific dimensions of cumulative effects assessment: toward improvements in guidance for practice. *Environ. Rev.*, **52**, 121029052013006.

Fordham, D.A., Resit Akçakaya, H., Araújo, M.B. *et al.* (2012). Plant extinction risk under climate change: are forecast range shifts alone a good indicator of species vulnerability to global warming? *Glob. Change Biol.*, **18**, 1357-1371.

Halpern, B.S. & Fujita, R. (2013). Assumptions, challenges, and future directions in cumulative impact analysis. *EcoSphere*, **4**, 1-11.

Halpern, B.S., Walbridge, S., Selkoe, K. *et al.* (2008). A global map of human impact on marine ecosystems. *Science*, **319**, 948-952.

Hawke, A. (2009). *The Australian Environment Act: Report of the Independent Review of the Environment Protection and Biodiversity Conservation Act 1999.* Department of the Environment, Water, Heritage and the Arts, Canberra, ACT.

Kareksela, S., Moilanen, A., Tuominen, S. & Kotiaho, J.S. (2013). Use of inverse spatial conservation prioritization to avoid biological diversity loss outside protected areas. *Conserv. Biol.*, **27**, 1294-1303.

Kujala, H., Moilanen, A., Araújo, M.B. & Cabeza, M. (2013). Conservation planning with uncertain climate change projections. *PLoS One*, **8**, e53315.

Lentini, P.E. & Wintle, B.A. (2015). Spatial conservation priorities are highly sensitive to choice of biodiversity surrogates and species distribution model type. *Ecography*, **38**, 1101-1111.

Mills, M., Nicol, S., Wells, J.A. *et al.* (2014). Minimizing the cost of keeping options open for conservation in a changing climate. *Conserv. Biol.*, **28**, 646-653.

Moilanen, A., Franco, A.M.A., Early, R.I., Fox, R., Wintle, B.A. & Thomas, C.D. (2005). Prioritizing multiple-use landscapes for conservation: methods for large multi-species planning problems. *Proc. R. Soc. B*, **272**, 1885-1891.

Moilanen, A., Leathwick, J.R. & Quinn, J.M. (2011). Spatial prioritization of conservation management. *Conserv. Lett.*, **4**, 383-393.

Moilanen, A., Pouzols, F.M., Meller, L. *et al.* (2014). *Zonation: spatial conservation planning methods and software. Version 4 user manual.* C-BIG Conservation Biology Informatics Group, Department of Biosciences, University of Helsinki, Helsinki, Finland.

Moilanen, A. & Wintle, B.A. (2006). Uncertainty analysis favours selection of spatially aggregated reserve networks. *Biol. Conserv.*, **129**, 427-434.

Moilanen, A. & Wintle, B.A. (2007). The boundary-quality penalty: a quantitative method for approximating species responses to fragmentation in reserve selection. *Conserv. Biol.*, **21**, 355-364.

Myers, N., Mittermeier, R.A., Mittermeier, C.G., Da Fonseca, G.A.B. & Kent, J. (2000). Biodiversity hotspots for conservation priorities. *Nature*, **403**, 853-858.

Mörtberg, U.M., Balfors, B. & Knol, W.C. (2007). Landscape ecological assessment: a tool for integrating biodiversity issues in strategic environmental assessment and planning. *J. Environ. Manage.*, **82**, 457-470.

Phillips, S.J., Anderson, R.P. & Schapire, R.E. (2006). Maximum entropy modeling of species geographic distributions. *Ecol. Model.*, **190**, 231-259.

Sebastián-González, E., Sánchez-Zapata, J.A., Botella, F., Figuerola, J., Hiraldo, F. & Wintle, B.A. (2011). Linking cost efficiency evaluation with population viability analysis to prioritize wetland bird conservation actions. *Biol. Conserv.*, **144**, 2354-2361.

Strimbu, B. & Innes, J. (2011). An analytical platform for cumulative impact assessment based on multiple futures: the impact of petroleum drilling and forest harvesting on moose (Alces alces) and marten (Martes americana) habitats in northeastern British Columbia. *J. Environ. Manage.*, **92**, 1740-1752.

Therivel, R. & Paridario, M.R. (2013). *The practice of strategic environmental assessment*. Routledge, Taylor & Francis, Hoboken.

WAPC. (2010). *Directions 2031 and beyond: metropolitan planning beyond the horizon*. Western Australian Planning Commission, Perth, WA.

WAPC. (2012). *Economic and employment lands strategy: non-heavy industrial*. Western Australian Planning Commission, Perth, WA.

Wintle, B.A., Bekessy, S.A., Venier, L.A., Pearce, J.L. & Chisholm, R.A. (2005). Utility of dynamic-landscape metapopulation models for sustainable forest management. *Conserv. Biol.*, **19**, 1930-1943.

Endnotes

1. http://www.environment.gov.au/protection/environment-assessments
2. http://www.dpc.wa.gov.au/Consultation/StrategicAssessment/Pages/Default.aspx
3. www.dmp.wa.gov.au/Geological-Survey/Basic-Raw-Materials-1411.aspx;
   http://www.planning.wa.gov.au/publications/6274.asp
4. www.dpc.wa.gov.au/Consultation/StrategicAssessment/Documents/01-Strategic-Assessment-Summary.pdf
5. Atlas of Living Australia (www.ala.org.au); NatureMap (https://naturemap.dpaw.wa.gov.au/)

18

Shoreline Armoring in an Inland Sea: Science-Based Recommendations for Policy Implementation

Megan N. Dethier[1], Jason D. Toft[2], & Hugh Shipman[3]

[1] Friday Harbor Laboratories and Biology Department, University of Washington, Friday Harbor, WA, 98250 USA
[2] School of Aquatic and Fishery Sciences, University of Washington, Seattle, WA 98195, USA
[3] Washington State Department of Ecology, Olympia, WA 98504, USA

Keywords

Armoring; policy; restoration; beaches; sediments; ecosystem goods and services; coastal management; seawalls; shoreline development.

Correspondence

Megan N. Dethier, Friday Harbor Laboratories and Biology Department, University of Washington, Friday Harbor, WA 98250, USA.
E-mail: mdethier@uw.edu

Editor

Christopher Brown

Abstract

Shoreline armoring can impact a variety of ecosystem functions, goods and services provided by beaches. Shoreline managers struggle to balance genuine need for armoring to protect infrastructure versus unacceptable losses of ecosystem functions—whether these be in beaches, sand dunes, or marshes. We use our recent research effort in the Salish Sea, Washington, as a case study to illustrate how highlighting the negative consequences of shoreline armoring to publicly important ecosystem functions may help to strengthen implementation of policy and prioritize restoration actions. We focus on two distinct mechanisms of armoring impact that link strongly to key beach functions, and recommend: (1) where armoring is clearly necessary, place or move it as high on the beach as possible. Armoring emplaced relatively low on the shore is more likely to affect a variety of ecosystem functions from forage fish spawning to beach recreation; (2) prioritize protection or restoration (armor removal) of feeder bluffs that are critical for sediment supply to the beach; this sediment is essential to the maintenance of beach functions. In addition, we recommend that nature-based alternatives to armoring be given preferential regulatory consideration and that outreach efforts clarify the advantages of these engineering methods.

Introduction

Anthropogenic modifications of coastlines are common worldwide, including abundant stabilization of the shoreline with various sorts of walls, referred to as armoring. Recent conservative estimates put the amount of armored shoreline in the continental U.S. at 14% (Gittman *et al.* 2015), with very high proportions near population centers. Estimates for the amount of hard engineering in Europe, the United States, Australia, and Asia are much higher, at more than 50% (Dafforn *et al.* 2015; Manno *et al.* 2016). However, only in the past few decades has there been exploration of the unintended negative consequences of these shoreline alterations. Armoring impacts a variety of ecosystem functions, goods, and services (EFGS) (Millennium Ecosystem Assessment 2005), the benefits gained by humans that are provided by beaches (Table 1). These may include "supporting" services such as primary production, "regulating" services such as mitigation of eutrophication, "provisioning" services such as shellfish, and "cultural" services such as recreation. However, impacts of armoring may not be apparent to the public because they are often very gradual or are invisible below the ocean surface, whereas the benefits of armoring in terms of property protection and shoreline aesthetics are obvious. These tradeoffs and the uncertainties inherent in quantifying impacts mean that regulations proposed to restrict armoring are readily resisted or weakened. We argue that we now know enough about negative environmental consequences of shoreline armoring in a variety of physical environments and thus we can make clear science-based recommendations for firmer implementation of stronger policies and regulations. A key tool for this effort may be linking EFGS, which the public and politicians can relate to, with decision-making (Ruckelshaus *et al.* 2015).

Table 1 Direct and indirect mechanisms by which shoreline armoring affects beach ecosystem functions, goods, and services

Functions, goods, and services	Mechanism			
	Encroachment	Loss of connectivity	Sediment impoundment	Wave reflection
Recreation, nonconsumptive: park use and outdoor education	Indirect − Indirect +	Indirect −	Indirect −	Indirect −
Recreation, consumptive: shellfish, seaweed, and fish		Indirect −	Indirect −	
Forage fish spawning: surf smelt and sand lance[a]	Direct −	Direct −	Direct −	Direct −
Trophic support: supply of insects, crustacea, and worms[b]	Direct −	Direct −	Indirect −	
Nutrient cycling: from marine and terrestrial wrack[c]	Direct −	Direct −	Indirect −	
Habitat provision: logs and wrack microhabitats[d]	Direct −	Direct −	Indirect −	Direct −
Groundwater filtering[e]	Direct −	Direct −	Indirect −	Indirect −
Resilience to sea-level rise[f]	Direct −	Indirect −	Direct −	Direct −

[a]Rice 2006; Penttila 2007; Quinn *et al.* 2012

[b]Dethier 1990; Heerhartz *et al.* 2015; Heerhartz & Toft 2015

[c]Dugan *et al.* 2008; Heerhartz *et al.* 2014

[d]Heerhartz *et al.* 2014; Dethier *et al.* 2016

[e]McIntyre *et al.* 2015

[f]Shipman 2010; Berry *et al.* 2014; Johannessen *et al.* 2014

Encroachment = covering the upper shore with armoring and thus eliminating natural habitats. Loss of connectivity = breaking linkages of materials, energy, and organisms between land and sea. Sediment impoundment = preventing sediment from eroding from banks and bluffs and reaching beaches. Wave reflection = causing storm waves to reflect off armoring rather than running gradually up a beach, leading to beach scour. Types of impact of each mechanism are given for each function; Blank = no known impact, − = negative, + = positive. A sampling of references for information on impacts is given, mostly from the Salish Sea. See Supporting Information for more details and references.

One of the challenges with quantifying impacts of armoring is that the mechanisms by which it alters shorelines are diverse, dependent on regional context (wave energy and geomorphology), and likely to manifest at different scales of space and time. In addition, while some direct impacts are documented, indirect impacts are often hypothesized but difficult to demonstrate. Recent reviews have summarized how armored shorelines can affect beach shape and hydrodynamic processes (Bernatchez & Fraser 2012; Nordstrom 2014), local biodiversity (Chapman & Underwood 2011; Gittman *et al.* 2016a), and accumulation of beach wrack along with the primary and secondary consumers that depend on it (Dugan *et al.* 2011). In marsh habitats, armoring may entirely cover and eliminate the marsh and all of its attendant functions (e.g., Bozek & Burdick 2005; Gittman *et al.* 2015). For high-energy sandy beaches, two clear impacts are impoundment of sand that would otherwise "feed" the beach, and prevention of shoreline retreat (natural beach migration with erosion and sea-level rise) (e.g., Berry *et al.* 2014), resulting in narrowing and coarsening of the beach. Other mechanisms of impact include loss of connectivity across the land-sea ecotone (Heerhartz *et al.* 2014), and hydrodynamic effects such as active erosion caused by wave reflection from seawalls (Ruggiero 2010).

Disentangling these mechanisms and ascribing cause-and-effect for indirect impacts can make it difficult to convince the public and regulators about the need for action. Agardy (2015) notes: "Even with strong bases for science-based actions... the unavoidable uncertainties are often used to prevent action and allow business-as-usual" (see also, Green *et al.* 2015; Zaucha *et al.* 2016). If the public observes change, perceives at least some of it as "bad," and becomes convinced (e.g., by knowledge brokers, Naylor *et al.* 2012) that human actions are causing it, then it is socially and politically easier to make progress toward un-doing the change. All of this must happen before management interventions such as removing armoring can gain momentum. An added difficulty is that geomorphological impacts tend to occur over years, making them "slow disasters," unlike fires whose impacts play out over hours or days; thus risk aggregation is slow, and incentives to act quickly are reduced (Moritz & Knowles 2016). Interventions are also unlikely to produce rapid results. Finally, issues of jurisdiction and governance are unusually complex at the land-sea border, so that any change in policy is likely to involve multiple agencies with different mandates (Zaucha *et al.* 2016).

Here, we use our research in Washington State as a case study for considering how discussion of EFGS may enable the strengthening of support for regulatory and restoration actions. Puget Sound, in the southern Salish Sea, is a fjord-like estuary where most beaches consist of a mix of sand and gravel; this is predominantly derived from the episodic erosion of glacial and interglacial deposits and is distributed by longshore transport (Shipman 2010). Wave regime and local geology are the primary drivers of modern beach geomorphology.

Because of a lack of research on armoring impacts along such mixed-sediment beaches, local regulators have until recently had little data on which to base efforts to strengthen or enforce regulations for restricting armoring. However, recent research summarized below, combined with numerous efforts toward public education on marine-conservation issues, leads us to believe that the public may be ready to hear arguments that shoreline armoring is damaging the marine resources of the area. There is extensive press (e.g., Hamel *et al.* 2015) about declining salmon, seabirds, and orca whales in the Salish Sea, and many long-term residents have anecdotes about shrinking beaches and fewer clams. However, these perceptions may be counterbalanced by fears about rising sea levels and the increased shoreline erosion that will likely result. Broad claims that all armoring has adverse impacts may get little traction (Russell-Smith *et al.* 2015), but the data now indicate specific circumstances where impacts may be greater than others (see below). We recognize that policy-makers must consider issues besides environmental impacts (Rose & Parsons 2015), but we now have an opportunity to make specific recommendations regarding regulations and restoration priorities that acknowledge tradeoffs and target the most serious issues.

Goods and services of Salish Sea beaches: why should people care?

If policies and regulations related to shoreline armoring are to change, success is more likely if we focus on the aspects of armoring that appear to have the greatest impacts on EFGS, and that affect the public personally (Zaucha *et al.* 2016). The ecosystem services that people relate to vary considerably among individuals and groups, so bringing diverse EFGS into the discussion may be helpful. The primary positive benefit of shoreline armoring is clear: it protects property and infrastructure from erosion caused by wave damage. The negative aspects relate primarily to characteristics of the beaches seaward of the armoring. As is true for beaches worldwide, those of the Salish Sea provide a variety of EFGS (Figure 1). Details about these functions and relevant references are given in Supporting Information and are summarized in Table 1. Beach EFGSs include high real estate prices for water access and views, intense and diverse recreational activities in public parks (Figure 1a), and ecological and geomorphic functions. Natural beaches are productive and supply food to nearshore food webs, including to juvenile salmon and ultimately to both humans and orca whales. They provide essential habitat for organisms that degrade marine and terrestrial detritus, and for terrestrial birds and mammals. They are the sole spawning habitat

for certain "forage fish" that are key elements in marine food webs. Natural beaches are geomorphically resilient, as they respond more flexibly to storm events and can shift landward to accommodate rising sea levels. Other functions are discussed in Supporting Information.

Recommendations: using science to improve management

Shoreline armoring impacts ecosystem functions through different mechanisms, affecting beaches both directly and indirectly (Table 1). In this section, we highlight two specific concerns where policy improvements could reduce impacts of armoring on the Salish Sea and in other regions. These include the waterward position of the seawall and the impact of erosion control on sediment supply.

Of the impact mechanisms detailed in Table 1, all except sediment impoundment are likely to be increasingly severe the lower the armoring is on the beachface. In the Salish Sea, we found a threshold in the elevation of armoring—about 0.5 m below local Mean Higher High Water—below which there is an abrupt drop in the number of beach logs and the amount of wrack that accumulates (Dethier *et al.* 2016). When structures extend below this elevation, there is no upper beach on which material can be retained between high tides (Figure 1b). Other beach biotas depend on these habitat elements. Juvenile fish such as salmon swimming alongshore prefer to do so in shallow water (presumably to avoid predation); where structures extend lower on the beach, there is less shallow water habitat at high tide. Fish that preferentially spawn high on the beachface find suitable habitat reduced or eliminated by structures built across the beach (Quinn *et al.* 2012). In addition, structures lower on the beach result in more frequent interaction with more energetic waves, increasing scour and even alongshore transport (Ruggiero 2010), impacting the amount and stability of appropriately sized spawning substrate. While new seawalls in Washington are required to be built as high on the beach as possible, many older structures extend below this elevation. A clear recommendation is that when older structures need to be replaced, they be relocated at least as high as the current allowable elevation. Restoration programs could offer funding and guidance to encourage the relocation or removal of structures that extend to lower beach elevations.

The second critical mechanism of impact in our case study area is sediment impoundment (Figure 1c, Table 1). On Puget Sound, bluff erosion is a significant source of beach sediment (Shipman 2010) and armoring prevents the replacement of fine sediment that is

Figure 1 Examples of EFGS or their losses on beaches in Puget Sound. (a) Recreational enjoyment in a natural area of Seahurst Park; (b) condominiums built on fill retained by a low-elevation concrete bulkhead; such armoring impacts EFGS by all four mechanisms in Table 1; (c) former feeder bluff with sediment impounded by armoring; and (d) stairs to the beach protected by a rock bulkhead. All photos by Hugh Shipman.

naturally winnowed from beaches by waves over time. Many ecological functions as well as recreational uses decline as beaches get coarser (Dethier *et al.* 2016); for example, forage fish, which are a key link in food webs up to the iconic orca whales, require a mix of sand and gravel to spawn on the upper beach (Penttila 2007). These potential impacts also lead to straightforward policy recommendations. While eroding banks and bluffs are widespread around the Salish Sea, much beach-building sand and gravel come from a limited subset of these, locally referred to as feeder bluffs (Figure 1c). These bluffs have been mapped (Shipman *et al.* 2014, https://fortress.wa.gov/ecy/publi cations/parts/1406016part2.pdf), providing a clear spatial basis for targeting preservation and restoration efforts. Concern about diminishing sediment supplies suggests creating policies for feeder bluffs that (1) prohibit

new seawalls, (2) discourage replacement or expansion of failed armor, and (3) incentivize removal of armor. For armored feeder bluffs, simply moving armoring higher up the shore does not restore sediment supply, so the focus must be on removal of the structure—i.e., a different response than for low-elevation impacts. Washington has had some success in reducing new seawalls on feeder bluffs, but the effort is challenging as these sites are often where erosion and landslide hazards are most severe. The state requires that new armor only be constructed where there is an imminent threat to existing upland development, but this can lead to complex geotechnical arguments between proponents and agency experts. This type of conflict between land owners and coastal managers suggests that there is a need for increased emphasis on preventing development above feeder bluffs in the first place to minimize future problems. Policies could involve instituting and enforcing large setbacks, creating incentives for the relocation of at-risk structures, and acquiring and preserving particularly high-value feeder bluffs.

In contrast, flexibility in regulations should be able to accommodate situations where armoring has fewer impacts, e.g., where little sediment supply is impounded or impacts are easier to mitigate. In some cases, stabilization structures can be kept small or may be designed so that they can be relocated after significant erosion events, retreating with the coastline (e.g., Hill 2015). Vegetation can be planted to reduce impacts from seawall construction on riparian areas. Steps to the beach are common and most may have limited impacts on EFGS, although such structures raise concerns if there is a chance that they will facilitate additional at-risk development or lead to a need for bank stabilization in the future (e.g., Figure 1d).

The framework for regulating armoring differs from state to state in the United States. In Washington State, management occurs primarily through the state's Shoreline Management Act (SMA) and Hydraulics Code, which together restrict the conditions and methods under which armoring can be constructed. The SMA is administered by local governments and addresses most shoreline activities, including stabilization structures. State Guidelines make it increasingly difficult to build new armoring except when there is an imminent threat to an upland structure, but restrictions on replacing existing structures are less strict. The Hydraulics Code is implemented by the Washington Department of Fish and Wildlife and is intended to reduce impacts on fish. Hydraulics Projects Approvals are required for any armoring structure and typically include conditions on the methods and timing of construction, but rarely can prohibit structures altogether.

The Puget Sound Partnership has identified both regulatory measures and restoration actions to reduce impacts (Puget Sound Partnership (PSP) 2014). Recent permit data indicate that the rate of new armoring appears to be decreasing, while the number of bulkheads being removed through restoration is increasing (Hamel et al. 2015). However, other analyses (Kinney et al. 2015) show that a significant proportion of armoring is either built without permits or is not constructed to permit specifications (e.g., elevation), indicating the need for more effective implementation of existing policies, including inspections and better enforcement. These actions require substantial political will and funding, which again speaks to the need for heightened awareness of impacts to EFGS of beaches.

In Washington and elsewhere, there is increasing interest in softer shoreline protection techniques, or "living shorelines," which use nature-based approaches (such as establishing dune grasses or oyster reefs) to reduce erosion and improve ecosystem functions (e.g., Hill 2015; Popkin 2015; Sutton-Grier et al. 2015; Gittman et al. 2016b). In the Salish Sea, softer designs to reduce erosion often include logs anchored into the upper shore to absorb wave energy, nourishment with sand and gravel, and planting of native vegetation to provide some of the shade and terrestrial-marine connectivity that is generally lost with armoring. These living designs also enhance recreation and restore many of the ecosystem functions listed in Table 1. On Puget Sound, the state's Shoreline Management Act requires that property owners examine the feasibility of such soft alternatives and can only consider conventional armoring as a last resort (Carman et al. 2010). Recent guidance on the design and construction of soft shoreline structures on Puget Sound (Johannessen et al. 2014) supports both property owners and government agencies in selecting better approaches, but implementation remains difficult because the effectiveness of these techniques is not well established. Additional guidance products and increased technical assistance are needed to educate contractors as well as homeowners not only about the benefits of softer techniques in terms of expense and complexity, but also long-term resilience and ecosystem functions. Naylor et al. (2012) and Popkin (2015) note the importance of changing not only regulations but also the permitting process to further incentivize property owners to opt for lower impact structures. Where it is not possible to avoid or remove armoring, current research in "ecological engineering" is exploring ways of adding habitat and biodiversity value to hard defenses, both in Washington (Cordell et al. 2017) and internationally (Naylor et al. 2012; Firth et al. 2013; Nordstrom 2014; Dafforn et al. 2015). Monitoring of soft-shore and ecological engineering projects and subsequent outreach on effective techniques are essential to provide the feedback that can encourage future efforts.

Communicating recommendations

There is an increasing body of literature on how to more effectively translate science into policy, actions, and decisions, including using the leverage of the ecosystem services approach (Ruckelshaus et al. 2015; Zaucha et al. 2016). This translation is needed to ensure that problem-focused research actually gets used by decision makers. Scientists are not always effective at communicating with diverse groups about such findings and recommendations (e.g., Rose & Parsons 2015). Knowledge brokers and guidance documents (Naylor et al. 2012) can improve our ability to engage and effect changes in attitudes in the wider community by delivering academic and applied science in a useful way to those who need it (Russell-Smith *et al.* 2015). Social and ecological information needs to be integrated, and tradeoffs explicitly acknowledged (Kittinger *et al.* 2014). Four main target groups for such outreach in our case study region are:

1) Scientists. This is readily accomplished with publications and regional professional meetings.
2) Managers. Efforts in this direction include nontechnical articles such as this one, presentations at workshops, and directly to agency groups.
3) The general public. Greater public awareness of the marine environment can improve acceptance of responsibility for conservation, increased pressure on politicians and regulators, and greater support for environmental initiatives including volunteering time (Morris *et al.* 2016). The Puget Sound region has an engaged public, and researchers can work directly with the numerous organizations that bridge the gap between science and the public. Regional groups include the Puget Sound Partnership (http://www.psp.wa.gov/), the Shore Friendly campaign (http://www.shorefriendly.org/), the Northwest Straits Commission (http://www.nwstraits.org/our-work/forage-fish/), and the Sound Waters Stewards (http://soundwaterstewards.org/). Research into social marketing is exploring incentives to remove armoring in cases when it is not actually needed to protect homes (http://wdfw.wa.gov/grants/ps_marine_nearshore/files/final_report.pdf).
4) Politicians. Links to this key group are indirect, probably coming most effectively from agency personnel and an active citizenry. The challenge is making the need for tightening restrictions on armoring more compelling than are counterarguments that protecting shoreline development justifies the potential cumulative impacts on coastal ecosystems. We can emphasize the monetary as well as human well-being values of natural beaches (Ruckelshaus *et al.* 2015), and the fact that there are alternatives to armoring that are both cost-effective and can improve EFGS.

Conclusions

Armoring a shoreline involves putting a static structure into a dynamic environment, where impacts and interactions are diverse and unpredictable. Any armoring can have impacts on beach EFGS and these impacts are likely to be cumulative, since relatively small actions are widespread and because effects of structures tend to increase over time. Shoreline defense structures are con-

troversial worldwide as shoreline managers struggle to balance genuine need for protection against unacceptable losses of EFGS—whether these be in marshes, sand dunes, riparian habitats, or beaches. While we have primarily discussed EFGS and armoring issues in the southern part of the Salish Sea, both the results of our research and the policy recommendations will apply elsewhere, although the specific mechanisms and issues may be different. Our policy recommendations, based on scientific research in the Salish Sea, can aid restoration decisions by focusing on how to minimize the loss of EFGS benefits. In the face of increasing levels of coastal urban growth and sea-level rise, there is great potential for restoration to not only enhance shoreline health but also better protect coastal communities using more natural approaches (Arkema *et al.* 2013). This new scientific information has already increased awareness of the tradeoffs associated with shoreline armoring among resource managers, property owners, and local governments. It provides a foundation that agencies can use to review shoreline projects and to support decisions about where and where not to armor.

Acknowledgments

The science behind this perspective was laboriously gathered by a large team of researchers from the University of Washington, Skagit River System Coop, Washington Department of Natural Resources, WA Department of Fish and Wildlife, WA Department of Ecology, and Tulalip and Swinomish tribes. The work of Sarah Heerhartz and Wendel Raymond was especially central. This research was funded in part by a grant from the Washington Sea Grant program, University of Washington, pursuant to National Oceanic and Atmospheric Administration Award No. R/ES-57. The views expressed herein are those of the authors and do not necessarily reflect the view of NOAA or any of its subagencies. Additional support was generously provided by the Washington Department of Fish and Wildlife Agreement #12-1249 as grant administrator for the U.S. Environmental Protection Agency.

References

Agardy, T. (2015). Tundi's Take: science uptake requires good delivery AND a receptive audience. *Mar. Eco. Manag.,* **8**(3), 5.

Arkema, K.K., Guannel, G., Verutes, G. *et al.* (2013). Coastal habitats shield people and property from sea-level rise and storms. *Nat. Clim. Change*, **3**, 913-918.

Bernatchez, P. & Fraser, C. (2012). Evolution of coastal defence structures and consequences for beach width trends, Quebec, Canada. *J. Coast. Res.*, **28**, 1550-1566.

Berry, A.J., Fahey, S. & Meyers, N. (2014). Boulderdash and beachwalls—the erosion of sandy beach ecosystem resilience. *Ocean Coast. Manag.*, **96**, 104-111.

Bozek, C.M. & Burdick, D.M. (2005). Impacts of seawalls on saltmarsh plant communities in the Great Bay Estuary, New Hampshire USA. *Wetl. Ecol. Manag.*, **13**, 553-568.

Carman, R., Taylor, K. & Skowland, P. (2010). Regulating shoreline armoring in Puget Sound. Pages 49-54 in H. Shipman, M.N. Dethier, G. Gelfenbaum, K.L. Fresh, R.S. Dinicola, editors. *Puget Sound shorelines and the impacts of armoring—proceedings of a state of the science workshop*, May 2009: U.S. Geological Survey Scientific Investigations Report 2010-5254.

Chapman, M.G. & Underwood, A.J. (2011). Evaluation of ecological engineering of "armoured" shorelines to improve their value as habitat. *J. Exp. Mar. Biol. Ecol.*, **400**, 302-313.

Cordell, J.R., Toft, J.D., Munsch, S.H. & Goff, M. Benches, beaches, and bumps: how habitat monitoring and experimental science can inform urban seawall design.419-436 In D. Bilkovic, M. Mitchell, J. Toft, M. La Peyre, editors. *Living shorelines: the science and management of nature-based coastal protection.*, February 2017, CRC Press. Boca Raton, FL

Dafforn, K.A., Glasby, T.M., Airoldi, L., Rivero, N.K., Mayer-Pinto, M. & Johnston, E.L. (2015). Marine urbanization: an ecological framework for designing multifunctional artificial structures. *Front. Ecol. Environ.*, **13**, 82-90.

Dethier, M.N. (1990) A Marine and Estuarine Habitat Classification System for Washington State. *Natural Heritage Program*, Washington Department of Natural Resources. 60 pp. Olympia, WA.

Dethier, M.N., Raymond, W.W., McBride, A.N. *et al.* (2016). Multiscale impacts of armoring on Salish Sea shorelines: evidence for threshold and cumulative effects. *Estuar. Coast. Shelf Sci.*, **175**, 106-117.

Dugan, J.E., Airoldi, L., Chapman, M.G., Walker, S.J., & Schlacher, T. (2011). Estuarine and coastal structures: environmental effects, a focus on shore and nearshore structures. Pp. 17–41 in Wolanski, E. and McLusky, D.S. (eds). *Treatise on Estuarine and Coastal Science*, Vol. **8**. Waltham: Academic Press. DOI: 10.1016/B978-0-12-374711-2.00802-0

Dugan, J.E., Hubbard, D.M., Rodil, I.F., Revell, D.L. & Schroeter, S. (2008). Ecological effects of coastal armoring on sandy beaches. *Mar. Ecol.*, **29**, 160-170.

Firth, L.B., Mieszkowska, N., Thompson, R.C. & Hawkins, S.J. (2013). Climate change and adaptational impacts in coastal systems: the case of sea defences. *Environ. Sci. Proc. Imp.***15(9)**: 1665-1670. DOI: 10.1039/c3em00313b

Gittman, R.K., Fodrie, F.J., Popowich, A.M. *et al.* (2015). Engineering away our natural defenses: an analysis of shoreline hardening in the US. *Front. Ecol. Environ.*, **13**, 301-307.

Gittman, R.K., Peterson, C.H., Currin, C.A., Fodrie, F.J., Piehler, M.F. & Bruno, J.F. (2016b). Living shorelines can enhance the nursery role of threatened estuarine habitats. *Ecol. Appl.*, **26**, 249-263.

Gittman, R.K., Scyphers, S.B., Smith, C.S., Neylan, I.P. & Grabowski, J.H. (2016a). Ecological consequences of shoreline hardening: a meta-analysis. *Bioscience*, **66**, 763-773.

Green, O.O., Garmestani, A.S., Allen, C.R. *et al.* (2015). Barriers and bridges to the integration of social-ecological resilience and law. *Front. Ecol. Environ.*, **13**, 332-337.

Hamel, N., Joyce, J., Fohn, M., editors. (2015). 2015 *State of the sound: report on the Puget Sound vital signs*. November 2015. 86 pp. www.psp.wa.gov/sos.

Heerhartz, S.M., Dethier, M.N., Toft, J.D., Cordell, J.R. & Ogston, A.S. (2014). Effects of shoreline armoring on beach wrack subsidies to the nearshore ecotone in an estuarine fjord. *Estuar. Coasts*, **37**, 1256-1268.

Heerhartz, S.M., Toft, J.D., Cordell, J.R., Ogston, A.S. & Dethier, M.N. (2015) Shoreline armoring in an estuary constrains wrack-associated invertebrate communities. *Estuar. Coasts*, **39**, 171-188.

Heerhartz, S.M. & Toft, J.D. (2015) Movement patterns and feeding behavior of juvenile salmon (*Oncorhynchus* spp.) along armored and unarmored estuarine shorelines. *Environ. Biol. Fishes*, **98**, 1501-1511. DOI 10.1007/s10641-015-0377-5

Hill, K. (2015). Coastal infrastructure: a typology for the next century of adaptation to sea-level rise. *Front. Ecol. Environ.*, **13**, 468-476.

Johannessen, J., MacLennan, A., Blue, A. *et al.* (2014). *Marine shoreline design guidelines*. Washington Department of Fish and Wildlife, Olympia, Washington. http://wdfw.wa.gov/publications/01583/.

Kinney, A., Francis, T. & Rice, J. (2015). Analysis of effective regulation and stewardship findings. Puget Sound Institute, University of Washington. https://www.eopugetsound.org/sites/default/files/features/resources/AnalysisOfEffectiveRegulationAndStewardshipFindings`FINAL`2015-12-14.pdf.

Kittinger, J.N., Koehn, J.Z., LeCornu, E. *et al.* (2014). A practical approach for putting people in ecosystem-based ocean planning. *Front. Ecol. Environ.*, **12**, 448-456.

Manno, G., Anfuso, G., Messina, E., Williams, A.T., Suffo, M. & Liguori, V. (2016). Decadal evolution of coastline armouring along the Mediterranean Andalusia littoral (South of Spain). *Ocean Coast. Manag.*, **124**, 84-99.

McIntyre, J.K., Davis, J.W., Hinman, C, Macneale, K.H., Anulacion, B.F., Scholz, N.L. & Stark, J.D. (2015) Soil bioretention protects juvenile salmon and their prey from the toxic impacts of urban stormwater runoff. *Chemosphere* **132**:213-219.

Millennium Ecosystem Assessment. (2005). *Ecosystems and human well-being*. Island Press, Washington, D.C.

Moritz, M.A. & Knowles, S.G. (2016). Coexisting with wildfire. *Am. Sci.*, **104**, 220-227.

Morris, R.L., Deavin, G., Donald, S.H. & Coleman, R.A. (2016). Eco-engineering in urbanised coastal systems: consideration of social values. *Ecol. Manag. Restor.*, **17**, 33-39.

Naylor, L.A, Coombes, M.A., Venn, O., Roast, S.D., Thompson, R.C. (2012). Facilitating ecological enhancement of coastal infrastructure: the role of policy, people and planning. *Environ. Sci. Policy*, **22**, 36-46.

Nordstrom, K.F. (2014). Living with shore protection structures: a review. *Estuar. Coast. Shelf Sci.*, **150**, 11-23.

Penttila, D. (2007). Marine forage fishes in Puget Sound. Puget Sound Nearshore Partnership Report No. 2007-03. Published by Seattle District, U.S. Army Corps of Engineers, Seattle, Washington.

Popkin, G. (2015). Breaking the waves. *Science*, **350**, 756-759.

Puget Sound Partnership (PSP). (2014). The 2014/2015 action agenda for Puget Sound. 770 pp. www.psp.wa.gov/action_agenda_center.php.

Quinn, T., Krueger, K., Pierce, K. *et al.* (2012). Patterns of surf smelt, *Hypomesus pretiosus*, intertidal spawning habitat use in Puget Sound, Washington State. *Estuar. Coasts*, **35**, 1214-1228.

Rice, C.A. (2006) Effects of shoreline modification on a northern Puget Sound beach: microclimate and embryo mortality in surf smelt (*Hypomesus pretiosus*). *Estuar. Coasts*, **29**, 63-71.

Rose, N.A. & Parsons, E.C.M. (2015). "Back off, man, I'm a scientist!" When marine conservation science meets policy. *Ocean Coast. Manag.*, **115**, 71-76.

Ruckelshaus, M., McKenzie, E., Tallis, H. *et al.* (2015). Notes from the field: lessons learned from using ecosystem service approaches to inform real-world decisions. *Ecol. Econ.*, **115**, 11-21.

Ruggiero, P. (2010). Impacts of shoreline armoring on sediment dynamics. Pages 179-186 in H. Shipman, M.N. Dethier, G. Gelfenbaum, K.L. Fresh, R.S. Dinicola, editors. *Puget Sound shorelines and the impacts of armoring—proceedings of a state of the science workshop*, May 2009: U.S. Geological Survey Scientific Investigations Report 2010–5254.

Russell-Smith, J., Lindenmayer, D., Kubiszewski, K., Green, P., Costanza, R. & Campbell, A. (2015). Moving beyond evidence-free environmental policy. *Front. Ecol. Environ.*, **13**, 441-448.

Shipman, H. (2010). The geomorphic setting of Puget Sound: implications for shoreline erosion and the impacts of erosion control structures. Pages 19-34 in H. Shipman, M.N. Dethier, G. Gelfenbaum, K.L. Fresh, R.S. Dinicola, editors. *Puget Sound shorelines and the impacts of armoring—proceedings of a state of the science workshop*, May 2009: U.S. Geological Survey Scientific Investigations Report 2010–5254.

Shipman, H., MacLennan, A. & Johannessen, J. (2014). Puget Sound feeder bluffs: coastal erosion as a sediment source and its implications for shoreline management. Shorelands and Environmental Assistance Program, Washington Department of Ecology, Olympia, Washington. Publ. #14-06-016.

Sutton-Grier, A.E., Wowk, K. & Bamford, H. (2015). Future of our coasts: the potential for natural and hybrid infrastructure to enhance the resilience of our coastal communities, economies, and ecosystems. *Environ. Sci. Policy*, **51**, 137-148.

Zaucha, J., Conides, A., Klaoudatos, D. & Norén, K. (2016). Can the ecosystem services concept help in enhancing the resilience of land-sea social-ecological systems? *Ocean Coast. Manag.*, **124**, 33-41.

19

Global Biodiversity Indicators Reflect the Modeled Impacts of Protected Area Policy Change

Brendan Costelloe[1,2,3], Ben Collen[4], E.J. Milner-Gulland[1], Ian D. Craigie[5], Louise McRae[2], Carlo Rondinini[6], & Emily Nicholson[1,7,8]

[1] Department of Life Sciences, Imperial College London, Silwood Park, Buckhurst Road, Ascot, Berkshire SL5 7PY, UK
[2] Institute of Zoology, Zoological Society of London, Regent's Park, London NW1 4RY, UK
[3] The Royal Society for the Protection of Birds, Potton Road, Sandy, Bedfordshire SG19 2DL, UK
[4] Centre for Biodiversity and Environment Research, Department of Genetics, Evolution and Environment, University College London, Gower Street, London WC1E 6BT, UK
[5] ARC Centre of Excellence for Coral Reef Studies, James Cook University, Townsville, QLD 4811, Australia
[6] Global Mammal Assessment Program, Department of Biology and Biotechnologies, Sapienza University of Rome, Viale dell'Università 32, Rome 00185, Italy
[7] School of Botany, University of Melbourne, VIC 3052, Australia
[8] Centre for Integrative Ecology, School of Life and Environmental Sciences, Deakin University, Burwood, Victoria 3125, Australia

Keywords
Biodiversity indicators; convention on biological diversity; Living Planet Index; protected areas; population decline; Red List Index.

Correspondence
Ben Collen, Centre for Biodiversity and Environment Research, Department of Genetics, Evolution and Environment, University College London, Gower Street, London WC1E 6BT, UK.
E-mail: b.collen@ucl.ac.uk

Abstract

Global biodiversity indicators can be used to measure the status and trends of biodiversity relating to Convention on Biological Diversity (CBD) targets. Whether such indicators can support decision makers by distinguishing among policy options remains poorly evaluated. We tested the ability of two CBD indicators, the Living Planet Index and the Red List Index, to reflect projected changes in mammalian populations in sub-Saharan Africa in response to potential policies related to CBD targets for protected areas (PAs). We compared policy scenarios to expand the PA network, improve management effectiveness of the existing network, and combinations of the two, against business as usual. Both indicators showed that more effective management would provide greater benefits to biodiversity than expanding PAs alone. The indicators were able to communicate outcomes of modeled scenarios in a simple quantitative manner, but behaved differently. This work highlights both the considerable potential of indicators in supporting decisions, and the need to understand how indicators will respond as biodiversity changes.

Introduction

Ambitious targets were agreed by the Convention on Biological Diversity (CBD, COP10 2010), in response to failure to meet their previous goal of reducing the rate of biodiversity loss by 2010 (Butchart et al. 2010). A mid-term review suggests that the prospects of achieving these targets are poor (Tittensor et al. 2014). Among failings that undermined attempts to reach the 2010 target was insufficient policy-specific scientific information to aid the decision-making process (Harrop 2011). Research has typically focused on measuring and predicting declines rather than evaluating the actions needed to

reverse them (Collen et al. 2011). Attention has recently turned to whether biodiversity indicators could be used proactively to generate predictions of different policy outcomes (Nicholson et al. 2012; Collen & Nicholson 2014; Visconti et al. 2015).

One of the main responses to biodiversity loss has been to establish protected areas (PAs; Jenkins & Joppa 2009). Two key measures of PAs' contribution to conservation are extent of coverage and effectiveness at conserving biodiversity. The CBD's latest target is for 17% of terrestrial areas and 10% of marine areas to be in effectively managed PAs by 2020 (COP10 2010). While PA coverage is measured, indicators of effectiveness remain undeveloped (Walpole *et al.* 2009). Coverage alone does not provide an accurate barometer of protection; there are many examples of ineffective "paper parks" (Craigie *et al.* 2010; Laurance *et al.* 2012).

In this study, we tested the ability of two global biodiversity indicators, the Living Planet Index (LPI) and the Red List Index (RLI), to convey the potential conservation outcome of the CBD's PA target in sub-Saharan Africa. The RLI and the LPI are two of the best-developed CBD indicators (Walpole *et al.* 2009). The RLI quantifies change in relative extinction risk based on changes in species' IUCN Red List categories (Butchart *et al.* 2004; Butchart *et al.* 2007). The LPI provides an aggregated measure of change in vertebrate abundance (Loh *et al.* 2005). Combined, the two may provide an indication of whether PAs are achieving two key objectives; preventing extinctions and reducing declines in common species.

We modeled the impact of continental-scale policies for African terrestrial PAs on large mammal abundance trends (Craigie *et al.* 2010) and the two indicators. We expanded on the preliminary analyses in Nicholson *et al.* (2012) by evaluating two CBD indicators, and by modeling the effects of combinations of three policies: protecting 10% of the earth's land area (CBD target in 2010), the updated target of 17% PA coverage by 2020, and improving management effectiveness (ME) within PAs, compared to a business as usual (BAU) scenario. We evaluated the ability of each indicator to detect species' responses to these policies and to inform choices between policies. The aim of the exercise was not to carry out detailed modeling of the drivers of change for African wildlife, nor to assess the effectiveness of PA designs, but to assess the feasibility of using existing CBD indicators to evaluate the likely effects of different policy actions on biodiversity trends.

Methods

Overall approach

We modeled the impacts of six policy scenarios on population trends for 53 large mammals species. Our approach was to:

(1) Estimate population size and recent trends in abundance of large mammals inside and outside PAs;
(2) Generate projected abundance trends for each species in response to six policy scenarios for PA coverage and effectiveness;
(3) Calculate the two indicators from generated abundance trends;
(4) Compare resultant indicator trends, and infer their ability to reflect actual trends.

Recent abundance and trends

We collated information on recent abundance, distribution, and population trends for 53 species of large mammal from 41 countries in sub-Saharan Africa. We collated country-level data on the population size of each species inside and outside of PAs (SI1), to estimate initial population sizes prior to policy implementation (SI2).

We estimated recent population trends (1970–2005) for species in PAs using data from Craigie *et al.* (2010). Data were available for most regions (as defined by the United Nations Geoscheme), but with differing numbers of observations and time frames. Trends were aggregated to the regional level because of a lack of country-level data, by calculating the geometric mean inter-annual change in population size for each species in each region (see SI5). Where data were not available for a region but the species was known to be present, trends were extrapolated from the most closely related species in that region. The resultant trend estimates per species and region (SI3) were used in the scenario analyses.

Model abundance trends in response to policy scenarios

We modeled six PA policy scenarios:

(1) BAU: continuation of current population trends in current PA network (2010), as documented in Craigie *et al.* (2010)
(2) Expand PA coverage to 10%: PAs were expanded to meet the 2010 CBD target in each country, recent population trends continued;
(3) Expand PA coverage to 17%: the 2020 CBD target, recent population trends continued;
(4) Improved ME: all declining populations in the 2010 PA network assumed to stabilize, and undergo small annual increases (>1.7%) due to effective management;
(5) Expand PA coverage to 10% and increase ME (as per scenario 4);
(6) Expand PA coverage to 17% and increase ME.

The impact of each policy scenario on each species was modeled for 30 years after implementation by projecting species' abundance trends from initial population sizes (SI2) using the annual rates of change for the relevant scenario (SI3). Each country was divided into populations inside PAs and outside PAs, which were subjected to different (but constant) rates of change.

Assumed trends outside PAs

There is a remarkable lack of research comparing population trends inside and outside PAs (Western *et al.* 2009), although several studies have shown PAs to be more effective at conserving biodiversity than unprotected areas (e.g., Struhsaker *et al.* 2005; Setsaas *et al.* 2007; Western *et al.* 2009). Lack of data beyond PA boundaries required an assumption: that all trends would be 25% worse *outside* PAs (non-PA) than *inside*; positive trends were decelerated by 25% and negative trends accelerated by 25%. Sensitivity analyses, where differences in trends were varied (no difference, 50%, 75% worse than in PAs), suggested that relative performance of policies was insensitive to the assumed value (Costelloe 2010).

Assumed impact of effective PA management on trends

No consistent data exist on the impact of effective management on mammal populations. However, PAs in Southern Africa are considered to be particularly effectively managed (Craigie *et al.* 2010), providing a benchmark. We assumed populations in effectively managed PAs experienced the same annual population trend as the average for Southern Africa (+1.7%). Those with a more positive annual trend than +1.7% kept their current trend.

Expansion of PAs

We used Marxan (Ball & Possingham 2000) to expand the current PA network up to 10% (scenarios 2 and 4) or 17% of each country (scenarios 3 and 5), with continental-level targets for suitable habitat under protection for each species (Rondinini *et al.* 2005). We assumed suitable habitat captured within new PAs would contain the same average population densities as existing PAs, and redistributed populations from outside PAs to inside PAs so that total abundance of the species within a country remained the same upon policy implementation (detailed methods in SI6).

Calculating indicators

The RLI measures overall extinction risk of sets of species and tracks changes in that risk (Butchart *et al.* 2004; Butchart *et al.* 2007). The index is a function of the proportion of species in each category at given points in time, and changes as the status of individual species improves or deteriorates. Population projections under each scenario were used to assign each species to a Red List threat category (IUCN 2001) at decadal intervals, using criteria A2 (population reduction) and C (population size, ignoring the subcriteria), based on the total modeled population size across the continent, starting from each species' 2010 Red List status (see SI1 and SI2). The RLI

was calculated for the 53 study species in each scenario every 10 years described in SI7 (Butchart *et al.* 2007).

The LPI is an aggregated measure of proportional change in abundance (Loh *et al.* 2005; Collen *et al.* 2009), specifically, the geometric mean change in abundance. The LPI was calculated for each scenario as per the method in Collen *et al.* (2009), described in detail in SI7. Each country comprised two populations: one aggregated population inside PAs, and one population outside PAs. We averaged the inter-annual rate of change for each species first at the regional level, to counter disproportionate impacts of trends in data rich areas; the LPI data for African mammals are biased toward Southern and Eastern African populations (Craigie *et al.* 2010). We performed a sensitivity analysis to test the effects on LPI trends of giving equal weighting across populations, by ignoring region and calculating the LPI with the average rate of population change for each species across all populations; see SI7 for details.

Results

In 2010, PAs comprised 2,709,082 km^2, an average of 13% of each sub-Saharan country (range 0%–36%). Fifteen of the 41 study countries has less than 10% of land under protection; 29 had less than 17% (IUCN and UNEP 2010; SI4). Under the PA expansion scenarios, total PA area across the continent increased to an average of 15% of each country (range 10%–36%) under the 10% PA coverage target, and by almost half to an average of 19% (17%–36%) per country under the 17% PA coverage (SI4). Baseline rates of population change before scenarios implementation varied substantially among regions and species, with greatest declines in West Africa, followed by Central and East Africa; on average, populations were increasing slightly in Southern Africa (SI3).

Both indicators predicted that improved ME would provide greater benefits to wildlife than PA expansion (Figure 1). The indicators showed only a small predicted increase from expansion without improved management, compared with BAU, due to on-going declines in most regions, and little difference between expanding PA coverage to 10% or 17%. PA expansion in conjunction with increased effectiveness gave little benefit above increasing the effectiveness of the existing network. The two indicators differentiated similarly between policies; however, there were differences in the overall trends they displayed. The RLI was stable or marginally increased under the improved management scenarios, and declined over the first decade before starting slow recovery under the scenarios without improved management. By contrast, the LPI declined at an attenuating rate, with

(a)

(b)

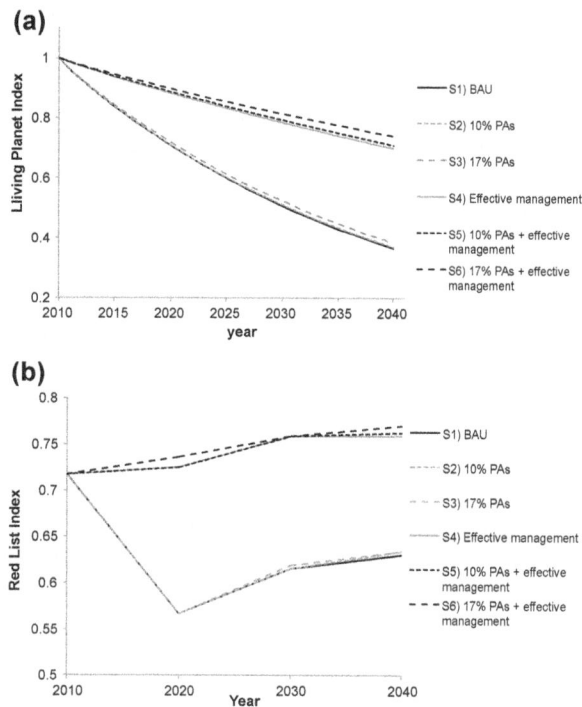

Figure 1 Comparative performance of scenarios for continental-wide indices, measured by (a) abundance trends (LPI) and (b) extinction risk trends (RLI). Scenarios 1, 2, 4, and 5 in (b) are shown in Nicholson *et al.* (2012).

(a)

(b)

(c)

Figure 2 Impact of projected changes under scenario 1 (BAU) for Tsessebe (*Damaliscus lunatus*) for (a) total population abundance, (b) extinction risk (RLI), and (c) abundance trend (LPI) where all populations were weighted equally (dashed line) or weighted by region (solid line).

more severe declines apparent for the scenarios without improved management.

The RLI showed an improvement in later years due to a changing ratio of abundance among regions. For illustration, we show a species-specific example under scenario 1, BAU (Figure 2). The Tsessebe (*Damaliscus lunatus*) started the simulations with a large East African population subject to a particularly strong annual decline (−13.4% p.a.) that drove the continent-wide population down at a rate sufficient to be classified Critically Endangered under Criterion A (89% decline over three generations, Figure 2a). Over time, the declining East African population comprised a smaller proportional share of the continent-wide population, with the previously smaller Southern African population increasing (positive trend +1.3% p.a.). By the final decade, the Southern African population comprised the majority of the total (albeit heavily depleted) population, resulting in a classification of Least Concern (Figure 2b). By contrast, the single-species LPI trend (i.e., geometric mean change in abundance of populations) closely mirrors the change in overall population size (Figure 2c), regardless of weighting procedure.

In some cases, the LPI was sensitive to weighting of trends in populations, particularly where trends

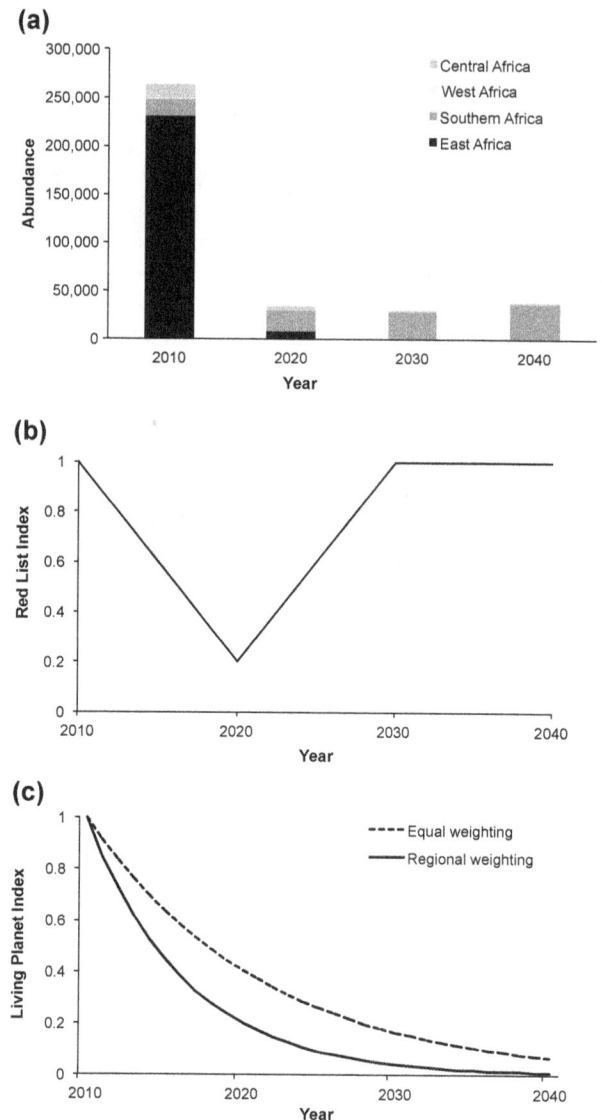

contrasted greatly. For illustration, under scenario 4, the African wild dog was predicted to decline steeply in smaller regional populations but increase in larger populations in Southern Africa; thus, the total abundance for the species increased (Figure 3a, no weighting). When changes were aggregated first at the regional level, the overall index declined (Figure 3a, regional weighting) because the modeled declines in the East, West, and Central regions were greater than projected increases in Southern Africa. Similarly, when using equal weighting across populations, the multi-species LPI showed a

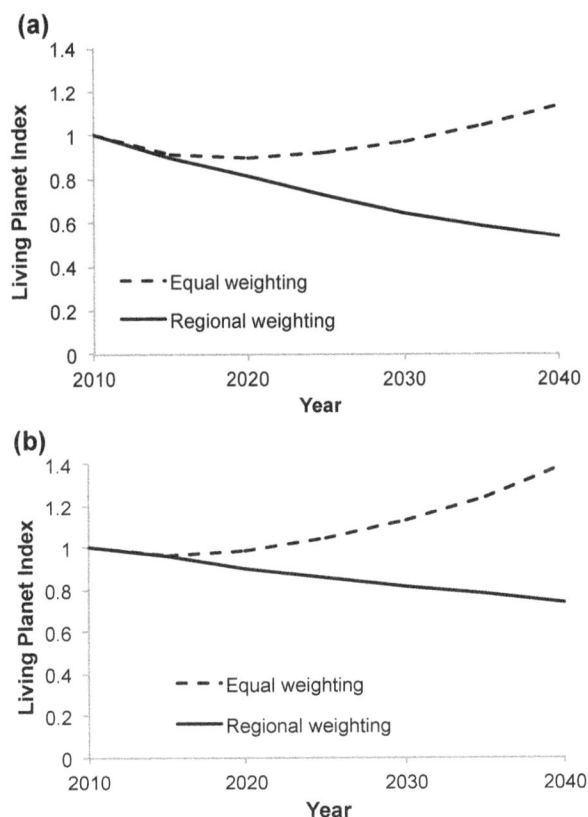

Figure 3 The sensitivity of the LPI to weighting of abundance trends when aggregating across populations for species, using two different weighting methods: where all populations were weighted equally (dashed line) or weighted by region (solid line). Results are shown for scenario 6, expand PA coverage to 17% of each country and increase ME of PAs, for (a) the African Wild dog (*Lycaon pictus*), and (b) all study species.

different trajectory to the regionally weighted index (Figure 3b).

Discussion

An indicator used to inform decision makers about the most effective policy must be able to discriminate among the predicted impacts of different policy options (Collen & Nicholson 2014). For our highly stylized model, both CBD indicators clearly demonstrated that policies to increase ME were more beneficial for biodiversity than those that only expanded PA coverage. Although both indicators ranked the scenarios in the same order, the difference in their behavior reflected the way they utilize underlying data. The RLI was less sensitive to abundance changes, in part because it measures extinction risk (which has a loose association with abundance; Purvis *et al.* 2000), and in part because of the coarse nature of the Red List categories (Mace *et al.* 2008); changes in Red

List status only occur when a species moves between classification thresholds. By contrast, the LPI detected broad scale population trends, and was sensitive to small and large changes in common and rarer species (Buckland *et al.* 2011).

The indicators also differ in the timeframes over which assessment occurs. The LPI is assessed annually and considers change in abundance relative to a reference year, whereas Red List assessments occur less frequently (modeled here as decadal intervals), with declines assessed over species-specific timeframes (10 years or 3 generations). Declines prior to the assessment window are therefore "forgotten" by the RLI, resulting in a shifting baseline, exemplified by projected changes in status for the Tsessebe. The impacts of shifting baselines may be considered a weakness of the RLI, however, the aim of the Red List is to evaluate extinction risk, in part a function of steep declines as well as absolute population size. These results show the potential risks of building an indicator from measures designed for other purposes, without examining the impacts on indicator behavior.

Our sensitivity analysis showed that the way in which species trends are aggregated is critical in determining overall aggregated trend. The LPI has been criticized for treating all proportional decreases in population size equally, regardless of absolute numbers (Pereira & Cooper 2006). The potential impacts of this on aggregated trends has been investigated, but absolute abundance data are rarely available (Collen *et al.* 2009). Giving equal weight to average trends across species and by region has merit when data are biased toward well-studied species and regions; weightings can be readily altered if data allow.

We focused on the ability of the indicators to reflect population trends under deterministic conditions with perfect knowledge. Real world data are imperfect and frequently biased, so the robustness of indicators to varying data quality and availability must be further explored. In this study, we did not evaluate the impacts of bias in sampling (all modeled species were represented in the indicators) or imperfect detection; both affect indicator behavior (Fulton *et al.* 2005; Branch *et al.* 2010; Nicholson *et al.* 2012).

We used trends to assess the impacts of policies, ignoring system dynamics. To do otherwise would be difficult for large numbers of species at the continental scale. We assumed trends remained the same inside expanded PAs, ignoring effects of size and suitability for each species, and that all study species fared worse outside PAs. Although our predictions appear relatively insensitive to assumed trends under effective management (Costelloe 2010), the assumption that all study species will respond equally is unlikely; in reality, species will respond differently to management, and their response will vary spatially depending on ecological processes, including species interactions (e.g., predator–prey and inter-predator), and the distribution of threats. A thorough exploration of impacts of management on species trends, and subsequently on biodiversity indicators, requires extensive analyses of data on threats and responses to management, and the modeling of species-specific responses. Such analyses would help identify the management actions required to halt declines.

It is unsurprising that improving ME was predicted to be more beneficial than expanding the coverage of PAs, given the assumptions in our relatively simplistic model, designed to examine indicator behavior rather than examine on-the-ground management options. On average 81% of species' populations were already in PAs, meaning that the PA expansion often saw relatively small increases in protected abundance, in contrast with considerable improvement in both indicators from a reduction in declines from better PA management. Ineffective management is cited as the main factor behind the rapid population declines in African PAs (Craigie *et al.* 2010). Our findings support assertions that shifting effort toward better management of existing PAs is preferable to simply annexing more land for protection (Jenkins & Joppa 2009), and raise doubts over the likely efficacy of CBD Target 11 in helping to meet the mission of "halting biodiversity loss" (COP10 2010). Indeed, the use of PA coverage as both a conservation target and a measure of conservation success, without measuring effectiveness, raises the risk of countries increasing paper parks to meet global targets with minimal biodiversity benefit. Goodhart's law states that when a measure becomes a target, it ceases to be a good measure; this may be a manifestation of that law for biodiversity indicators (Newton 2011).

The different behavior of the indicators tested, based on their underlying structure, show the importance of using multiple indicators to measure complementary aspects of biodiversity change (Purvis & Hector 2000). Both indicators are used to engage policy makers and the public by communicating simply complex measures of biodiversity change (Jones *et al.* 2011); our results show that they may also be used to summarize and communicate projected changes and potential policy outcomes. Although our model assumptions were simplistic, our analyses show how evidence-based modeling can allow the causal relationships between policy actions, biodiversity change, and indicators of change to be better understood (Nicholson *et al.* 2012; Collen & Nicholson

2014). Then indicators can start to tell us how we can best conserve biodiversity, not simply that we are failing to do so.

Acknowledgments

This article was a product of a workshop funded by the NERC Centre for Population Biology, which also supported B. Costelloe we thank the workshop participants for their insightful discussion and input to the design of the study. B. Collen was funded by grants from the Rufford Foundation and WWF International; B. Collen and E.N. were funded by the ARC Centre of Excellence for Environmental Decisions; E.N. was funded by a Marie Curie Fellowship and a Centenary Research Fellowship, and I.D.C. was funded through NERC studentship NER/S/A/2006/14094 with CASE support from UNEP-WCMC. This article is a product of Imperial College's Grand Challenges in Ecosystems and the Environment initiative.

References

Ball, I.R. & Possingham, H.P. (2000). Marxan version 1.8.3; http://www.ecology.uq.edu.au/marxan.htm.

Branch, T.A., Watson, R., Fulton, E.A., *et al.* (2010). The trophic fingerprint of marine fisheries. *Nature* **468**, 431-435.

Buckland, S.T., Studeny, A.C., Magurran, A.E., Illian, J.B. & Newson, S.E. (2011). The geometric mean of relative abundance indices: a biodiversity measure with a difference. *Ecosphere* **2**, art100.

Butchart, S.H.M., Akçakaya, H.R.,Chanson, J., *et al.* (2007). Improvements to the Red List Index. *PLoS One* **2**, e140.

Butchart, S.H.M., Stattersfield, A.J., Bennun, L.A., *et al.* (2004). Measuring global trends in the status of biodiversity: Red List Indices for birds. *PLoS Biol.* **2**, e383.

Butchart, S.H.M., Walpole, M., Collen, B., *et al.* (2010). Global biodiversity: indicators of recent declines. *Science* **328**, 1164-1168.

Collen, B., Loh, J., Whitmee, S., *et al.* (2009). Monitoring change in vertebrate abundance: the living planet index. *Conserv. Biol.* **23**, 317-327.

Collen, B., McRae, L., Deinet, S., *et al.* (2011). Predicting how populations decline to extinction. *Philos. Trans. R. Soc. B Biol. Sci.* **366**, 2577-2586.

Collen, B. & Nicholson, E. (2014). Taking the measure of change. *Science* **346**, 166-167.

COP10 (2010). Conference Of The Parties To The Convention On Biological Diversity. Decision X/2, Strategic Plan for

Biodiversity 2011–2020. Convention On Biological Diversity.

Costelloe, B.T. (2010). The power of global biodiversity indicators to predict future policy outcomes. MSc in Conservation Science. Imperial College, London; http://www.iccs.org.uk/wp-content/thesis/consci/2010/Costelloe.pdf.

Craigie, I.D., Baillie, J.E.M., Balmford, A., *et al*. (2010). Large mammal population
declines in Africa's protected areas. *Biol. Conserv.* **143**, 2221-2228.

Fulton, E.A., Smith, A.D.M. & Punt, A.E. (2005). Which ecological indicators can robustly detect effects of fishing? *ICES J. Mar. Sci.* **62**, 540-551.

Harrop, S.R. (2011). 'Living In Harmony With Nature'? Outcomes of the 2010 Nagoya Conference of the Convention on Biological Diversity. *J. Environ. Law* **23**, 117-128.

IUCN (2001). IUCN Red List Categories and Criteria: Version 3.1. IUCN Species Survival Commission, Gland, Switzerland.

IUCN & UNEP (2010). *The World Database on Protected Areas (WDPA)*. UNEP-WCMC, Cambridge, UK.

Jenkins, C.N. & Joppa, L. (2009). Expansion of the global terrestrial protected area system. *Biol. Conserv.* **142**, 2166-2174.

Jones, J.P.G., Collen, B., Atkinson, G., *et al*. (2011). The why, what and how of global biodiversity indicators beyond the 2010 target. *Conserv. Biol.* **25**, 450-457.

Jones, K.E., Bielby, J., Cardillo, M., *et al*. (2009). PanTHERIA: a species-level database of life history, ecology, and geography of extant and recently extinct mammals. *Ecology* **90**, 2648-2648.

Laurance, W.F., Useche, D.C., Rendeiro, J., *et al*. (2012). Averting biodiversity collapse in tropical forest protected areas. *Nature* **489**, 290-294.

Loh, J., Green, R.E., Ricketts, T., *et al*. (2005). The Living Planet Index: using species population time series to track trends in biodiversity. *Philis. Trans. R. Soc. B*, **360**, 289-295.

Mace, G.M., Collar, N.J., Gaston, K.J., *et al*. (2008). Quantification of extinction risk: IUCN's system for classifying threatened species. *Conserv. Biol.*, **22**, 1424-1442.

Newton, A.C. (2011). Implications of Goodhart's Law for monitoring global biodiversity loss. *Conserv. Lett.*, **4**, 264-268.

Nicholson, E., Collen, B., Barausse, A., *et al*. (2012). Making robust policy decisions using global biodiversity indicators. *PLoS One*, **7**, e41128.

Pereira, H.M. & Cooper, H.D. (2006). Towards the global monitoring of biodiversity change. *Trends Ecol. Evol.*, **21**, 123-129.

Purvis, A., Gittlemann, J., Cowlishaw, G., & Mace, G. M. (2000). Predicting extinction risk in declining species. *Proc. Roy. Soc. Lon. B*, **276**, 1947-1952.

Purvis, A. & Hector, A. (2000). Getting the measure of biodiversity. *Nature*, **405**, 212-219.

Rondinini, C., Stuart, S. & Boitani, L. (2005). Habitat suitability models reveal shortfall in conservation planning for African vertebrates. *Conserv. Biol.*, **19**, 1488-1497.

Setsaas, T.H., Holmern, T., Mwakalebe, G., Stokke, S. & Roskaft, E. (2007). How does human exploitation affect impala populations in protected and partially protected areas? A case study from the Serengeti Ecosystem, Tanzania. *Biol. Conserv.*, **136**, 563-570.

Struhsaker, T.T., Struhsaker, P.J. & Siex, K.S. (2005). Conserving Africa's rain forests: problems in protected areas and possible solutions. *Biol. Conserv.*, **123**, 45-54.

Tittensor, D.P., Walpole, M., Hill, S.L.L., *et al*. (2014). A mid-term analysis of progress toward international biodiversity targets.*Science*, **346**, 241-244.

Visconti, P., Bakkenes, M., Baisero, D., *et al*. (2015). Projecting global biodiversity indicators under future development scenarios. *Conserv. Lett.*, in press.

Walpole, M., Almond, R.E.A., Besançon, C., *et al*. (2009). Tracking progress toward the 2010 biodiversity target and beyond. *Science*, **325**, 1503-1504.

Western, D., Russell, S. & Cuthill, I. (2009). The status of wildlife in protected areas compared to non-protected areas of Kenya. *PLoS One*, **4**, e6140.

20

Design Features and Project Age Contribute to Joint Success in Social, Ecological, and Economic Outcomes of Community-Based Conservation Projects

Jeremy S. Brooks

School of Environment and Natural Resources, The Ohio State University, 2021 Coffey Rd., Columbus, OH, 43212, U.S.A.

Keywords

Biodiversity conservation; capacity; community conservation; conservation and development; sustainable development; synergies; tradeoffs .

Correspondence

Jeremy S. Brooks, School of Environment and Natural Resources, The Ohio State University, 2021 Coffey Rd., Columbus, OH 43212, USA
E-mail: brooks.719@Osu.edu

Editor

Krister Andersson

Abstract

Community-based conservation (CBC) seeks to align various ecological, economic, and social goals. While a number of comparative analyses have examined the factors associated with successful outcomes in each of these domains, far fewer studies have explored joint success across domains. Understanding when and how CBC improves multiple outcomes can generate more sustainable and socially acceptable policies and programs. Here, I use a comparative database of 136 CBC projects identified from a systematic literature review to assess which aspects of national socio-economic and political context, community-characteristics, and project design features are associated with win–win outcomes. Using multivariate logistic regressions within a multilevel analysis and model-fitting framework, I show that capacity building, local participation, environmental education, and project age contribute to win–win outcomes. These results hold across various national and local contexts and resource domains and suggest that general project design features can contribute to joint success in CBC projects.

Introduction

Governments, donors, and nongovernment organizations have devoted significant resources to integrative conservation approaches. Between 1980 and 2008 nearly 75% of the $18 billion in international biodiversity aid was devoted to such projects (Miller 2014). These integrative approaches, including comanagement, ecotourism, and integrated conservation and development (hereafter referred to as community-based conservation [CBC]), typically involve some combination of devolving control over resources, engaging local communities, and linking conservation with economic development. CBC is based on the premise that socio-economic benefits and community engagement can alleviate poverty, improve social cohesion, increase support for conservation, and reduce threats to biodiversity (Borgerhoff Mulder and Coppolillo 2005). Thus, the criteria by which conservation projects are evaluated has expanded, requiring conservation planners and practitioners to be attentive to economic, social,

and ecological outcomes as well as potential synergies and tradeoffs among them.

The proliferation of CBC, coupled with the set of ambitious win–win objectives outlined in the recently announced global sustainable development goals (SDGs) (United Nations General Assembly 2015), make the question of when and how win–win outcomes emerge even more critical (Adams et al. 2004). Despite the positive view portrayed in the SDGs, the consensus about CBC is that tradeoffs are more common than synergies (McShane et al. 2011). However, few studies have explicitly and systematically examined the factors associated with joint success in multiple outcomes (but see Chhatre and Agrawal 2009; Persha et al. 2010; Persha et al. 2011).

This study heeds calls for quantitative, comparative analyses of CBC to provide insights into the conditions under which win–win outcomes can be achieved (e.g. Agrawal and Redford 2006; Persha et al. 2011). I use a large, comparative database of CBC projects identified from a systematic literature review to assess which

Conservation project outcomes

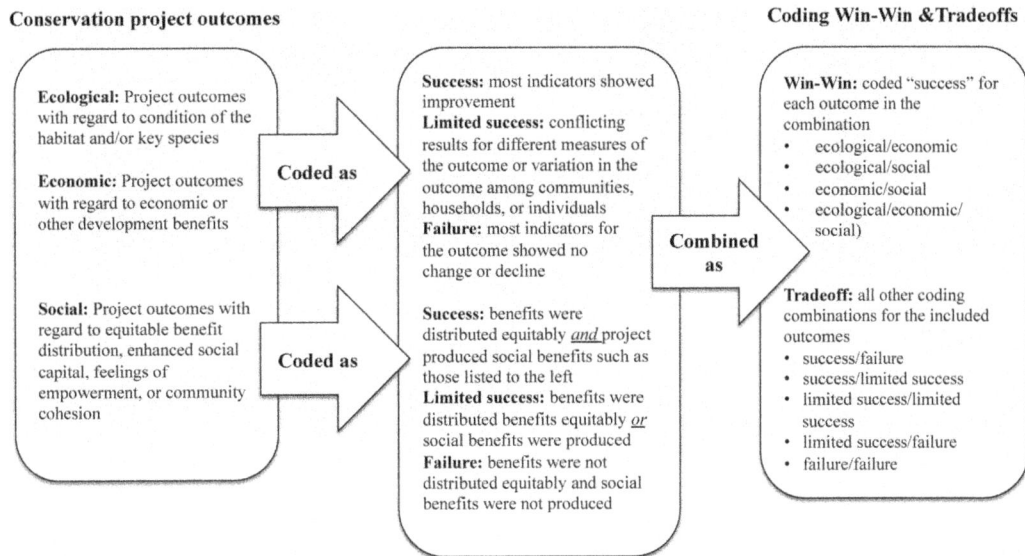

Coding Win-Win &Tradeoffs

Ecological: Project outcomes with regard to condition of the habitat and/or key species

Economic: Project outcomes with regard to economic or other development benefits

Coded as

Social: Project outcomes with regard to equitable benefit distribution, enhanced social capital, feelings of empowerment, or community cohesion

Coded as

Success: most indicators showed improvement
Limited success: conflicting results for different measures of the outcome or variation in the outcome among communities, households, or individuals
Failure: most indicators for the outcome showed no change or decline

Success: benefits were distributed equitably *and* project produced social benefits such as those listed to the left
Limited success: benefits were distributed benefits equitably *or* social benefits were produced
Failure: benefits were not distributed equitably and social benefits were not produced

Combined as

Win-Win: coded "success" for each outcome in the combination
- ecological/economic
- ecological/social
- economic/social
- ecological/economic/social)

Tradeoff: all other coding combinations for the included outcomes
- success/failure
- success/limited success
- limited success/limited success
- limited success/failure
- failure/failure

Figure 1 Coding of outcome variables and description of how they were combined for measures of synergies and tradeoffs. The combined outcome variables make up the four dependent variables for this study: ecological/economic, ecological/social, economic/social, ecological/economic/social. See supplementary material for additional details on measurement of project outcomes.

aspects of national socio-economic and political context, community-characteristics, and project design features are associated with win–win outcomes, where "win–win" is defined as success in two or more of the outcomes measured (ecological, economic, social) and "tradeoffs" are defined as some combination of success, limited success, or failure (see Figure 1 and supplementary material text). The sample includes data from 136 CBC projects nested within 40 countries.

Win–Win Outcomes and Tradeoffs

While win–win outcomes have been elusive (McShane et al. 2011), there is evidence that they are possible (Chhatre and Agrawal 2009; Persha et al. 2011; Miller et al. 2012) and there is a general sense of how ecological, economic, and social outcomes are linked. For instance, economic benefits (economic) should incentivize the protection or sustainable use of species, resources, or habitats (ecological) (Abbot et al. 2001), which may be more likely with direct payments (Ferraro and Kiss 2002) or indirect use (e.g. ecotourism) (Gössling 1999), than with direct use in highly market-integrated communities with incentives for overharvesting (Brewer et al. 2012).

Similarly, protection or sustainable use of resources (ecological) should not exacerbate existing inequalities or come at the expense of equitable access (social). Positive outcomes may emerge with high levels of local participation, which can ensure that conservation rules

are suited to local ecological and cultural conditions (Ostrom 1990) and enhance social capital, empower community members, and overcome existing inequalities (Campbell et al. 2010). These social benefits may further generate positive ecological outcomes and may be more likely where rights, responsibilities, and sufficient resources are devolved to local communities (Ribot et al. 2006) that face low barriers to collective action (Agrawal 2001).

Further, equitable benefit sharing (economic and social) can empower community members, incentivize further participation, and enhance human capital, thus leading to additional economic and social benefits. This result may be more likely in nations or communities with good governance (Nelson 2010), when market access enables linkages with external decision-makers (Persha and Andersson 2014), or in smaller, homogeneous communities (Dasgupta and Beard 2007).

However, these simplified representations hide the challenges associated with aligning CBC outcomes, which include (i) potential changes in outcomes over time (e.g. if harvest rates become unsustainable or market value fluctuates), (ii) the potential value trading off ecological impacts for economic success in the short-term in order to generate the support necessary for ecological success over the long term (Miller et al. 2012), and (iii) tradeoffs among different measures within an outcome domain (Brown 2004; McShane et al. 2011). These complexities, and the limited quantitative analysis of the factors associated with win–win outcomes, have inhibited our understanding of the conditions under which such outcomes emerge .

This study builds on previous work that examined individual outcomes of CBC projects and found that participation, engagement with communities, capacity building and secure tenure were important for ecological and economic success (Brooks et al. 2012). The goal of this article is to identify the factors associated with win–win scenarios across ecological, economic and social outcomes. Following Brooks et al. (2012) I use a set of predictor variables derived from theory and observations about the conditions that support or hinder collective action and conservation behavior to provide insights into when win–win outcomes might emerge (see Tables 1 and 2 and supplementary material text for further justification of variables). These variables are grouped into the following three categories.

(1) National socio-economic and political context can affect levels of corruption, trust in actors and institutions, and the functioning of economic markets (Tallis et al. 2008; Nelson 2010).
(2) Community characteristics affect the likelihood of collective action and conservation behavior through mechanisms related to local economic opportunities, incentives or disincentives for behaviors, patterns of interactions among community members and stakeholders, and secure access to resources (Ostrom 1990; Agrawal 2001).
(3) Project design features structure involvement in decision-making, investments in human capital, the degree of resource use permitted, and the provision and distribution of benefits.

In addition to variables in these categories, I controlled for project age, the first author's disciplinary background, and the conservation status of the ecoregion in which the project was located (see supplementary material text for justification of these control variables).

Methods and Analysis

A systematic review of the CBC literature was conducted using online databases and the Advancing Conservation in a Social Context digital library (www.tradeoffs. org/app/Public/Catalog) (see supplementary material text for search details). These searches added 74 projects to an initial sample of 62 projects from previous reviews (Brooks et al. 2006; Waylen et al. 2010). Studies were included if they were published in the primary or gray literature and were the most recent of multiple sources on the same project, addressed a CBC intervention with conservation as the primary aim, measured at least two

of the outcomes that were the focus of a previous study (Brooks et al. 2012), and were missing information for no more than one-third of independent variables.

Using a coding protocol modified from previous reviews (Brooks et al. 2006; Waylen et al. 2010), up to sixty-five pieces of information were collected for each project. Only data related to variables in the three aforementioned categories are presented here. Coding was done by the author and another researcher (see supplementary material for coding procedures) and coding decisions were based solely on the information presented in articles in the sample. To reduce the amount of missing information, the corresponding authors of papers in the sample were e-mailed a questionnaire containing a subset of topics in the coding protocol (for use of these data see supplementary material text).

Following Brooks et al. (2012), the analysis proceeded in two steps. Bivariate analyses were conducted using 2D contingency tables for categorical predictors and logistic regression for continuous predictors. The Goodman–Kruskal γ-statistic summarizes the association between predictors and outcomes and was used as a test statistic for Monte Carlo significance tests. Multiple testing was controlled for by adjusting significance levels using q values (Storey 2002) to obtain approximate control of the false-discovery rate. The P values obtained from the contingency tables and the regression models were supplied to the q value software (available at http://www.bioconductor.org/packages/release/bioc/html/qvalue.html).

Multivariate analyses were conducted after imputing missing values (Little and Rubin 2002). Five unique datasets were created using the MICE package in R. All predictor and uncombined outcome variables were used to impute missing values (missing values were not imputed for outcomes). After imputing missing values, some predictors were combined to reduce the number of variables (see Table 1 and Table S7 for the correlation matrix for predictors). Best-fitting logistic regression models for each dependent variable (Figure 1) were selected using a forward, stepwise Akaike Information Criterion (AIC) procedure. The best-fitting model was then fit in lrm and the robcov function was used to calculate robust SEs to account for clustering of projects within countries. Estimates for the five imputed datasets were averaged and pooled SEs were calculated (Gelman and Hill 2007).

Results

The sample size for each combination of outcomes ranged from 64 (ecological & economic; ecological, economic & social) to 118 (economic & social) (see Figure 2). Tradeoffs were more common than synergies across all

Table 1 Measurement and source of national-level predictors

	National Political Context
Governance	World Bank scores for quality of governance based on measures of: voice in politics, government stability, government effectiveness, regulatory quality, rule of law, and corruption. Single value obtained from the first factor score from a principal components analysis of the six measures. Data collected from (http://info.worldbank.org/governance/wgi/sc_country.asp#). Scores taken from the year closest to the date of research.
Rights	Values for political rights and civil liberties from the FreedomHouse database. Single value obtained from the first factor score from a principal components analysis of the two measures. Data collected from (http://www.freedomhouse.org/report-types/freedom-world). Scores taken from the year closest to the date of research.
	National Socio-economic Context
HDI	Human Development Index score from the 2009 UNDP trends dataset accessed throughThe United Nations Development Programme (http://hdr.undp.org/en/statistics/). Scores taken from the year closest to the date of research.
Gini	Gini inequality coefficient obtained ranked by country from data compiled from 2000 to 2007 (http://www.wri.org/project/earthtrends/). One score for each country.

outcome combinations and were most common for the triad of ecological, economic, and social outcomes. Synergies were most common between ecological and economic outcomes (41% of projects) and least common between economic and social outcomes (14% of projects).

Many of the significant associations in the bivariate analysis were project design variables including local participation, engagement with local cultural traditions, investments in institutional capacity, and environmental education (see Table 3). For national context, higher HDI and good governance were associated with win–win outcomes for multiple combinations. Population size was the only community characteristic associated with win–win outcomes for more than one combination. Finally, project age was associated with win–win outcomes in all four combinations.

The multivariate analysis indicates that capacity building, project age, participation, and environmental education are particularly important for win–win outcomes (see Figures 3a–d and Table S1 for full results). The alignment of ecological and social outcomes is most likely when the project builds the capacity of individuals and institutions and with greater local participation in project design and implementation. Economic and social synergies are more likely for older projects and when the project provides environmental education. Ecological and economic synergies are more likely for older projects and when the project helps build the capacity of individuals and institutions. Finally, all three outcomes are more likely to align in older projects.

Discussion

Our results support the consensus that tradeoffs are more common than synergies in CBC (Brown 2004;

McShane et al. 2011). Win–win results were most common for ecological and economic outcomes and least common for economic and social outcomes. Thus, economic returns do not have to come at the expense of conservation goals, but it may be difficult to distribute economic benefits equitably and manage them in a way that does not disrupt social dynamics.

The results suggest that several project design features and project age are associated with win–win outcomes across diverse national contexts, resource domains, and project types. Importantly, these project features can be incorporated in nearly any context. Thus, while project success may be context dependent (Adams et al. 2004), these factors may provide a foundation for aligning success across outcomes.

Capacity building remained in the best-fitting model for all outcome combinations and was significantly associated with win–win outcomes for two combinations. Scholars have noted the importance of capacity building in the context of CBC (Campbell et al. 2010), and numerous reviews have emphasized the importance of individual skills development (Salafsky et al. 2001; Tallis et al. 2008) and strengthening institutional capacity (Agrawal and Benson 2011; Persha et al. 2011).

Understanding the mechanisms through which capacity building produces joint success is difficult because our combined measure includes individual-level training (e.g. biological monitoring, business skills, record keeping, equipment use) and institutional capacity (e.g. strengthening rural organizations and institutions, conflict negotiations, creating management committees). Increasing human capital through individual capacity building can increase pride, self-confidence, and a sense of belonging (Scanlon and Kull 2009) and provide skills

Table 2 Description, measurement, and coding of project design and community characteristics predictors. Some predictors used in the multivariate analysis are a combination of variables that were used in the bivariate analysis (see Table S1), as indicated in the description below

	Community characteristics
Market access	The degree communities are market integrated including wage labor, selling and purchasing goods, and remoteness, three categories: low, moderate, high.
Threat	Threats to local resources and/or the protected area (e.g., logging, hunting, land clearance). Up to three noted for each project and coded by motivation (subsistence = 1, mixed = 2, commercial = 3). Sum of the three threats divided into three categories: low (1,2), moderate (3-5), high (6-9).
Local Institutions	Comprised of Supportive local culture (do local traditions and beliefs support conservation: unsupportive, mixed, supportive) and Effective local government (quality of preexisting local governance institutions: ineffective, mixed, effective). Local institutions combines variables above. Three categories: low (both ineffect./unsupp. or one ineffect./unsupp. and one mixed), moderate (one is ineffect./unsupp. and one is effect./supp.), high (both effect./supp. or one effect./ supp. and one mixed).
Tenure	Control or ownership over primary resources targeted by project. Three categories: no community, mixed, community (entity within community including private or communal ownership).
Charisma	Presence of charismatic individual/group of individuals facilitating project. Coded as: no, yes.
Pop. size[a]	Size of human population, three categories: low (<5000), moderate (5001–50000), high (>50000). Adapted from Waylen et al. (3).
Pop. heterogen.[b]	Ethnical/cultural diversity of community. Two categories: low (one ethnic/cultural group present, or < 33% and > 67% of one group), high (multiple ethnic groups, >33% and <67% of community of one group and/or the author notes disharmony based on caste, class, or ethnic divisions).
	Project Design Features
Participation	Comprised of Impetus (whether impetus for project came from the community), Establishment (level of community involvement in project establishment, and Daily management (level of community involvement in daily project management. Coded as: no or low =1, some =2, joint or complete involvement=3. Participation combines variables above. Three categories: no/low (3,4), moderate (5,6), high (7-9).
Engagement	Comprised of Approach local culture (whether project engaged with local cultural traditions and beliefs), and Approach local institutions (whether the project engaged with local institutions and/or leaders). Each coded as: no engagement/conflict=1, mixed=2, engaged=3. Engagement combines variables above. Three categories: no/low (2,3), moderate (4), high (5,6).
Protectionism	IUCN ranking for protected area associated with the project. Two categories[b]: Strict Nature Reserve/National Park, Other (national monument, habitat/species management area, protected landscape, managed resource area, no protected area).
Resource use	Constraints on resource use, coded into three categories: protected, regulated, unregulated.
Benefits provision	Economic/development benefits provided by project and type of resource use[c]. Four categories: ecotourism (indirect use of targeted species/habitat), CBC (community efforts to minimize resource use), compensation/substitution (prohibition or minimized use of targeted resource but other benefits provided), enhancement (increasing marketable use of the targeted resource).
Capacity	Comprised of Capacity skills (did project build skills intended to aid development or conservation efforts), and Capacity institutions (were local institutions relating to environmental management reinforced or built as a result of the project). Coded as: yes, no. Capacity combines variables above. Two categories: no (neither skills nor institutions), yes (skills, institutions or both).
Envt. education	Whether environmental education was a component of the project. Coded as: no, yes.
	Controls
Project Age[c]	Number of years the project has been running.
Author discipline	Affiliation of first author, four categories: biological sciences, social sciences, interdisciplinary science or department, employed by an NGO.
Ecoregion status	Conservation status of ecoregion(s) in project area[d]. When multiple exist only lowest status value is coded, three categories: critically endangered, vulnerable, relatively stable.

[a]We collapsed Waylen et al.'s (2010) seven-level variable to three categories to reduce the number of predictors in the models. It was important to avoid categories with low representation in the analysis, so, we considered the structure of the data while constructing the three categories.

[b]Modified from: Oldekop JA, Bebbington AJ, Brockington D, & Prieziosi RF (2010) Understanding the lessons and limitations of conservation and development. Conserv Biol 24(2):461-469.

[c]Year of project initiation was subtracted from the year research for the project was conducted. When the year in which research was initiated was not reported, we subtracted from the publication year the mean number of years between the initiation of research and the publication year for all other studies in the sample.

[d]See: Olson DM & Dinerstein E (1998) The global 200: A representation approach to conserving the Earth's most biologically valuable ecoregions. Conserv Biol 12(3):502-515.

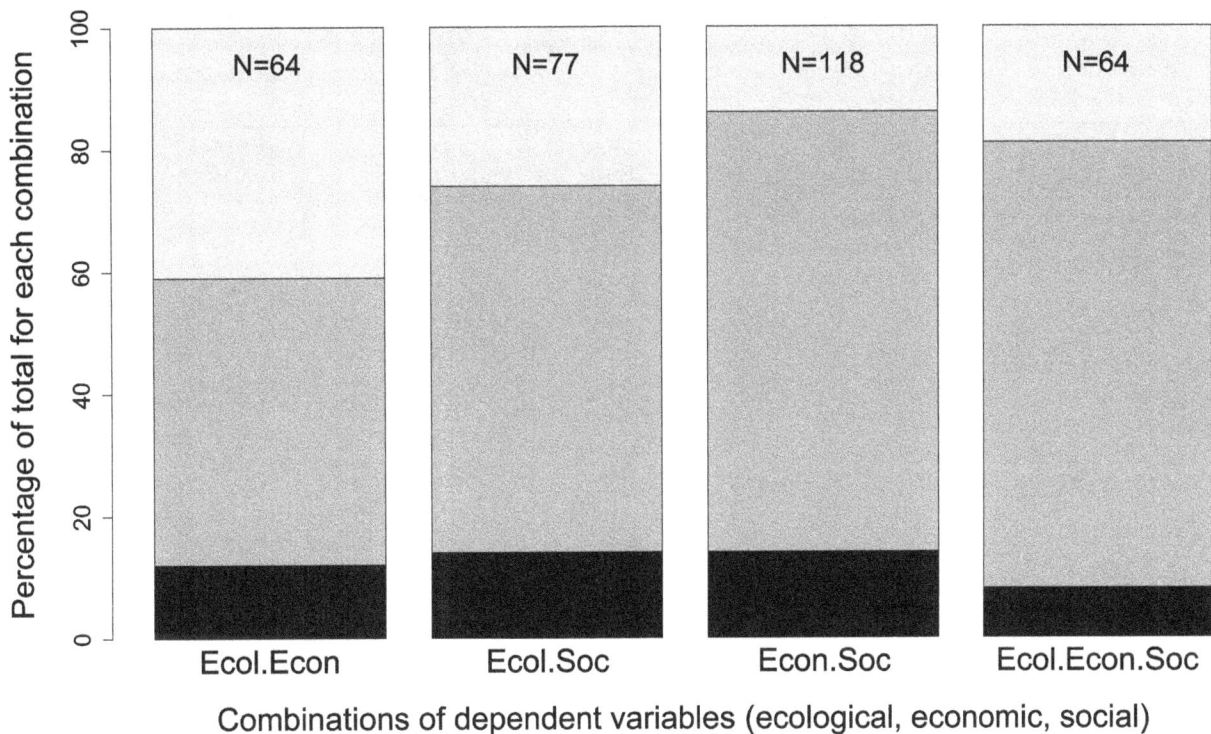

Figure 2 Percentage of synergies (light grey), tradeoffs (dark grey) and dual failures (black) for each combination of outcomes. Dual failures where included as tradeoffs for the analyses.

that facilitate participation and help individuals harness economic opportunities. Thus, individual capacity building can be a component of more holistic approaches to poverty reduction (Agrawal and Redford 2006) that go beyond income generation by strengthening human capital and improving human capabilities (Sen 1999). If the human capital generated by capacity building is applied across social, ecological, and economic components of the project, then it may provide a foundation for synergies and long-term CBC success.

The results of the bivariate analysis suggest, however, that institutional capacity building is a better predictor of win–win outcomes. Strengthening institutional capacity may help foster interactions among community members that are important for multiple outcomes. For instance, informal institutions to improve communication and mobilize community members were needed in one community in Taiwan to support voluntary patrolling efforts (Tai 2007). These informal communication networks strengthened social capital (social outcome) and lead to more formal institutions that protected wildlife habitat (ecological outcome) that was also a prime tourist destination (economic outcome).

Local participation and environmental education were also important, albeit to a lesser degree. Several reviews

have found local participation and autonomy to be crucial for independent CBC outcomes (Padgee et al. 2006; Brooks et al. 2012), as well as for synergies between outcomes (Chhatre and Agrawal 2009; Persha et al. 2011). Locally derived management decisions are thought to be better suited to local ecological dynamics and to lead to better compliance (Ostrom 1990). In addition, social outcomes may be enhanced when participation can lead to feelings of empowerment, pride, and buy-in from community members (Campbell et al. 2010). Further, locally crafted rules about benefits distribution may be better adapted to local socio-economic and cultural conditions. Thus participation may enhance ecological and social outcomes in distinct ways, or indirectly improve ecological conditions through improved social dynamics (Miller et al. 2012). For instance, Scanlon and Kull (2009) provide evidence that CBC helped create a shared identity among communities in Namibia, which played a role in reducing hunting.

Environmental education was associated with joint economic and social success, but the lack of detailed descriptions of environmental education makes it difficult to determine how. Educational programs may be structured as forums that bring community members together, thereby contributing to social gains. These

Table 3 Results of the bivariate analysis showing coefficients from proportional odds logistic regressions (National Context variables and CTR-Years project running) and Goodman–Kruskal gamma values from two-way contingency tables (all other variables), indicating significant relationships between predictors and outcomes. This analysis was conducted on the dataset that included missing values (not imputed data). Only significant values are shown. Some variables were combined for the multivariate analysis (second column) as described in Table 2

Domain and variable name	Combined variable name	Ecological & social	Economic & social	Ecological & economic	Ecol., econ. & soc.
National socio-economic & political context					
Governance	–	0.50	–	0.41	–
Rights	–	–	–	–	–
HDI	–	–	4.84	5.21	5.66
Gini	–	–	0.06	–	–
Community characteristics					
Market access	–	–	–	–	–
Threat	–	–	–	–	–
Supportive local culture	Local institutions	0.54	–	–	–
Effective local govt.	Local institutions	–	–	–	–
Tenure	–	0.31	–	–	–
Charisma	–	–	–	-0.27	–
Population size	–	0.35	0.58	–	0.57
Population heterogeneity	–	–	–	–	–
Project design					
Impetus	Participation	0.64	0.56	–	–
Establishment	Participation	0.61	0.44	0.35	0.46
Daily management	Participation	0.48	0.68	0.47	0.53
Approach local culture	Engagement	0.71	0.80	0.96	1.00
Approach local institutions	Engagement	0.68	–	0.66	–
Protectionism	–	–	–	–	–
Resource use	–	–	–	–	–
Provision benefits	–	–	–	0.38	–
Capacity skills	Capacity	–	–	0.46	–
Capacity institutions	Capacity	0.64	0.53	0.58	0.66
Environmental educ.	–	0.54	0.73	–	0.71
Controls					
Ecoregion status	–	–	0.35	–	–
Author discipline	–	–	–	–	–
Years project running	–	0.11	0.10	0.13	0.14

educational programs may inform communities about important ecosystem services as well as the potential economic value of conservation, which could contribute to positive economic outcomes. For instance, Becker et al. (2005) describe efforts to inform communities about the ecological and economic value of premontane moist forests in Ecuador. Education programs increased awareness of ecological importance and economic value, and also lead to involvement in environmental monitoring efforts, which enhanced community relationships.

Finally, project age was associated with win–win outcomes for all combinations in the bivariate analysis, and three of the combinations in the multivariate analysis. Previous studies have noted that success in CBC projects and conservation enterprises may take years to emerge (Salafsky et al. 2001), which may be a function

of the complex nature of social-ecological systems and the trial-and-error process through which institutions co-evolve with local conditions (Rammel et al. 2007). Interestingly, projects older than twelve years had a significantly greater proportion of synergies than younger projects, although this association may be due to the small sample size for projects of that age and could be a function of publication bias (see supplementary material text).

Importantly, the causal link between project age and win–win outcomes is unclear. While win–win outcomes may result from projects being given sufficient time to mature, it is plausible that projects that show early signs of success across multiple outcomes receive more financial and/or community support and are thus more likely to persist.

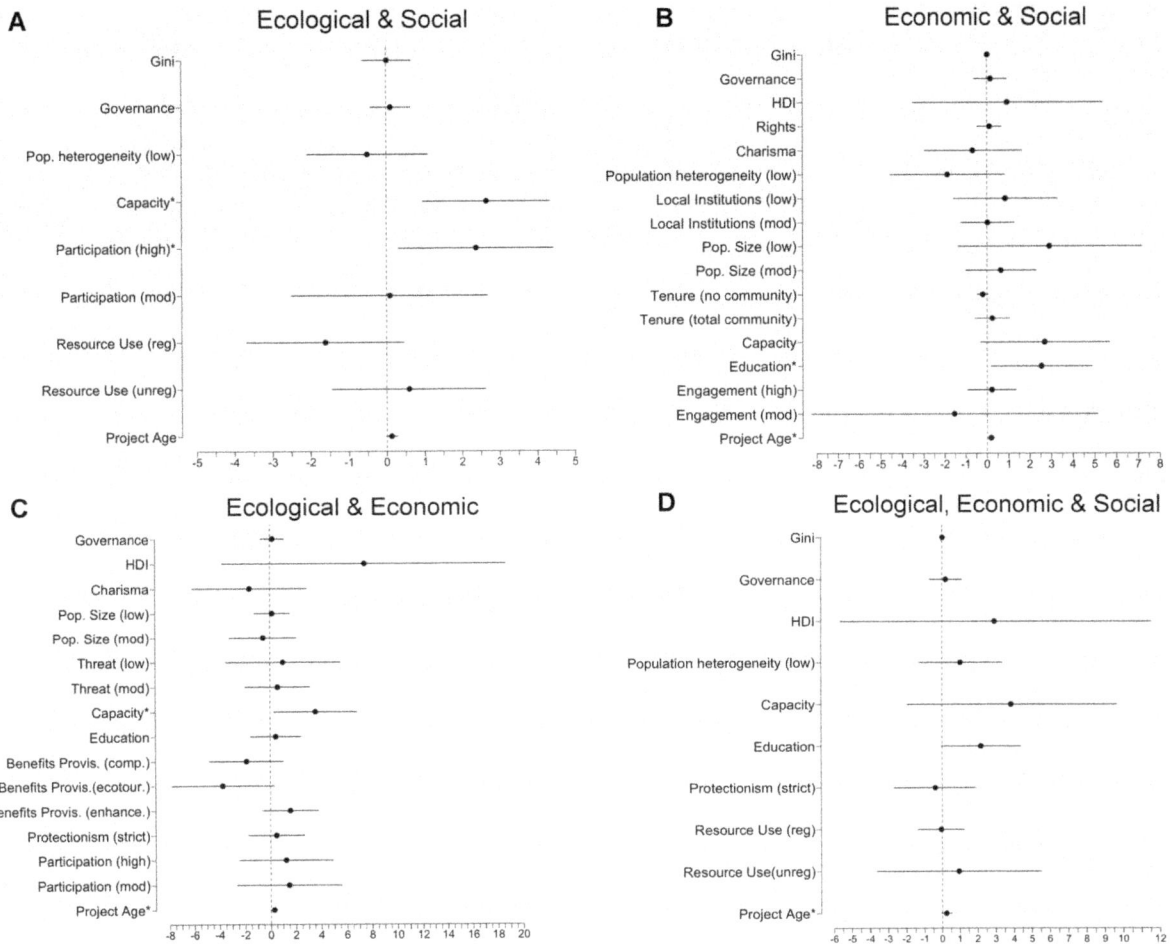

Figure 3 (a)–(d) Plots of the pooled coefficients (dots) and 95% confidence intervals (bars) (X-axis) for variables in the reduced-fit model for each outcome combination as selected by forward, stepwise AIC. Asterisks denote variables that are significantly associated with win–win outcomes for that combination. See supplementary material text for a discussion of the interpretation of the values for the pooled coefficients that are provided in Table S1 and are represented by the dots above.

The importance of project age and design features does not mean that community characteristics and national context are unimportant. It is well established that particular community characteristics can contribute to sustainable resource management (Agrawal 2001), and Tallis et al. (2008) present evidence that leads them to hypothesize that strong governance is important for win–win outcomes. Further, our bivariate analysis suggests that governance and HDI may contribute to win–win outcomes. However, the effect of national context may be mediated by project design or community characteristics in ways that we cannot discern through the exploratory analyses conducted here. A larger dataset and/or specific hypotheses for a smaller number of predictors would be needed to explore interactions with national context variables (e.g. Miller et al. 2015).

In addition, with this analysis I cannot make claims about direct synergies or tradeoffs between outcomes (Agrawal and Benson 2011). Joint success indicates neither a causal relationship between outcomes nor that the same underlying factor contributed to success in multiple outcomes. Similarly, the absence of win–win outcomes does not necessarily indicate direct tradeoffs whereby an increase in one outcome leads to a decline in another outcome. Instead outcomes may be related through complex feedbacks (Miller et al. 2012) or may be affected by independent factors. Longitudinal studies that indicate whether and how outcomes change over time and provide information on mechanisms underlying those changes are important for understanding causal relationships between outcomes. In addition, quasi-experimental designs (Ferraro and Hannauer 2014) and qualitative reviews that examine linkages between key variables and outcomes across scales can also provide important insights (Robbins et al. 2015).

Finally, several potentially important predictors, such as local ecological conditions, supportive national policies, and economic heterogeneity, were excluded from the analysis due to insufficient information in the papers in our sample (see supplementary material text for additional omitted variables). These variables and others could also explain conservation success or failure and should be included in future work.

Conclusion

This study suggests that capacity building, local participation, environmental education, and project age are associated with joint success in CBC. These features may strengthen a community's ability to participate in project management, capitalize on opportunities, respond to change, and adapt over time. While win–win outcomes are possible and it is important to explore when, where, and why they emerge, it is evident that tradeoffs are more common across all pairing examined here. Therefore, I reiterate calls (McShane et al. 2011) for project organizers to explicitly discuss the potential for tradeoffs in order to temper expectations and prepare communities to anticipate and manage conflicts that arise in the course of a CBC project.

Acknowledgments

This project was funded by the Beckman Institute, University of Illinois, Urbana-Champaign and the Ohio Agricultural Research and Development Center. I thank Kerry Waylen for her efforts with research design and coding, Monique Borgerhoff Mulder for guidance and feedback on early drafts, Peter Brosius at the Center for Integrative Conservation Research, University of Georgia for facilitating the initial research, Mark Grote for statistics advice, Shandel Brown, Lauren Weisenfluh, and David Kim for assistance with research, and two anonymous reviewers for very helpful suggestions.

Supporting Information

Table S1a-d. Results of logistic regression after stepwise, forward AIC model selection for each of the four combinations of outcome variables.

Table S2. Projects and associated publications included in the study

Table S3. Summary statistics for all independent and dependent variables for raw data before imputing missing values

Table S4. Correlation matrix of Spearman's r values for independent variables used in the multivariate analysis for all five imputed datasets.

Figure S1. Proportion of win-win outcomes by project age for each combination. The sample size for each age grouping in each combination of outcomes is provided within the bars.

References

Abbot, J.I.O., Thomas, D.H.L., Gardner, A.A., Neba, S.E. & Khen, M.W. (2001). Understanding the links between conservation and development in the Bamenda highlands, Cameroon. *World Dev.*, **29**, 1115-1136.

Adams, W., Aveling, R., Brockington, D. et al. (2004). Biodiversity conservation and the eradication of poverty. *Science*, **306**, 1146-1149.

Agrawal, A. (2001) Common property institutions and sustainable governance of resources. *World Dev.*, **29**, 1649-1672.

Agrawal, A. & Benson, C. (2011) Common property theory and resource governance institutions: strengthening explanations of multiple outcomes. *Environ. Conserv.*, **38**, 199-210.

Agrawal, A., & Redford, K. (2006) Poverty, development and biodiversity conservation: shooting in the dark. Working Paper No 26. Wildlife Conservation Society.

Becker, C.D., Agreda, A., Astudillo, E., Costantino M., & Torres P. (2005) Community-based monitoring of fog capture and biodiversity at Loma Alta, Ecuador enhance social and institutional cooperation. *Biodivers. Conserv.*, **14**, 2695-2707.

Borgerhoff Mulder, M., & Coppolillo, P. (2005) *Conservation: linking ecology, economics and culture*. Princeton University Press, Princeton, NJ.

Brown, K. (2004) Addressing trade-offs in forest landscape restoration. In S. Mansourian, D. Vallauri, editors. *Forest Restoration in landscapes*. Springer-Verlag, New York, pg 59-62.

Brewer, T.D., Cinner, J.E., Green, A., & Pressey, R.L. (2012) Effects of human population density and proximity to markets on coral reef fishes vulnerable to extinction by fishing. *Conserv. Biol.*, **27**, 443-452.

Brooks, J.S., Franzen, M.A., Holmes, C.M., Grote, M., & Borgerhoff Mulder, M. (2006) Testing hypotheses for the success of different conservation strategies. *Conserv. Biol.*, **20**, 1528-1538.

Brooks, J.S., Waylen, K.A., & Borgerhoff Mulder, M. (2012) How national context, project design, and local community characterisics influence success in community-based conservation projects. *Proc. Natl. Acad. Sci. U.S.A.*, **109**, 21265-21270. doi:10.1073/pnas.1207141110.

Campbell, B.M., Sayer, J.A., & Walker, B. (2010) Navigating trade-offs: working for conservation and development outcomes. *Ecol. Soc.*, **15**, 16-20.

Chhatre, A., & Agrawal, A. (2009) Trade-offs and synergies between carbon storage and livelihood benefits from forest commons. *Proc. Natl. Acad. Sci. U.S.A.*, **106**, 17667-17670.

Dasgupta, A., & Beard, V.A. (2007) Community drive development, collective action, and elite capture in Indonesia. *Dev. Change* **38**, 229-249.

Ferraro, P.J., & Hannauer, M.M. (2014) Advances in measuring the environmental and social impacts of environmental programs. *Ann. Rev. Environ. Resour.*, **39**, 495-517.

Ferraro, P.J., & Kiss, A. (2002) Direct payments to conserve biodiversity. *Science*, **298**, 1718-1719.

Gelman A., & Hill, J. (2007) *Data analysis using regression and multilevel/hierarchical models*. Cambridge Univeristy Press, New York, NY.

Gössling, S. (1999) Ecotourism: a means to safeguard biodiversity and ecosystem functions? *Ecol. Econ.*, **29**, 303-320.

Little, R.J.A., & Rubin, D.B. (2002) *Statistical analysis with missing data*, 2nd ed. John Wiley, New York, NY.

McShane, T., Hirsch, P., & Trung, T.C. et al. (2011) Hard choices: making trade-offs between biodiversity conservation and human well-being. *Biol. Conserv.*, **144**, 966-972.

Miller, B.W., Caplow, S.C., & Leslie, P.W. (2012) Feedbacks between conservation and social-ecological systems. *Conserv. Biol.*, **26**, 218-227.

Miller, D.C. (2014) Explaining global patterns of international aid for linked biodiversity conservation and development. *World Dev.*, **59**, 341-359.

Miller, D.C., Minn, M., & Sinsin, B. (2015) The importance of national political context to the impacts of international conservation aid: evidence from the W National Parks of Benin and Niger. *Environ. Res. Lett.*, **10**, 1-12.

Nelson, F., editor. (2010) *Community rights, conservation, and contested land*. Earthscan, Washington, D.C.

Ostrom, E. (1990) *Governing the commons: the evolution of institutions for collective action*. Cambridge University Press, Cambridge.

Padgee, A., Kim, Y. & Daugherty, P.J. (2006) What makes community forestry management successful: a meta-study from community forests throughout the world. *Soc. Natur. Resour.*, **19**, 33-52.

Persha, L., Agrawal, A. & Chhatre, A. (2011) Social and ecological synergy: local rulemaking, forest livelihoods, and biodiversity conservation. *Science*, **331**, 1606-1608.

Persha, L., & Andersson, K. (2014) Elite capture risk and mitigation in decentralized forest governance regimes. *Global Environ. Change*, **24**, 265-276.

Persha, L., Fischer, H., Chhatre, A., Agarwal, A., & Benson, C. (2010) Biodiversity conservation and livelihoods in human-dominated landscapes: forest commons in South Asia. *Biol. Conserv.*, **143**, 2918-2925.

Rammel, C., S, S., & Wilfing, H. (2007) Managing complex adaptive systems—a co-evolutionary perspective on natural resource management. *Ecol. Econ.*, **63**, 9-21.

Ribot, J., Agrawal, A., & Larson, A.M. (2006) Recentralizing while decentralizing: how national governments reappropriate forest resources. *World Dev.*, **34**, 1864-1886.

Robbins, P., Chhatre, A., & Karanth, K. (2015) Political ecology of commodity agroforests and tropical biodiversity. *Conserv. Lett.*, **8**, 77-85.

Salafsky, N., Cauley, H., & Balachander, G. et al. (2001) A systematic test of an enterprise strategy for community-based biodiversity conservation. *Conserv. Biol.*, **15**, 1585-1595.

Scanlon, L.J., & Kull, C.A. (2009) Untangling the links between wildlife benefits and community-based conservation at Torra Conservancy, *Namibia. Dev. S. Africa*, **26**, 75-93.

Sen, A.K. (1999). *Development as freedom*. Alfred A. Knopf, Inc., New York.

Storey, J.D. (2002) A direct approach to false discovery rates. *J. R. Stat. Soc. Ser. B–Stat. Methodol.*, **64**, 479-498.

Tai, H.-S. (2007) Development through conservation: an institutional analysis of indigenous community-based conservation in Taiwan. *World Dev.*, **35**, 1186-1203.

Tallis, H., Kareiva, P., Marvier, M., & Chang, A. (2008) An ecosystem services framework to suppport both practical conservation and economic development. *Proc. Nat. Acad. Sci. U.S.A.*, **105**, 9457-9464.

United Nations General Assembly. (2015) *General Assembly Draft Resoluton, Transforming our World: the 2030 Agenda for Sustainable Development*. New York, NY.

Waylen, K.A., Fischer, A., McGowan, P.K., Thirgood, S.J., & Milner -Gulland, E.J. (2010) *The effect of local cultural context on community-based conservation interventions: evaluating ecological, economic, attitudinal, and behavioural outcomes*. Systematic Review No 80. Collaboration for Environmental Evidence, Birmingham, U.K.

21

Unexpected Management Choices When Accounting for Uncertainty in Ecosystem Service Tradeoff Analyses

Reniel B. Cabral[1], Benjamin S. Halpern[1,2,3], Christopher Costello[1], & Steven D. Gaines[1]

[1] Bren School of Environmental Science and Management, University of California, Santa Barbara, CA 93106, USA
[2] Imperial College London, Silwood Park Campus, Buckhurst Road, Ascot SL57PY, UK
[3] National Center for Ecological Analysis and Synthesis, 735 State St. Suite 300, Santa Barbara, CA 93101, USA

Keywords
Bioeconomic model; decision making; ecosystem-based management; ecosystem services; multisector planning; second best theory; spatial planning; tradeoff analysis; uncertainty.

Correspondence
Reniel B. Cabral and Benjamin S. Halpern, Bren School of Environmental Science and Management, University of California, Santa Barbara, CA 93106, USA.
E-mail:rcabral@ucsb.edu; halpern@nceas.ucsb.edu

Editor
Richard Zabel

Abstract

Resource management and conservation increasingly focus on ecosystem service provisioning and potential tradeoffs among services under different management actions. Application of bioeconomic approaches to tradeoffs assessment is touted as a way to find win-win outcomes or avoid unnecessary stakeholder conflict. Yet, nearly all assessments to date have ignored inherent uncertainties in the provision and valuation of services. We incorporate uncertainty into the ecosystem services analytical framework and show how such inclusion improves optimal decision making. In particular, we show: (1) "suboptimal" solutions can become optimal when uncertainties are accounted for; (2) uncertainty paradoxically makes stakeholders value conservation despite their lack of preference for it; and (3) substantial losses or missed gains in ecosystem service provisioning can be incurred when uncertainty is ignored. Our results highlight the urgency of accounting for uncertainties in ecosystem services in tradeoff assessments given the widespread use of this approach by government agencies and conservation organizations.

Introduction

Communities face increasingly complex and uncertain decisions about how to effectively and efficiently manage natural resources, in particular as new uses of lands and oceans move into places already allocated for other purposes. These changes increase the number and diversity of stakeholders who benefit from, and often whose welfare depends on, how resources are managed. As such, environmental management decisions frequently involve tradeoffs among different ways that people use and value ecosystems. To better articulate the nature of tradeoffs among services and identify solutions that may reduce conflict and promote win-win solutions, decision makers are increasingly turning to an ecosystem services (ESs) decision framework (Polasky *et al.* 2008; White *et al.* 2012;

Lester *et al.* 2013), which models the potential supply and value of services from ecosystems under different management schemes (Tallis *et al.* 2012). The economic theory underpinning this approach has been around for decades, but it has only recently begun to gain significant traction in the context of resource management and conservation[1].

Recent advances in modeling ES have extended the tradeoff framework to many classes of problems, including conservation planning (Klein *et al.* 2013), reserve network design (Costello *et al.* 2010; Halpern *et al.* 2011; Rassweiler *et al.* 2012), land use regulation (Polasky *et al.* 2008; Goldstein *et al.* 2012; Johnson *et al.* 2012), habitat conversion (Barbier *et al.* 2008; Zavalloni *et al.* 2014), and resource use permitting (Kim *et al.* 2012; White *et al.* 2012). One of the general outcomes from these diverse studies is that modeling the full set of possible management actions commonly identifies solutions that provide more value to more stakeholder groups, thereby potentially reducing conflict, than would normally arise in less comprehensive policy discussions (Rassweiler *et al.* 2014). Although there is a great potential value in scientific forecasts of diverse options, one of the challenges is that such forecasts have inherent uncertainty. To date, the theory and application of bioeconomic assessments of these tradeoffs have largely ignored such uncertainty (Nicholson *et al.* 2009; Johnson *et al.* 2012; Grêt-Regamey *et al.* 2013).

Uncertainty arises from inevitable limits to scientific knowledge about the natural world and the effects of humans' actions on it, often making outcomes hard to predict in isolation, and even more so in combination. Uncertainty also stems from unavailability of high-quality datasets required to parameterize and inform predictive models. Moreover, how individuals and organizations respond to information adds an additional layer of uncertainty. Although previous studies have considered the impact of uncertainty on environmental management and the provision of ES and conservation of biodiversity (Doyen & Béné 2003; Lande *et al.* 2003; Grafton & Kompas 2005; Regan *et al.* 2005b; Halpern *et al.* 2006; McCarthy & Possingham 2007; Johnson *et al.* 2012), it is rare for analyses of ES tradeoffs to incorporate uncertainties, either in the natural or the human systems that determine service provision and valuation. Ignoring uncertainty creates a number of potential challenges, such as misinformed losses and gains from implementing a policy, miscalculation about actual risks, and misunderstanding about the implications of stakeholder preferences. Furthermore, second best theory (Lipsey & Lancaster 1956) predicts that the optimal strategy in the first-best world (with perfect information) would no longer be the optimal in a second-best world (with uncertainties or imperfect information), while the suboptimal strategy in the first-best world would perform better than the optimal strategy in a second-best world. Here, we address several unresolved questions about how uncertainty affects decision making around ES provision and value, namely: (1) what are the potential losses or gains in total service provision when uncertainty is ignored?, (2) does uncertainty and associated risk tolerance of stakeholders affect the choice of optimal management solutions?, and (3) will stakeholder preferences for different services change given uncertainty?

Methods

We use two case studies to illustrate the incorporation of uncertainty into the ES tradeoff framework and the effect of uncertainty on optimal decision making. Our case studies examine (1) tradeoffs between converting mangroves to shrimp aquaculture versus their preservation as nursery habitat for fisheries (see Supplementary Information) and (2) tradeoffs between conservation (fish biomass) and yield in a stochastic fish population in a region deciding whether to create a marine protected area (MPA). In both cases, we explore a range of possible management actions, focusing on how uncertainty in underlying models affects optimal actions.

Tradeoff between mangroves' nursery function and aquaculture production

We model the nursery function (N) of a mangrove patch as a function of the distance from the watercourse (x, in unit of 100 m) using an exponential decay model,

$$N(x) = e^{-k_N x}. \tag{1}$$

k_N dictates the rate of decay in the nursery function with distance. We assume that the nursery function can have two states, $k_N = \{0.1, 0.15\}$. On average, mangroves' nursery functioning is highest and least variable near the watercourse.

Clean water is a limiting factor for aquaculture productivity. As an aquaculture pond's distance from the watercourse increases, so does the chance of reduced productivity and level of variability in productivity (Binh *et al.* 1997). We assume the productivity loss function for aquaculture to have a similar form as that of the nursery function ($A(x) = e^{-k_A x}$), with aquaculture production having two possible states, $k_A = \{0, 0.3\}$. $k_A = 0$ implies that all patches have the same aquaculture suitability of 1, while $k_A > 0$ implies a declining aquaculture productivity with distance from the watercourse.

We assume that a favorable state ($k_N = 0.1$ and $k_A = 0$) occurs with probability $p = 0.5$ and an unfavorable state ($k_N = 0.15$ and $k_A = 0.3$) occurs with probability $p = 0.5$. On average, an approximately 10% reduction in aquaculture productivity is expected at $x = 2$ and an approximately 40% reduction can occur at $x = 10$.

The area adjacent to the watercourse is suitable for both aquaculture and nurseries. However, the conversion of mangrove patches near the watercourse reduces the nursery functioning of mangroves. Furthermore, aquaculture structures can interfere with the spatial connectivity of the nursery function in a nonlinear manner, i.e., establishing aquaculture structures adjacent to the watercourse reduces the accessibility of the interior mangroves, thus reducing the contribution of interior mangroves to the total nursery functioning of the system (see Supplementary Information).

We consider four heuristic scenarios/strategies for establishing aquaculture ponds: (1) single-sector decisions where aquaculture stakeholders optimize production by deploying ponds close to the watercourse (▲); (2) single-sector decisions where conservation/wild fisheries stakeholders optimize nursery function by directing aquaculture establishment away from the watercourse (■); (3) random deployments of aquaculture ponds (●); and (4) a block strategy where mangroves are grouped together from the smallest connected mangroves to the largest connected mangroves, and then groups of mangroves are converted into aquaculture ponds, from smallest to largest (◆).

Tradeoff between fish yield and fish biomass in a stochastic fish population

We consider a Ricker (1954) difference equation model to describe the dynamics of a hypothetical fish population. The total population biomass at time $t+1$ (i.e., B_{t+1}) is a function of the population biomass at the previous time step, i.e., $B_{t+1} = f(B_t)$ where

$$f(B_t) = B_t e^{r\left(1 - \frac{B_t}{K}\right)}. \tag{2}$$

The parameters r and K represent the population's growth rate and carrying capacity, respectively.

Given any positive finite values of K and r, Equation (2) is always positive. Additionally, f is positive for any positive finite stochastic values of K and r. Therefore, Equation (2) is an ideal model for investigating the effect of harvesting on the population given various magnitudes of population stochasticity. The nonzero equilibrium point of Equation (2) is globally/asymptotically stable for $r<2$. We use $r = 0.5$ in our model.

We assume a constant-yield fishery where a fixed target yield (Y) is set. Given harvesting, the population model (Equation (2)) becomes

$$f(B_t) = B_t e^{r\left(1 - \frac{B_t}{K}\right)} - Y. \tag{3}$$

The maximum sustainable yield (MSY) occurs when $f'(B^*) = 1$, where $B^* = B_{MSY}$ is the steady-state population biomass at MSY. Hence,

$$\left(1 - \frac{r B_{MSY}}{K}\right) e^{r\left(1 - \frac{B_{MSY}}{K}\right)} = 1. \tag{4}$$

There is no analytic solution to Equation (4). Given $r = 0.5$ and $K = 1,000$, B_{MSY} and Y_{MSY} were derived numerically: $B_{MSY} = 467.7971$ and $Y_{MSY} = 142.6161$.

Since the point at MSY is unstable (Brauer & Sánchez 1975; Beddington & May 1977; Roughgarden & Smith 1996; Murray 2007), given population variability, it is necessary to engineer the system to achieve an optimal sustainable yield. We limit the engineering to MPA es-

tablishment. We assume that an MPA reduces the fishing ground and thus the yield. We simply assume that yield at MSY is reduced in proportion to the size of the MPA, i.e., $Y = (1–\text{MPA\_size})*Y_{MSY}$, where the MPA size is from 0 (no MPA) to 1 (no fishing allowed).

We introduced a uniformly distributed noise $\upsilon(a,b)$ with zero mean and a symmetric limit of a and b to the carrying capacity of the population, i.e., $f(B_t) = B_t e^{r(1 - \frac{B_t}{K \pm \upsilon(a,b)})} - Y$. We ran variability from 0 to $\pm0.99K$ in steps of $0.01K$. For each instance of variability, we derive the size of the MPA needed to achieve an optimal sustainable yield. We ran the population for 1,000,000 time steps at each instance of variability.

ES tradeoff, utility, and indifference curve

Production theory, which deals with the production of goods and services given resource input, has been applied successfully to ES tradeoff assessments (e.g., White *et al.* 2012; Lester *et al.* 2013). ES can exhibit tradeoffs or synergistic relationships whose nature can be identified easily by plotting the production of services under different policy scenarios on a Cartesian coordinate system whose axes represent the assessed services. An indifference curve is a convenient graphical representation of stakeholders' preferences within the service production space. A curve contains the possible combinations of ES values (ES bundles) where stakeholders are indifferent about any of the combinations. Multiple, nonintersecting indifference curves represent the sets of indifferent ES bundles; an arrow is usually used to indicate the direction of the most preferred sets of ES bundles. The selected ES bundle will depend on both the production of ES and the stakeholder's preference.

The above framework assesses expected services produced by diverse policy options, and then uses these outcomes to make several broad management recommendations. First, outer-bound solutions are the best-case options (points a–f in Figure 1A); the choice between these options depends on stakeholders' relative preferences for different services, represented by indifference curves (Figure 1A). Second, interior points are suboptimal (e.g., option "g") and thus poor decisions, as one can improve the outcome of at least one service at no cost to other services. Policy option "c" has the highest utility in our example; it is positioned at the third indifference curve and the rest are positioned below it (Figure 1A). Third, the shape of the curve (concave in Figure 1A) is used to define the nature of the tradeoff between services (in this case, weak to moderate). Yet, all of these management guidelines ignore the potential influence of uncertainty in modeling these management outcomes (Figure 1B).

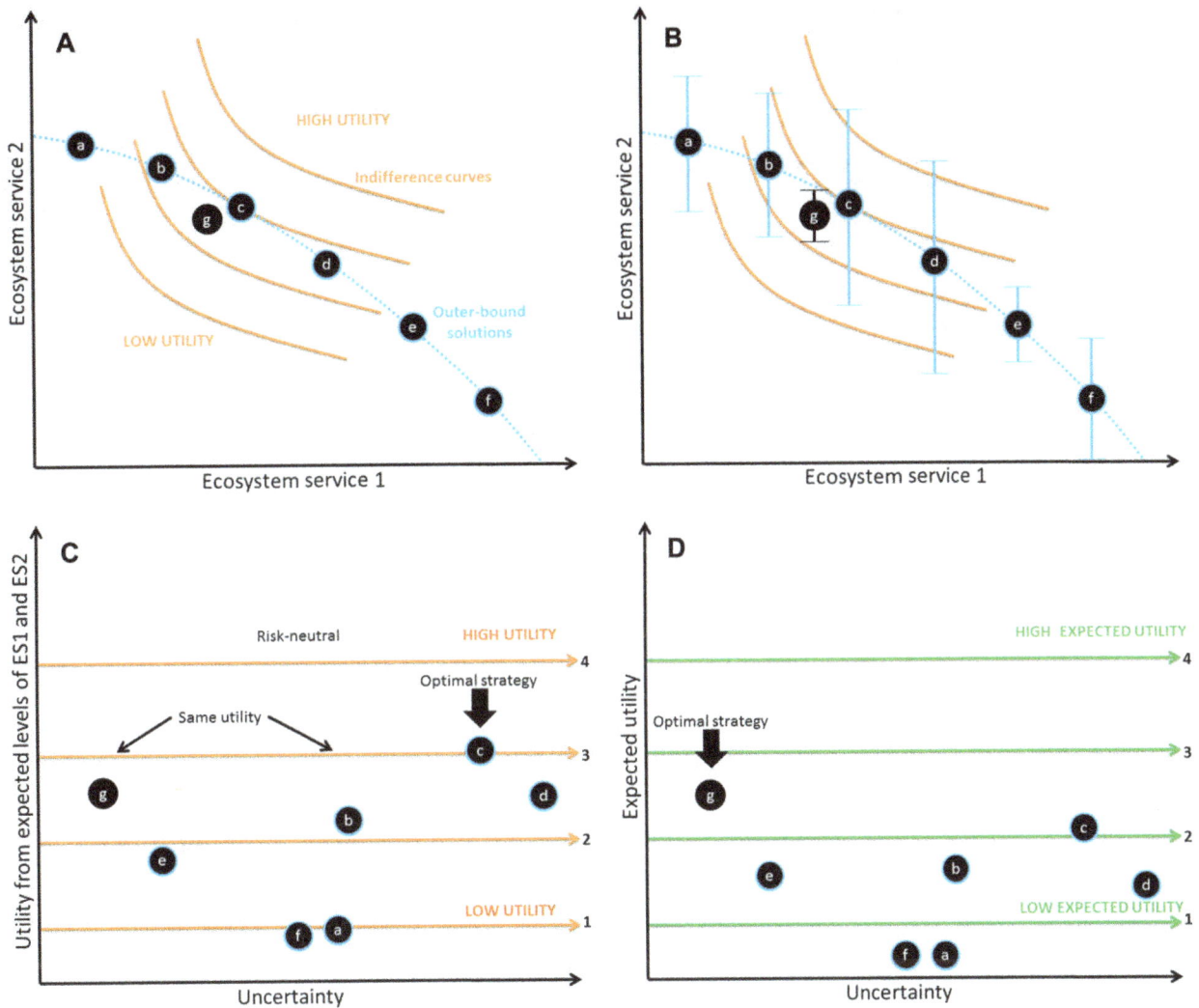

Figure 1 Ecosystem services (ES) tradeoff analysis with uncertainty. (A) Conventional representation of policy options in ES tradeoff analysis. (B) Policy outcomes with uncertainty. Uncertainty is only in ES 2. (C–D) Proposed graphical representation (utility v. uncertainty) for analyzing policy options with uncertainty under the ES tradeoff analysis framework. Points within a single line have the same utility. (C) A stakeholder that does not account or care for risk would choose policy option "c." (D) Example where accounting for risk makes a risk-averse stakeholder chooses policy option "g" instead of "c."

Some insights about the role of uncertainty in ES trade-off evaluation can be gained from plotting the policy options into the utility versus uncertainty axes (Figure 1B–D). For example, a stakeholder who does not account for or care about uncertainty might choose policy "c" as expected (Figure 1C), i.e., the expected utility is constant at any level of uncertainty. A risk-averse stakeholder would instead try to minimize the chance of an undesirable outcome, or avoid any risk at all, and would account for uncertainty (e.g., Mangel 2000) leading to the choice of policy "g" (Figure 1D). In other words, a risk-averse stakeholder would rationally choose policy option "g," an interior solution, over policy "c."

We formalize the above idea by using the concept of utility and expected utility (Von Neumann & Morgenstern 1944; Varian 2009). Utility is a numerical description of a stakeholder's preferences. For the aquaculture-nursery function tradeoff, we use a common utility function (U) of the form

$$U = \left(\sum_i A_i\right)^\alpha + \left(\sum_i N_i\right)^\alpha, \qquad (5)$$

where $N_i = N(x_i)$ and $A_i = A(x_i)$ for patch i distance x_i. The summation is evaluated throughout the 1,125 mangrove grid cells: areas that can be conserved or converted into aquaculture ponds. The parameter α describes

Table 1 Expected utility of two bundles of services produced by two different policies for different levels of risk-aversion (α) (see Figure 2 for the bundles' positions)

α	0.1	0.2	0.3	0.4	0.5	0.6	0.7	0.8	0.9	1.0
ES bundle 1	3.57	6.56	11.94	21.77	39.78	72.85	133.65	245.60	452.04	833.18
ES bundle 2	3.63	6.70	11.99	21.82	39.71	72.31	131.74	240.16	438.02	799.29

Note: $\alpha = 1$ is a risk-neutral stakeholder. The lower the α, the more risk-averse the stakeholder is. Shaded boxes indicate optimal solution for different α.

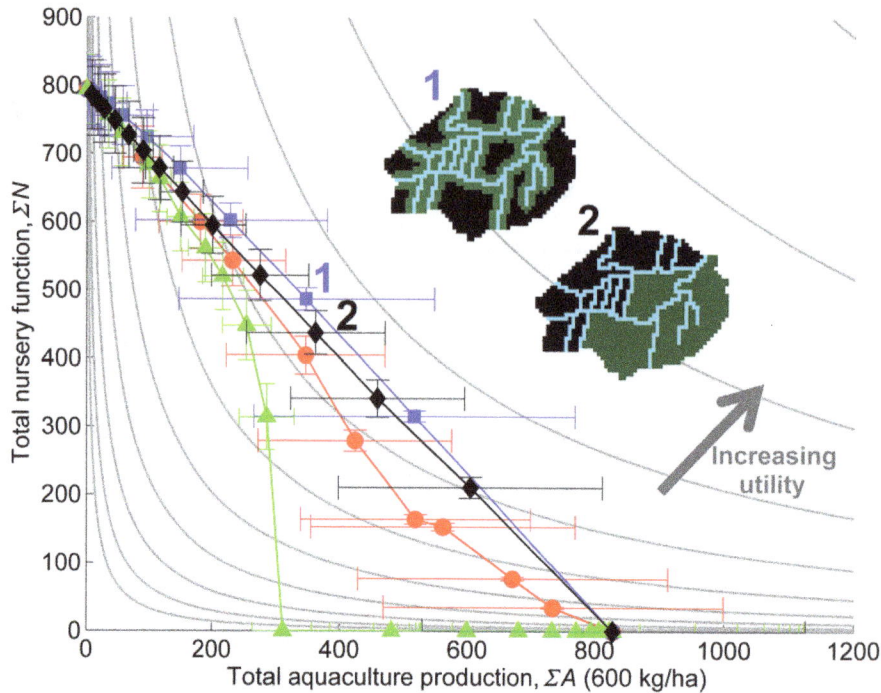

Figure 2 Tradeoff curves between the aquaculture production and nursery function of a mangrove forest in Viet Nam for variable heuristic decisions: (■) preferentially establishing ponds away from the mangrove edge, (●) random location of ponds, (▲) preferentially establishing ponds near the mangrove edge, and (◆) deploying ponds by blocks. Inset: Policy design for options "1" and "2" with black as ponds, green as mangroves, and blue as watercourse. $\alpha = 0.1$ is used for the indifference curves.

the risk preference of a stakeholder. $\alpha = 1$ is a risk-neutral stakeholder because U increases linearly with $\sum A$ and $\sum N$. A $0 < \alpha < 1$ describes a risk-averse stakeholder, i.e., the marginal gain in U decreases at increasing values of $\sum A$ and $\sum N$ (U is concave).

For any given configuration of aquaculture development, two states can occur (favorable state ($\sum_i A_i^+, \sum_i N_i^+$) and unfavorable state ($\sum_i A_i^-, \sum_i N_i^-$)). These give rise to favorable utility U_+ with probability p_+ and an unfavorable utility U_- with probability $(1 - p_+)$. The expected utility of the ES bundle is

$$EU = p_+ U_+ + (1 - p_+) U_-. \qquad (6)$$

Note that for a risk-averse stakeholder, EU is less than the utility of the mean production values, i.e., $EU < U'$, where $U' = \left(\frac{\sum_i A_i^+ + \sum_i A_i^-}{2}\right)^{\alpha} + \left(\frac{\sum_i N_i^+ + \sum_i N_i^-}{2}\right)^{\alpha}$; the opposite holds for a risk-loving stakeholder.

Results

Across each potential management strategy for mangrove conversion to shrimp ponds, incorporation of uncertainty in the underlying biophysical model determining service production leads to differences in predicted service provision of up to 60% (Figure 2). The worst-case scenario (▲) results when stakeholders neglect the interactions of the two ESs, acting in favor of a single sector (in this case, aquaculture). Optimizing aquaculture production requires ponds to be built along habitat edges. When all edge patches are converted into ponds, the nursery functioning of all interior mangroves vanishes. Uncertainty accumulates as total aquaculture production increases; uncertainty in aquaculture is highest when the total aquaculture production is highest. However, the uncertainty in the nursery function is highest instead when aquaculture ponds are preferentially placed along habitat

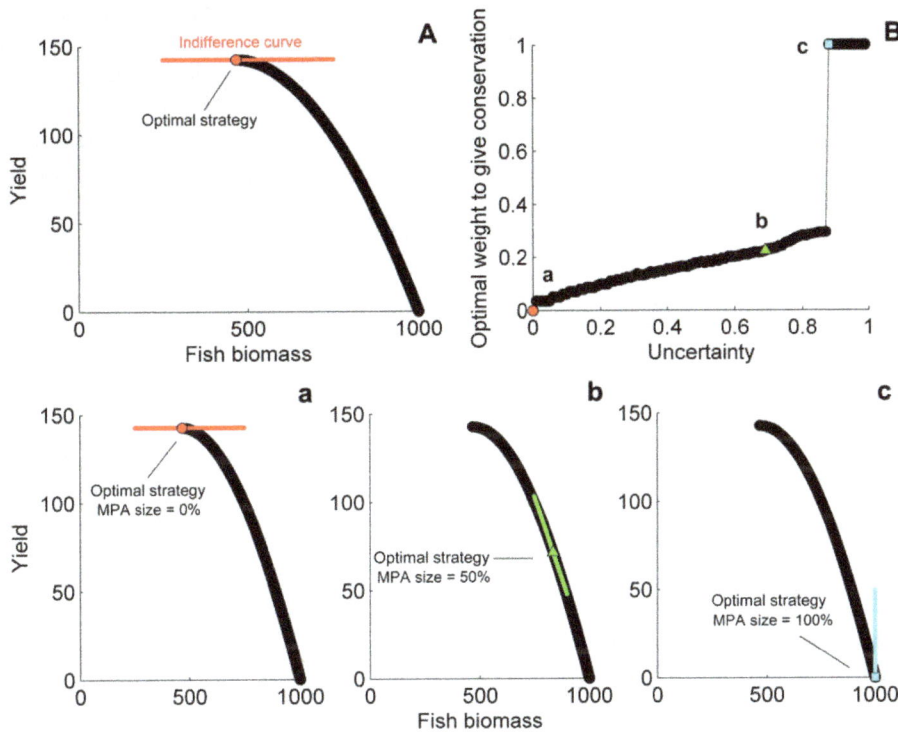

Figure 3 Tradeoff between conservation (fish biomass) and yield. (A) Fish biomass versus yield for variable size of marine protected area (MPA). The highest yield corresponds to Y_{MSY} (with no MPA). MPA has the effect of reducing yield. Closing the entire fishery resulted to zero yield and biomass at carrying capacity ($K = 1,000$). Stakeholders are assumed to have zero preference for conservation (horizontal indifference curve). (B) Effect of uncertainty on stakeholder's preference. With increasing uncertainty, stakeholders' preference appeared to shift toward conservation even though they have zero preference for it, i.e., (a) at zero uncertainty, the optimal strategy is to fish at MSY; (b) with uncertainty, the optimal strategy is to establish MPAs or lower fishing effort; and (c) at sufficiently high uncertainty, the optimal strategy is to not fish the stock at all. Uncertainty pertains to the level of variability in K, which ranges from a, $b = 0$ (uncertainty = 0) to $\pm 0.99K$ (uncertainty = 0.99) (see Methods).

edges, because ponds block interior mangroves and this limits water flow into the interior mangroves (Figure 2).

Based on tradeoff assessments without uncertainty, the optimal policy solution lies along the outer-bound (frontier) solutions (■, marked by "1," Figure 2). However, accounting for uncertainty makes a risk-averse stakeholder chooses an interior solution (marked by "2," see Table 1), because it has higher expected utility than any of the outer-bound solutions. The two solutions vary considerably in outcome. One involves utilizing the entire mangrove forest by converting mangrove patches that are away from the watercourse into aquaculture ponds, while the other involves converting mangroves to ponds by blocks, hence retaining other blocks of mangroves in their pristine state (Figure 2, inset). The optimal solution changes from an interior to an outer-bound as the stakeholder becomes less risk-averse, i.e., $\alpha \geq 0.5$ (Table 1).

In the second case, the stakeholder is assumed to care only about fish yield with no interest in conservation (i.e., horizontal indifference curve, weight toward conserva-

tion = 0, Figure 3A). Optimizing yield implies that the stakeholder would fish the stock at MSY and would not consider establishing an MPA. However, the fish population is unstable at MSY, and population variability may push the population to rapid declines. When considering uncertainty in model evaluation, the stakeholder should prefer to establish an MPA, or to lower fishing effort, as fishing at MSY with high variability can ultimately lead to population collapse. Optimizing catch without accounting for uncertainties would result to losses or missed gains for both catch and biomass, and these losses depend on the level of population stochasticity (Figure 4). Higher uncertainty implies a less optimal sustainable catch limit and higher fish biomass.

Indeed, the outcome of uncertainty is similar to an outcome where the stakeholder has more preference for conservation (question 3, see Figure 3). Incorporating uncertainty makes stakeholder behave "as if" they had strong biological or conservation preferences, even when they have no preference for conservation at all.

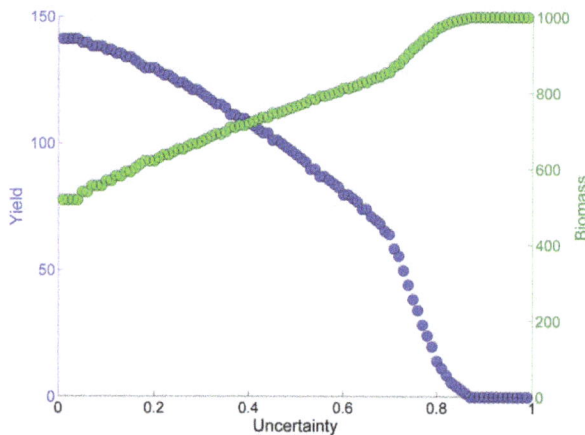

Figure 4 Losses and missed gains given various levels of uncertainties. Green indicates the mean lowest level of fish biomass that can be retained in the fishery, while blue shows the mean highest yield that can be obtained when uncertainties are accounted for.

Sufficiently large population variability results in the case where the stakeholder will support a strongly conservationist strategy and will fish the stock very little, as more intensive harvesting will only lead to population collapse (Figure 3B). Note that in general, MSY will not be the optimal harvest level when the fishery is concerned with profit (Gordon 1954; Scott 1955).

Discussion

Most ES tradeoff assessments to date have focused on outer-bound solutions as optimal, yet there is inherent uncertainty in the ability to achieve those solutions (Regan *et al.* 2005a; Kareiva *et al.* 2011; Rassweiler *et al.* 2014). We showed that interior solutions can achieve preferred outcomes, especially when accounting for stakeholder risk tolerance. This finding is consistent with the second best theory (Lipsey & Lancaster 1956), where the optimal solution in a world with perfect information becomes suboptimal in a world with uncertainties.

Incorporating uncertainty in tradeoff analysis also allows the visualization of hidden conservation bias in the indifference curve, as illustrated by the case where purely production-oriented stakeholders, regardless of their risk tolerance, would need to consider conservation more as uncertainty increases. Our result reinforces the precautionary approach to fisheries management of considering both fisheries yield and conservation by setting target production lower than MSY (Roughgarden & Smith 1996) or establishing MPAs to manage both fishing effort (Hastings & Botsford 1999) and uncertainty in fisheries (Hsieh *et al.* 2006).

Ignoring or underestimating uncertainty risks foregoing potential value that could be derived from ES (if the mean is underestimating value) or collapsing systems or setting expectations that are too high (if the mean is overestimated). In general, incorporating uncertainties into ES tradeoff analyses has three main advantages. First, risks and uncertainties in tradeoffs of policy options are explicitly illustrated and quantified; therefore, the costs and benefits of ignoring (or reducing) uncertainties are demonstrated. Second, uncertainties have the paradoxical effect of making stakeholders value conservation even though their preference is solely to maximize gain from an extractive ES. This hidden conservation bias in indifference curves cannot be visualized by traditional treatment of uncertainties. Such an effect may also arise with ESs that are tightly coupled and may be common in nature. And finally, the ES decision framework appeared to be a convenient and practical tool for communicating tradeoffs and risks.

Although it is widely known that uncertainty affects decision making, management within the ES framework has yet to embrace this reality. So far, we have explored a few types of uncertainties; there are broad classes of uncertainties in decision making focused on natural resource management that warrant attention (Regan *et al.* 2002). Factors including irrational behavior of stakeholders, discount rates, and market imperfection could influence the conservation outcome and interact with other ecosystem components. While this work also focuses on uncertainties in the context of provision and valuation of ES, another gap that merits attention is the flows and ultimate use of ES. Alternative methods based on Bayes' theorem, which quantify uncertainties in ES supply and demand given some level of available data, have been proposed (Villa *et al.* 2014). To address the complexities in the types and nature of these uncertainties, which themselves merit further evaluation, the framework presented here is general and can be used to account for other uncertainties in human systems, biophysical systems, and their interactions.

Acknowledgments

We thank K. Siegel for the language edit and the reviewers for the comments and suggestions that improved our work. The publication fee was supported by the UCSB Open Access Fund.

References

Barbier, E.B., Koch, E.W., Silliman, B.R. *et al.* (2008). Coastal ecosystem-based management with nonlinear ecological functions and values. *Science*, **319**, 321-323.

Beddington, J.R. & May, R.M. (1977). Harvesting natural populations in a randomly fluctuating environment. *Science*, **197**, 463-465.

Binh, C.T., Phillips, M.J. & Demaine, H. (1997). Integrated shrimp-mangrove farming systems in the Mekong delta of Vietnam. *Aquac. Res.*, **28**, 599-610.

Brauer, F. & Sánchez, D.A. (1975). Constant rate population harvesting: equilibrium and stability. *Theor. Popul. Biol.*, **8**, 12-30.

Costello, C., Rassweiler, A., Siegel, D., Leo, G.D., Micheli, F. & Rosenberg, A. (2010). The value of spatial information in MPA network design. *Proc. Natl. Acad. Sci.*, **107**, 18294-18299.

Doyen, L. & Béné, C. (2003). Sustainability of fisheries through marine reserves: a robust modeling analysis. *J. Environ. Manage.*, **69**, 1-13.

Goldstein, J.H., Caldarone, G., Duarte, T.K. *et al.* (2012). Integrating ecosystem-service tradeoffs into land-use decisions. *Proc. Natl. Acad. Sci.*, **109**, 7565-7570.

Gordon, H.S. (1954). The economic theory of a common-property resource: the fishery. *J. Polit. Econ.*, **62**, 124-142.

Grafton, R.Q. & Kompas, T. (2005). Uncertainty and the active adaptive management of marine reserves. *Mar. Policy*, **29**, 471-479.

Grêt-Regamey, A., Brunner, S.H., Altwegg, J. & Bebi, P. (2013). Facing uncertainty in ecosystem services-based resource management. *J. Environ. Manage.*, **127**, S145-S154.

Halpern, B.S., Regan, H.M., Possingham, H.P. & McCarthy, M.A. (2006). Accounting for uncertainty in marine reserve design. *Ecol. Lett.*, **9**, 2-11.

Halpern, B.S., White, C., Lester, S.E., Costello, C. & Gaines, S.D. (2011). Using portfolio theory to assess tradeoffs between return from natural capital and social equity across space. *Biol. Conserv.*, **144**, 1499-1507.

Hastings, A. & Botsford, L.W. (1999). Equivalence in yield from marine reserves and traditional fisheries management. *Science*, **284**, 1537-1538.

Hsieh, C., Reiss, C.S., Hunter, J.R., Beddington, J.R., May, R.M. & Sugihara, G. (2006). Fishing elevates variability in the abundance of exploited species. *Nature*, **443**, 859-862.

Johnson, K.A., Polasky, S., Nelson, E. & Pennington, D. (2012). Uncertainty in ecosystem services valuation and implications for assessing land use tradeoffs: an agricultural case study in the Minnesota River Basin. *Ecol. Econ.*, **79**, 71-79.

Kareiva, P., Tallis, H., Ricketts, T.H., Daily, G.C. & Polasky, S. (2011). *Natural capital: theory and practice of mapping ecosystem services*. Oxford University Press Inc., New York.

Kim, C.-K., Toft, J.E., Papenfus, M. *et al.* (2012). Catching the right wave: evaluating wave energy resources and potential compatibility with existing marine and coastal uses. *PLoS ONE*, **7**, e47598.

Klein, C.J., Tulloch, V.J., Halpern, B.S. *et al.* (2013). Tradeoffs in marine reserve design: habitat condition, representation, and socioeconomic costs. *Conserv. Lett.*, **6**, 324-332.

Lande, R., Engen, S. & Sæther, B.-E. (2003). *Stochastic population dynamics in ecology and conservation*. Oxford University Press Inc., New York.

Lester, S.E., Costello, C., Halpern, B.S., Gaines, S.D., White, C. & Barth, J.A. (2013). Evaluating tradeoffs among ecosystem services to inform marine spatial planning. *Mar. Policy*, **38**, 80-89.

Lipsey, R.G. & Lancaster, K. (1956). The general theory of second best. *Rev. Econ. Stud.*, **24**, 11-32.

Mangel, M. (2000). Irreducible uncertainties, sustainable fisheries and marine reserves. *Evol. Ecol. Res.*, **2**, 547-557.

McCarthy, M.A. & Possingham, H.P. (2007). Active adaptive management for conservation. *Conserv. Biol.*, **21**, 956-963.

Murray, J.D. (2007). *Mathematical biology: I. An introduction*. 3rd edition. Springer, New York.

Nicholson, E., Mace, G.M., Armsworth, P.R. *et al.* (2009). Priority research areas for ecosystem services in a changing world. *J. Appl. Ecol.*, **46**, 1139-1144.

Polasky, S., Nelson, E., Camm, J. *et al.* (2008). Where to put things? Spatial land management to sustain biodiversity and economic returns. *Biol. Conserv.*, **141**, 1505-1524.

Rassweiler, A., Costello, C., Hilborn, R. & Siegel, D.A. (2014). Integrating scientific guidance into marine spatial planning. *Proc. R. Soc. Lond. B Biol. Sci.*, **281**, 20132252.

Rassweiler, A., Costello, C. & Siegel, D.A. (2012). Marine protected areas and the value of spatially optimized fishery management. *Proc. Natl. Acad. Sci.*, **109**, 11884-11889.

Regan, H.M., Ben-Haim, Y., Langford, B. *et al.* (2005a). Robust decision-making under severe uncertainty for conservation management. *Ecol. Appl.*, **15**, 1471-1477.

Regan, H.M., Colyvan, M. & Burgman, M.A. (2002). A taxonomy and treatment of uncertainty for ecology and conservation biology. *Ecol. Appl.*, **12**, 618-628.

Regan, T.J., Burgman, M.A., McCarthy, M.A. *et al.* (2005b). The consistency of extinction risk classification protocols. *Conserv. Biol.*, **19**, 1969-1977.

Ricker, W.E. (1954). Stock and recruitment. *J. Fish. Res. Board Can.*, **11**, 559-623.

Roughgarden, J. & Smith, F. (1996). Why fisheries collapse and what to do about it. *Proc. Natl. Acad. Sci.*, **93**, 5078-5083.

Scott, A. (1955). The fishery: the objectives of sole ownership. *J. Polit. Econ.*, **63**, 116-124.

Tallis, H., Lester, S.E., Ruckelshaus, M. *et al.* (2012). New metrics for managing and sustaining the ocean's bounty. *Mar. Policy*, **36**, 303-306.

Varian, H.R. (2009). *Intermediate microeconomics: a modern approach*. 8 edition. W. W. Norton & Company, New York.

Villa, F., Bagstad, K.J., Voigt, B. *et al.* (2014). A methodology for adaptable and robust ecosystem services assessment. *PLoS ONE*, **9**, e91001.

Von Neumann, J. & Morgenstern, O. (1944). *Theory of games and economic behavior*. Princeton University Press, Princeton, New Jersey.

White, C., Halpern, B.S. & Kappel, C.V. (2012). Ecosystem service tradeoff analysis reveals the value of marine spatial planning for multiple ocean uses. *Proc. Natl. Acad. Sci.*, **109**, 4696-4701.

Zavalloni, M., Groeneveld, R.A. & van Zwieten, P.A.M. (2014). The role of spatial information in the preservation of the shrimp nursery function of mangroves: a spatially explicit bio-economic model for the assessment of land use trade-offs. *J. Environ. Manage.*, **143**, 17-25.

22

Cultural Ecosystem Services in Protected Areas: Understanding Bundles, Trade-Offs, and Synergies

Judith M. Ament[1,2,3], Christine A. Moore[1,4], Marna Herbst[5], & Graeme S. Cumming[1,6]

[1] Percy FitzPatrick Institute, DST/NRF Centre of Excellence, University of Cape Town, Rondebosch, Cape Town 7701, South Africa
[2] Institute of Zoology, Zoological Society of London, Regent's Park, London NW1 4RY, UK
[3] Centre for Biodiversity and Environment Research, Department of Genetics, Evolution and Environment, University College London, London WC1E 6BT, UK
[4] School of Geography and the Environment, University of Oxford, South Parks Road, Oxford OX1 3PY, UK
[5] South African National Parks, Scientific Services, Private Bag X1021, Phalaborwa 1390, South Africa
[6] ARC Centre of Excellence for Coral Reef Studies, James Cook University, Townsville, QLD 4811, Australia

Keywords

Cultural ecosystem services; protected areas; South Africa; bundles, trade-offs, synergies; social preferences.

Correspondence

Judith M. Ament, Percy FitzPatrick Institute, DST/NRF Centre of Excellence, University of Cape Town, Rondebosch, Cape Town 7701, South Africa.
E-mail: judith.ament@ioz.ac.uk

Abstract

The concept of ecosystem services (ES) provides a potentially useful tool for decision-making in natural area management. Provisioning and regulating ES often occur in "bundles" that are cohesive because of coprovisioning or codependence. We asked whether individual preferences for cultural benefits also define service bundles. Data from a large survey of visitor preferences ($n = 3,131$ respondents) from all 19 South African National Parks indicated five bundles of cultural ecosystem services: (1) "natural history," (2) "recreation," (3) "sense of place," (4) "safari experiences," and (5) "outdoor lifestyle." Trade-offs and synergies between bundles of services depended on the ecosystem providing them and on alignment between demand for services and the supply of particular service bundles in specific ecosystems. Our results show that identifying demand for multiple services can both help us to understand why people visit and value protected areas, and better inform the management choices that influence service provision.

Introduction

Over the last 50–60 years, Conservation Biology has worked through several different framings of the relationships between people and nature. The observation that natural areas supply ecosystem goods and services (ES) to people has provided a useful link between ecosystems and human well-being (De Groot *et al.* 2002; MA 2005). The practicalities of quantifying and modeling this link are still, however, a work in progress (Carpenter *et al.* 2009; Seppelt *et al.* 2011; Reyers *et al.* 2013; Mace 2014). Explicitly connecting change in ecosystems to human well-being requires a comprehensive approach that considers both tangible and intangible benefits (Russell *et al.* 2013).

Many tangible ecosystem benefits are readily quantified through economic measures, such as the costs of water purification or the market values of food and fuel. Intangible benefits, or cultural ecosystem services (CES), are harder (but not impossible) to measure using approaches that recognize the difficulties of aggregating human values and deliberately maintain a plurality of perspectives and epistemologies (Chan *et al.* 2012b; Satz *et al.* 2013). Assessments of CES now cover topics such as recreation (e.g., Driver & Knopf 1977; Chan *et al.* 2006), culture and heritage (e.g., Tengberg *et al.* 2012; Nahuelhual *et al.* 2014), sense of place (e.g., Trentelman 2009; Ardoin *et al.* 2012), and mental health (e.g., Bratman *et al.* 2012; 2015), and promise to contribute to more resilient strategies for ecosystem management (Chan *et al.* 2012a). So far, however, incorporation of CES into decision-making—from landscape management to international policy—has been minimal in comparison to more tangible ES, such as food provision and climate regulation, despite continuous recognition of the value of CES and the instrumental role they play in securing public support for the protection of ecosystems (Daniel *et al.* 2012; Wolff *et al.* 2015). Satz *et al.* (2013) discuss a wide range of reasons for why CES are often ignored in decision-making processes, including problems such as the marginalization of rural communities whose decisions may be heavily influenced by cultural values, the difficulties (incommensurability) of comparing economic and cultural values, the interconnected nature of different benefits, failings in deliberative processes, and the perception that CES are "luxury goods" relative to more tangible benefits.

Research on ES has recognized that different services often occur together in "bundles" (Cumming & Peterson 2005). Service bundles have previously been described as cohesive because of either coprovisioning (one ecosystem provides several services) or codependence (one service requires another) (Bennett *et al.* 2009). For example, tangible ES, including some CES (tourism, deer hunting, nature appreciation, summer cottages, and forest recreation), have previously been quantified and compared directly from maps to explore the concepts of clustering in ES and ES bundles (Raudsepp-Hearne *et al.* 2010). There is, however, a third way of describing service bundles, which has not been considered in great depth: that is, based on the preferences of stakeholders (Martín-López *et al.* 2012; Klain *et al.* 2014). Bundles defined by user preference are particularly relevant in the context of cultural services, where understandings of human perceptions of the environment can benefit support for and resilience of environmental policy and strategies (Martín-López *et al.* 2007; Asah *et al.* 2014) and can improve CES indicator quality (Hernández-Morcillo *et al.* 2013).

Analysis of ES bundles is important for making decisions about trade-offs between multiple services more effective and financially defensible (Nelson *et al.* 2009), but has focused primarily on the supply of provisioning and regulating services (e.g., Maes *et al.* 2012; Qiu & Turner 2013), or recreational aspects of cultural services (Raudsepp-Hearne *et al.* 2010; Turner *et al.* 2014; Queiroz *et al.* 2015). The idea that potentially antagonistic bundles of human preferences may exist has been explored in the literature on tourist travel motivations (e.g., Bieger & Laesser 2002; Dolnicar & Grun 2007), with several examples for natural areas (Uysal *et al.* 1994; Tao *et al.* 2004; van der Merwe & Saayman 2008), but has only recently been considered under the ES framework (Burkhard *et al.* 2012; Wolff *et al.* 2015).

In protected areas where profit generation influences conservation success and depends on tourist numbers (Mayer *et al.* 2010; Clements *et al.* 2016), understanding

human preferences is especially relevant. If cultural service bundles exist, then park managers may have to choose between providing a balance of service bundles or favoring a particular bundle (Rodriguez *et al.* 2006). In either case, the starting point for recognizing trade-offs and choosing strategies is to describe the relevant bundles of CES.

We used a large data set of tourist interviews from South African national parks to test (1) whether tourist demands for CES fall into distinct categories and hence, whether CES can be captured in distinct "bundles" with different recipient groups; and (2) whether trade-offs and synergies between these bundles emerge within ecosystems with different characteristics.

Methods

Data collection

Data were collected between February 2013 and May 2015 by means of tourist questionnaires distributed to all gates and reception desks of the 19 South African national parks (SANParks) (Figure 1). In addition, research assistants visited all parks on sampling trips to encourage visitors to complete the self-explanatory questionnaires at campsites, restaurants, and picnic spots. The sample population was limited to adults of 16 years or older. The survey comprised a brief explanation about the research, questions on demographic details of the respondents, and 30 Likert-type questions asking visitors to rate their appreciation of different aspects of protected areas on a five-point scale (1 = strongly disagree, 2 = disagree, 3 = neutral, 4 = agree, and 5 = strongly agree). These questions were indicators of six subcategories of cultural ES (MA 2005): aesthetic ($n = 7$), cultural and heritage ($n = 2$), education ($n = 2$), recreation ($n = 11$), social ($n = 4$), and spiritual and religious ($n = 4$) (Table 1). Other subcategories were excluded from the study, as they were found too difficult to translate into indicator questions.

Data analysis

After capture, all years of data were pooled. The total sample included responses from 4,093 individuals, from which 3,131 complete responses were retained for analysis. Analysis consisted of two stages. We first investigated demand for CES in protected areas by subcategory through assessment of the services most appreciated (i.e., ticked "strongly agree") by visitors to the national parks.

Second, to understand the patterns of service demand by individual visitor and park, we assessed correlations between appreciation of services in an exploratory

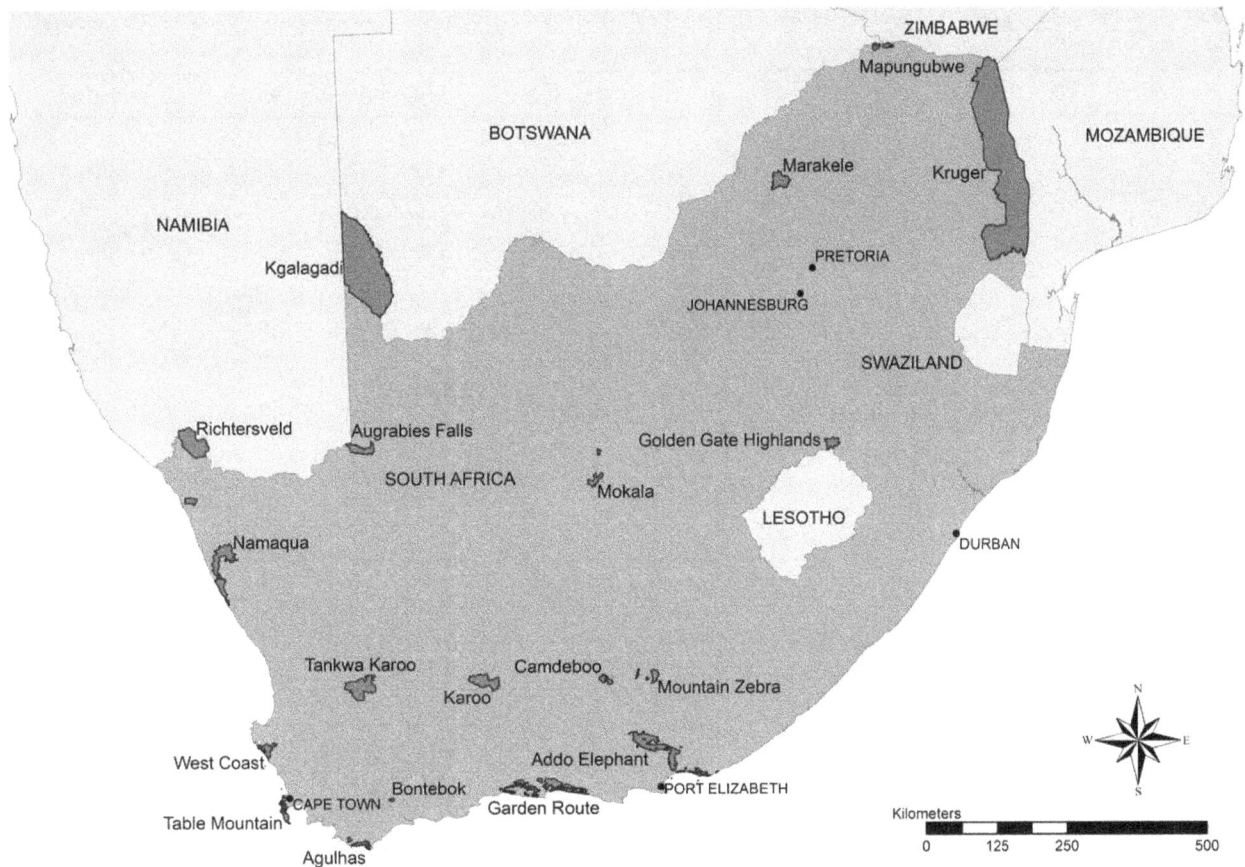

Figure 1 Map of South African national parks from which data originated: Addo Elephant NP ($n = 171$), Agulhas NP ($n = 120$), Augrabies Falls NP ($n = 115$), Bontebok NP ($n = 130$), Camdeboo NP ($n = 118$), Garden Route NP ($n = 276$), Golden Gate Highlands NP ($n = 145$), Kruger NP (North: $n = 141$; South: $n = 325$; private: $n = 53$; and unspecified: $n = 75$), Karoo NP ($n = 112$), Kgalagadi NP ($n = 177$), Mapungubwe NP ($n = 104$), Marakele NP ($n = 139$), Mokala NP ($n = 94$), Mountain Zebra NP ($n = 152$), Namaqua NP ($n = 37$), Richtersveld NP ($n = 131$), Table Mountain NP (Boulders Beach: $n = 76$; Cable Way: $n = 88$; Cape Point: $n = 91$; and Silvermine: $n = 145$), Tankwa Karoo NP ($n = 57$), and West Coast NP ($n = 59$).

factor analysis (R "Stats" package, *factanal* function, and varimax rotation), interpreting survey responses (i.e., 1–5) as numerical interval data. Factor analysis is a powerful statistical procedure capable of uncovering the structure in service demand (i.e., bundling survey questions that vary together) without enforcing a priori ideas about the clustering of visitor preferences. We used visual (scree plot) and analytical (parallel analysis) methods to determine the number of factors to extract (seven), and after inspection of meaningfulness, the first five were retained for interpretation. We calculated Spearman's rank correlation coefficients of factor loadings to assess trade-offs and synergies between factors, and Bartlett factor scores to investigate patterns in tourist preferences in different parks. Demographic predictors of factor scores were assessed with two-way ANOVA's followed by post-hoc Tukey tests (R "multcomp" package). All analyses were performed in R version 3.2.2 (R Core Team 2014).

Results

Demand for individual cultural ecosystem services

Demographic characteristics of respondents were well balanced (see Supplementary Information, Figures S1–S4 and Table S1). Respondents were generally very positive in their responses to survey statements, recording high mean and median responses for nearly all services (Table 1). "Relaxation" received highest mean response ($\mu = 4.59$), closely followed by "refreshing the spirit" ($\mu = 4.52$). The statement gauging the importance of "feeling closer to God" in protected areas generated a strongly bimodal response distribution ($\mu = 3.45$, $\sigma = 1.47$). Service demand was further spread over all subcategories of CES (Figure 2). The top quintile of most highly demanded services ($n = 6$) comprised four different subcategories (spiritual

Table 1 Mean and median responses of protected area visitors to preference statements on cultural ecosystem services. Responses were recorded on a five-point Likert scale

Survey statement	Indicator variable	Mean response ($\pm$ SD)	Median response
Aesthetic			
Looking at big mammals	Big mammals	4.43 ($\pm$ 0.72)	5—Strongly agree
Sitting, enjoying the view	View	4.38 ($\pm$ 0.70)	4—Agree
Looking at birds	Birds	4.11 ($\pm$ 0.94)	4—Agree
Looking at flowers	Flowers	3.72 ($\pm$ 1.02)	4—Agree
Looking at reptiles	Reptiles	3.52 ($\pm$ 1.09)	4—Agree
Trying to identify plants	Plants	3.38 ($\pm$ 1.10)	3—Neutral
Looking for and or listening to frogs	Frogs	3.06 ($\pm$ 1.13)	3—Neutral
Cultural and heritage			
The experience reminds me of my childhood	Childhood	3.48 ($\pm$ 1.21)	4—Agree
It helps me to understand my culture and or history	Culture and history	3.31 ($\pm$ 1.11)	3—Neutral
Educational			
Learning more about nature	Learning	4.25 ($\pm$ 0.75)	4—Agree
Doing guided tours	Guided tours	3.18 ($\pm$ 1.16)	3—Neutral
Recreational			
I enjoy camping	Camping	3.85 ($\pm$ 1.26)	4—Agree
Doing game drives	Game drives	4.11 ($\pm$ 0.99)	4—Agree
Cooking or braaing	Cooking	3.89 ($\pm$ 1.05)	4—Agree
Taking photographs	Photography	4.39 ($\pm$ 0.82)	5—Strongly agree
Hiking or climbing	Hiking/climbing	3.72 ($\pm$ 1.09)	4—Agree
Reading and/or writing	Reading/writing	3.52 ($\pm$ 1.10)	4—Agree
Driving off road, four wheel driving, or dirt biking	Off-roading	3.28 ($\pm$ 1.32)	3—Neutral
Swimming, surfing, or doing other watersports	Swimming	3.09 ($\pm$ 1.26)	3—Neutral
Boating or canoeing	Boating	2.98 ($\pm$ 1.24)	3—Neutral
Sunbathing	Sunbathing	2.51 ($\pm$ 1.24)	2—Disagree
Fishing	Fishing	2.33 ($\pm$ 1.26)	2—Disagree
Social			
It is a way to spend time with my family friends	Family/friends	4.24 ($\pm$ 0.90)	4—Agree
Hanging out at the campsite or chalet	Hanging out	3.40 ($\pm$ 1.19)	4—Agree
Talking to other visitors	Talking	3.31 ($\pm$ 1.00)	3—Neutral
Having a party with my friends	Partying	2.59 ($\pm$ 1.30)	3—Neutral
Spiritual and religious			
It refreshes my spirit	Spirit	4.52 ($\pm$ 0.71)	5—Strongly agree
It makes me feel closer to God	God	3.45 ($\pm$ 1.47)	4—Agree
It helps me to relax	Relaxation	4.59 ($\pm$ 0.67)	5—Strongly agree
I like to get away from modern conveniences	Away	4.09 ($\pm$ 0.97)	4—Agree

and religious, recreational, aesthetic, and social) (MA 2005).

Bundles of cultural ecosystem services

Exploratory factor analysis identified five bundles of CES that cumulatively explained 35.3% of variance in survey responses, with low cross-loadings between bundles. All but four CES loaded strongly (factor loading>0.35) and uniquely or semiuniquely on one bundle ("learning" loaded strongly on two bundles) (Table 2).

The first CES bundle contained biodiversity and natural history-type services ("learning" and all aesthetic values except "big mammals" and "enjoying the view")

and explained 9.3% of total variation in survey responses. One social service ("partying") was strongly negatively correlated to this bundle. Women and older people scored significantly higher on this bundle than men and younger people (Figure S5a). The second bundle also explained 9.3% of variation and identified strong synergies between most recreational activities ("boating," "swimming," "sunbathing," "partying," "hiking," "fishing," and "off-road driving"). This bundle showed evidence of a trade-off with "bird-watching" and was strongly associated with women, younger people, and people visiting from North America (Figure S5b). The third bundle explained 7.4% of variation in responses and revealed synergies between a combination of services

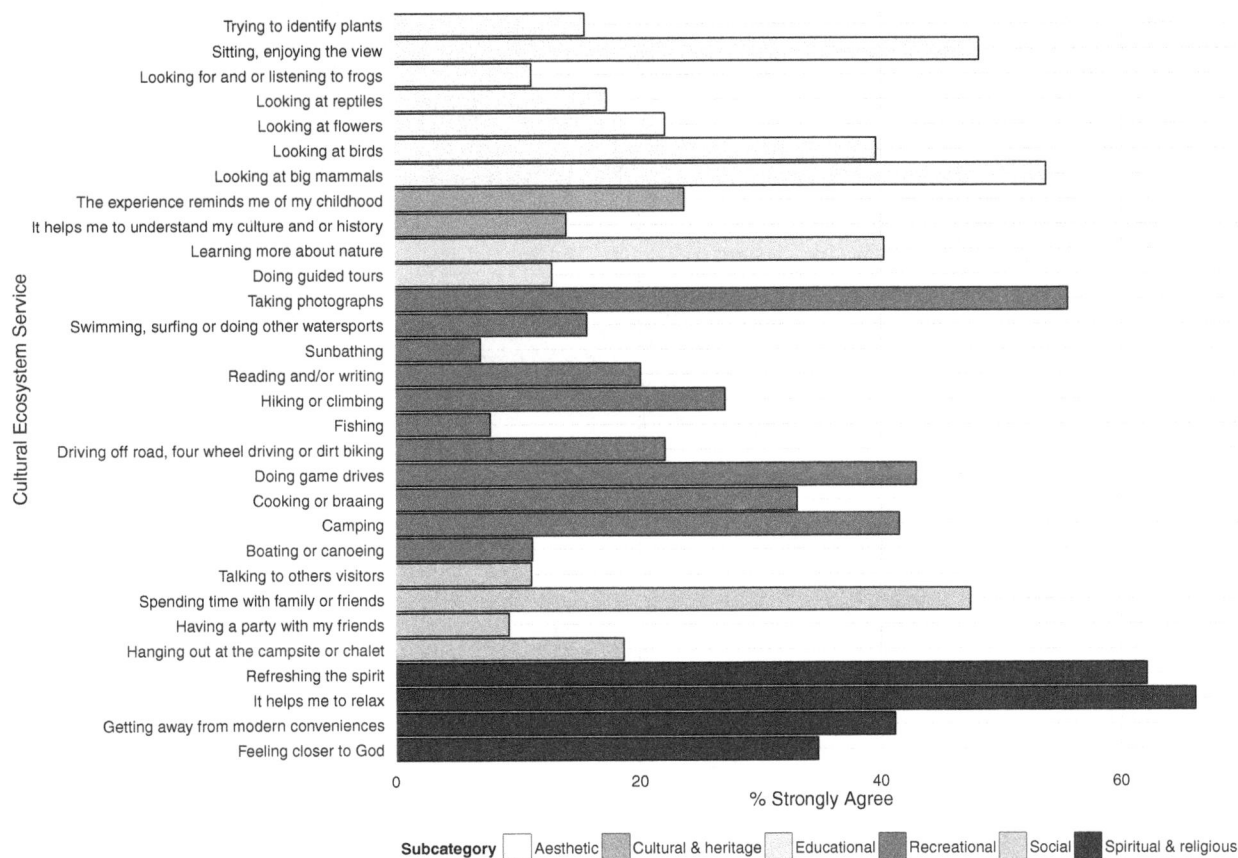

Figure 2 Most highly demanded cultural ecosystem services by subcategory, as defined in the Millennium Ecosystem Assessment (MA 2005). Service demand is distributed over all subcategories, with five of six categories receiving high rates of service demand (>40% strongly agree).

that concern visitors' emotional connections to these ecosystems, which, in conjunction, might be understood as sense of place. This bundle highlights several clusters of intangible benefits, such as psychological ("refreshing the spirit," "relaxation," and "feeling closer to God"), socio-cultural ("spending time with family and friends" and "reliving childhood memories"), and experiential ("camping" and "getting away from modern conveniences"). This bundle did not involve strong trade-offs and was associated with women, middle-aged people, South Africans, and Africans in general (Figure S5c). Fourth was a bundle explaining just 5.5% of variation in survey responses, but with a very clear and unique interpretation: enjoying a "safari experience." Indicator variables loading strongly onto this bundle were "viewing big mammals," "doing game drives," "taking photographs," and "learning." "Feeling closer to God" was negatively correlated with this bundle. This bundle was associated with women, younger people, Europeans and North Americans, and visitors during school holidays (Figure S5d). The last bundle of CES included strong re-

lationships between sedentary, low-key activities ("cooking," "reading/writing," and "hanging out at the campsite or chalet"). This bundle explained 3.8% of variation and reflected the value of protected areas in providing a space for enjoying an "outdoor lifestyle." Men, people with less formal education, and South Africans scored higher on this bundle (Figure S5e).

Trade-offs and synergies between bundles of services

Correlations of factor loadings between bundles were mostly negative, with the strongest trade-off existing between natural history and recreation ($\rho = -0.583$, $P<0.001$) (Table 3). Other trade-offs with recreation existed with safari experiences ($\rho = -0.507$, $P<0.01$) and "sense of place" ($\rho = -0.410$, $P<0.05$). Natural history was further positively correlated with safari experiences ($\rho = 0.371$, $P<0.05$), but negatively with an outdoor lifestyle ($\rho = -0.478$, $P<0.01$).

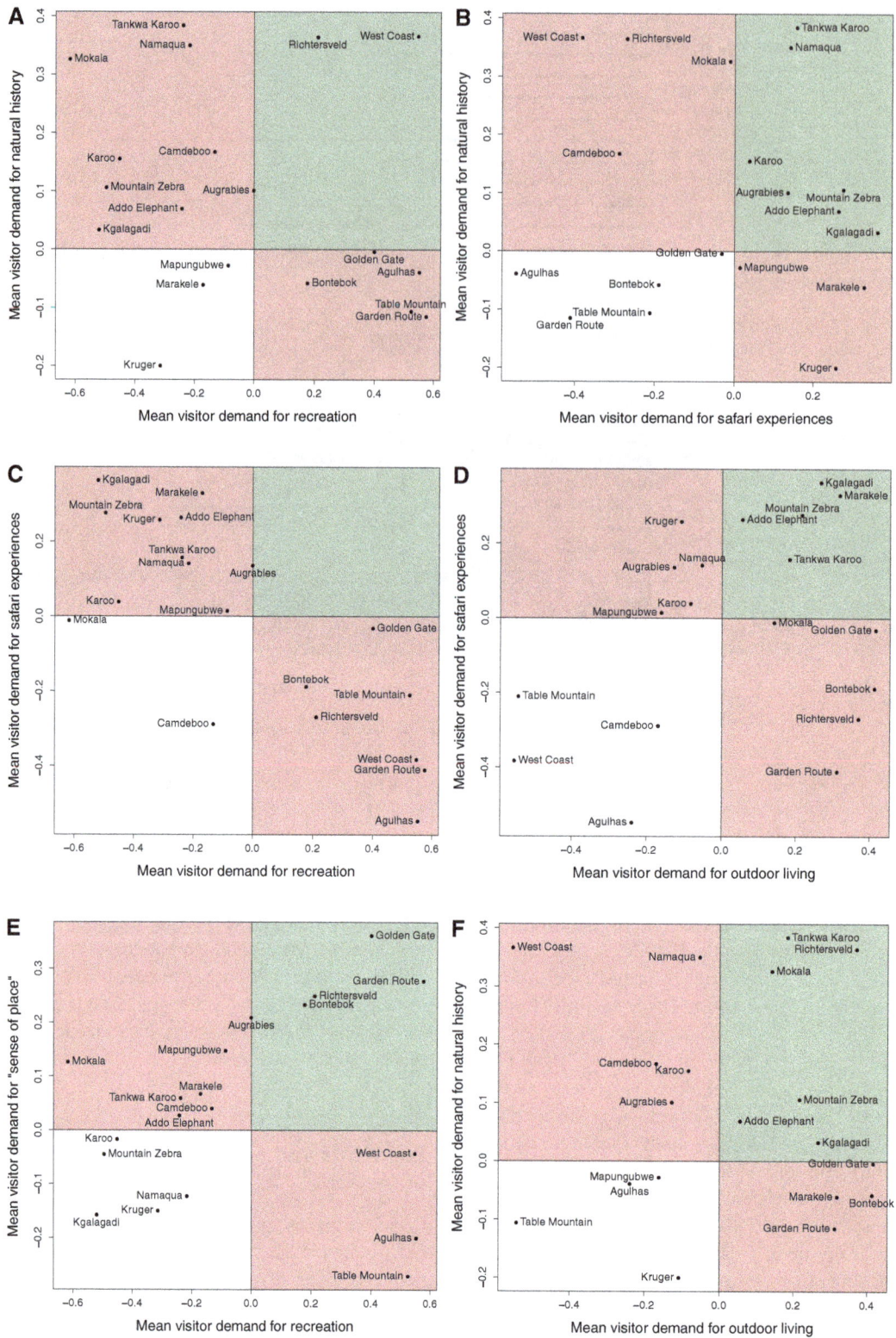

Figure 3 Trade-offs and synergies between bundles of cultural ecosystem services in South African national parks. Mean park scores on service bundles are calculated from pattern matrix of tourist preferences. Parks in green shading exhibit synergies between particular bundles; parks in the red tradeoffs.

Table 2 Loadings of individual cultural ecosystem services on bundles (factors) of ecosystem services

Indicator variable	Bundle 1: natural history	Bundle 2: recreation	Bundle 3: sense of place	Bundle 4: safari experience	Bundle 5: outdoor living
Flowers	**0.760**	0.089	0.084	0.141	0.061
Plants	**0.750**	−0.030	0.108	0.116	0.089
Birds	**0.632**	−0.149	0.124	0.258	0.041
Frogs	**0.574**	0.139	0.092	0.153	0.011
Learning	**0.474**	−0.080	0.223	**0.415**	−0.044
Reptiles	**0.354**	0.122	0.062	0.263	−0.063
Boating	0.116	**0.781**	0.083	0.049	0.024
Swimming	−0.019	**0.738**	0.124	0.064	0.058
Sunbathing	−0.072	**0.600**	0.040	0.030	0.119
Partying	−0.166	**0.508**	0.059	0.031	0.192
Hiking/climbing	0.217	**0.446**	0.258	0.099	−0.005
Fishing	0.149	**0.453**	0.078	0.099	0.160
Off-roading	−0.029	**0.368**	0.063	0.202	0.180
Spirit	0.188	−0.051	**0.661**	0.180	−0.030
Relaxation	0.062	0.012	**0.562**	0.225	0.057
Away	0.136	0.069	**0.462**	0.111	0.068
Family/friends	−0.056	0.185	**0.460**	0.087	0.182
God	0.086	−0.051	**0.447**	0.112	0.217
Camping	0.091	0.157	**0.434**	0.018	0.288
Childhood	0.167	0.188	**0.388**	0.010	0.136
Big mammals	0.160	0.009	0.140	**0.619**	0.039
Game drives	0.158	0.066	0.057	**0.509**	0.116
Photography	0.173	0.073	0.097	**0.409**	0.046
Cooking	0.056	0.165	0.287	0.141	**0.639**
Hanging out	−0.050	0.267	0.222	0.103	**0.441**
Reading/writing	0.269	0.126	0.167	0.081	**0.354**
View	0.244	0.140	0.260	0.333	0.209
Guided tours	0.082	0.303	0.059	0.322	−0.026
Culture and history	0.320	0.127	0.221	0.125	0.046
Talking	0.132	0.195	0.098	0.167	0.153

Note: Gray shading indicates strong loading (> 0.35) of ecosystem service on particular factor (i.e., service bundle).

Table 3 Relationships between bundles of ecosystem services (factors) expressed by Spearman's rank correlation coefficients of factor loadings

	Bundle 1: natural history	Bundle 2: recreation	Bundle 3: sense of place	Bundle 4: safari experience	Bundle 5: outdoor living
Natural history	1.000				
Recreation	−0.583***	1.000			
Sense of place	0.035	−0.410*	1.000		
Safari experience	0.371*	−0.507**	−0.119	1.000	
Outdoors lifestyle	−0.478**	0.264	0.158	−0.478**	1.000

$n = 30$; * $p < 0.05$; ** $p < 0.01$; *** $p < 0.001$.

Analysis of mean factor scores in individual parks revealed distinct differences of CES demands between parks. Most interestingly, three significant trade-offs between bundles of CES uncovered in the correlation matrix were found to materialize in preference patterns in South African national parks. These were between natural history and recreation (Figure 3A), between natural history and safari (Figure 3B), and between safari and recreation (Figure 3C). These trade-offs are discussed below. Other significant trade-offs did not materialize in specific parks (Figure 3D–F).

Discussion

We found that the most-valued cultural services of protected areas were spread over all CES subcategories (MA

2005). Despite the lack of agreement on what constitutes "sense of place" (Trentelman 2009; Ardoin *et al.* 2012), even this category seemed to emerge as a combination of socio-cultural, psychological, and experiential aspects recognized as dimensions of sense of place (Ardoin 2006; Lewicka 2011), illustrating the diversity and complexity of the valuation of this and other CES (Hernández-Morcillo *et al.* 2013; Hausmann *et al.* 2015).

Factor analysis identified a total of five bundles of CES in South African national parks: (1) "natural history," (2) "recreation," (3) "sense of place," (4) "safari experiences," and (5) "outdoor lifestyle," confirming that visitors to protected areas have distinct travel motivations. These bundles were largely in agreement with bundles of visitor motivations previously identified for a small subset of our park sample (Kruger & Saayman 2010) and they complement the investigation of more tangible ES for South Africa (Egoh *et al.* 2008).

Three trade-offs between CES bundles were found to materialize in our sample of South African national parks. First, most parks showed trade-offs between natural history and recreation (Figure 3A). Parks in which this trade-off was strong (Tankwa-Karoo, Namaqua, and Mokala) had negative scores for recreation and positive scores for natural history. Tankwa-Karoo and Namaqua are located in the Succulent Karoo biome, a fragile dryland Conservation International biodiversity hotspot (CEPF 2001), which could easily be damaged if high-impact recreational activities such as off-road driving or mountain biking were permitted. In the Richtersveld and the West Coast, however, these bundles were in synergy. The Richtersveld Cultural and Botanical Landscape is a UNESCO world heritage site, where high demands for natural history uncovered in our study may reflect the rich biodiversity of this region, while its longstanding tradition of communal land comanagement (UNESCO 2009) and varied activities—fishing, water sports, and four-wheel driving—underscores the identity of the park as a cultural landscape. Similar synergies were present in West Coast National Park, which is in the Cape Floristic Region, the smallest of six recognized floral kingdoms of the world and an area of extraordinarily high diversity and endemism. This park also offers a variety of recreational activities—swimming, hiking, and mountain biking, for instance—again illustrating the potential for synergies between recreation and biodiversity when managed accordingly. Interestingly, Table Mountain National Park (TMNP) as a whole did not demonstrate this synergy, despite the park's location within the Cape Floristic Region and its appointment as a World Heritage Site within this region. TMNP comprises four different managerial sections, however, which all have very different identities (Table Mountain and Cape Point are mainly hiking and scenic points, Silvermine is a recreational area popular for swimming and cycling, while Boulders Beach is home to a colony of endangered African Penguins). These section-specific characteristics may prevent the park on a whole from generating a clear demand profile (see Figure S6 for trade-offs by park section).

Second, most parks that generated high demands for safari experiences disclosed synergies with natural history, recognizing wildlife safaris as a means to observe biodiversity. Interestingly, this synergy was not found in Kruger NP, where demands for natural history were low (Figure 3B). Like Table Mountain, Kruger comprises two distinct managerial sections: the South, managed more commercially; and the North, set aside for wilderness experiences. The two areas have different biophysical characteristics that may prevent the demand for natural history from materializing in the park as a whole (see Figure S6 for trade-offs by section).

Third, trade-offs between CES demands for safari experiences and recreation were ubiquitous among all parks (Figure 3C). All of the parks that scored high on demands for safari experiences (Kgalagadi, Marakele, Mountain Zebra, Addo-Elephant, and Kruger) were parks that contained some or all of the big five (African lion, African leopard, African elephant, Black and/or White rhinoceros, and Cape buffalo). Parks of this type necessarily cannot offer a wide range of recreational activities, and invariably, these parks scored low on demands for them. Two other parks containing some of the big five (Karoo and Mapungubwe) generated low demand for safari experiences, signaling opportunities for better marketing of these services in these parks. In parks that scored high on demands for recreation (West Coast, Agulhas, Table Mountain, Garden Route, and Golden Gate Highlands), opportunities for activities are abundant due to the absence of large and/or dangerous wildlife, which simultaneously reduces the realistic availability of safari-type services and explains the low demand for those services in these areas. Safari experiences reflect the large mammal fauna of Africa, and seeing these animals in their natural habitat has parallels to other region-specific ("bucket list") services, such as seeing Komodo Dragons or scuba diving on the Great Barrier Reef.

Finally, four parks in our sample had intermediate scores on all bundles (Karoo, Camdeboo, Mapungubwe, and Augrabies), indicating that either they have distinct CES demand bundles that were not addressed by our questionnaire, or that they present opportunities for more adequate profiling and marketing.

Through a detailed multipark, multiservice approach, we were able to show that trade-offs and synergies between bundles of CES do not only arise as a function of heterogeneous landscapes (Burkhard *et al.* 2012; Turner *et al.* 2014; Queiroz *et al.* 2015), but can be understood purely from differences in social preferences for these CES, indicating that people visiting different protected areas may seek different bundles of services. This information, in combination with the general alignment of visitor CES demands with the availability of service bundles in specific parks, has strong practical implications for protected area management, particularly where protected area viability depends on economic returns from tourism. Management actions seeking to amplify a particular kind of CES, or improve access to that CES, are more likely to achieve their goals if they align with the specific properties of local ecosystems and locally specific tourist demand. Parks with natural availability of, and thus high demands for, natural history-type CES could increase their economic viability through greater investment in educational and viewing resources, such as species lists, bird hides, and vegetation maps; while parks with greater capabilities to deliver recreational-type CES may invest in activities on offer (e.g., horseback riding and mountain bike tours) or equipment hire (e.g., fishing equipment and bicycles). When managing ecosystems purely from an ecological perspective, managers may unknowingly make choices that lead to counter-productive trade-offs for visitors (e.g., reintroducing big five species into parks that visitors appreciate for their recreational opportunities). In addition to realizing that trade-offs exist in the CES protected areas, can provide protected area managers in South Africa should be aware of the currently skewed visitor demographics (notably, ethnic composition of visitors does not match that of the nation as a whole, Supplementary Information, S1–S5) and aim to broaden the appeal of national parks for people from ethnic and socioeconomic backgrounds who are currently underrepresented in the visitor base.

Finally, our results provide a mechanism that explains why successful revenue-generation approaches in one protected area do not necessarily translate well to other areas that may have different identities and from which visitors may seek different kinds of CES. Business models for revenue generation in South African national parks must clearly be tailored to fit individual locations and customer bases.

Acknowledgments

The authors thank current and previous laboratory members for useful discussions about this work, and SANParks for permission to undertake surveys in national parks. The study was supported by a CPRR grant from the Natural Research Foundation of South Africa, a James S. McDonnell Foundation complexity scholar award to G.S.C., and the DST/NRF Centre of Excellence at the Percy FitzPatrick Institute.

Supporting Information

Figure S1: Age distribution of survey population per park.

Figure S2: Origin of survey population per park. RSA, Republic of South Africa; RoA, Rest of Africa; Eur, Europe; NA, North America; SA, South America; Asia, Asia; and Oc, Oceania.

Figure S3: Gender distribution of survey population per park.

Figure S4: Education level of survey population per park. HS, high school; UG, undergraduate degree; PG, postgraduate degree; and other, other.

Figure S5: Effect of gender, formal education, age, continent of origin (n.b. "RoA," Rest of Africa; "Oc," Oceania; "Eur," Europe; "RSA," South Africa; and "N. Am," North America), and season of visit (n.b. "Peak," school holidays) on factor scores for natural history (a), recreation (b), sense of place (c), safari experiences, and outdoor lifestyle (e).

Figure S6: Trade-offs and synergies between bundles of cultural ecosystem services in the four managerial sections of Kruger National Park (KNP) and Table Mountain National Park (TMNP). Absence of clear signals of visitor demand in these parks on a whole may be explained by the great disparities between visitor types in different subsections.

Table S1: Self-reported race group of respondents.

References

Ardoin, N.M. (2006). Toward an interdisciplinary understanding of place: lessons for environmental education. *Can. J. Environ. Educ.*, **11**, 112-126.

Ardoin, N.M., Schuh, J.S. & Gould, R.K. (2012). Exploring the dimensions of place: a confirmatory factor analysis of data from three ecoregional sites. *Environ. Educ. Res.*, **18**, 583-607.

Asah, S.T., Guerry, A.D., Blahna, D.J. & Lawler, J.J. (2014). Perception, acquisition and use of ecosystem services—human behavior, and ecosystem management and policy implications. *Ecosyst. Serv.*, **10**, 180-186.

Bennett, E.M., Peterson, G.D. & Gordon, L.J. (2009). Understanding relationships among multiple ecosystem services. *Ecol. Lett.*, **12**, 1394-1404.

Bieger, T. & Laesser, C. (2002). Market segmentation by motivation: the case of Switzerland. *J. Travel Res.*, **41**, 68-76.

Bratman, G.N., Hamilton, J.P. & Daily, G.C. (2012). The impacts of nature experience on human cognitive function and mental health. *Ann. N. Y. Acad. Sci.*, **1249**, 118-136.

Bratman, G.N., Hamilton, J.P., Hahn, K.S., Daily, G.C. & Gross, J.J. (2015). Nature experience reduces rumination and subgenual prefrontal cortex activation. *Science*, **112**, 8567-8572.

Burkhard, B., Kroll, F., Nedkov, S. & Müller, F. (2012). Mapping ecosystem service supply, demand and budgets. *Ecol. Indic.*, **21**, 17-29.

Carpenter, S.R., Mooney, H.A., Agard, J. *et al.* (2009). Science for managing ecosystem services: beyond the millennium ecosystem assessment. *Science*, **106**, 1305-1312.

Critical Ecosystem Partnership Fund (CEPF). (2001). Ecosystem profile: the cape floristic region. South Africa. Available from: http://www.cepf.net/Documents/final.capefloristicregion.ep.pdf. Accessed 20 April 2016.

Chan, K.M.A., Guerry, A.D., Balvanera, P. *et al.* (2012a). Where are cultural and social in ecosystem services? A framework for constructive engagement. *BioScience*, **62**, 744-756.

Chan, K.M.A., Satterfield, T. & Goldstein, J. (2012b). Rethinking ecosystem services to better address and navigate cultural values. *Ecol. Econ.*, **74**, 8-18.

Chan, K.M.A., Shaw, M.R., Cameron, D.R., Underwood, E.C. & Daily, G.C. (2006). Conservation planning for ecosystem services. *PLoS Biol.*, **4**, 2138-2152.

Clements, H.S., Baum, J. & Cumming, G.S. (2016). Money and motives: an organizational ecology perspective on private land conservation. *Biol. Conserv.*, **197**, 108-115.

Cumming, G. & Peterson, G.D. (2005). Ecology in global scenarios. Pages 45–70 in S.R. Carpenter, P.L. Pingali, E.M. Bennett, M.B. Zurek, editors. *Ecosystems and human well-being*. Ecosystems and Human Well-Being, Washington, D.C.

Daniel, T.C., Muhar, A., Arnberger, A. *et al.* (2012). Contributions of cultural services to the ecosystem services agenda. *Science*, **109**, 8812-8819.

De Groot, R.S., Wilson, M.A. & Boumans, R. (2002). A typology for the classification, description and valuation of ecosystem functions, goods and services. *Ecol. Econ*, **41**, 393-408.

Dolnicar, S. & Grun, B. (2007). Cross-cultural differences in survey response patterns. *Int. Market. Rev.*, **24**, 127-143.

Driver, B.L. & Knopf, R.C. (1977). Personality, outdoor recreation, and expected consequences. *Environ. Behav.*, **9**, 169-193.

Egoh, B., Reyers, B., Rouget, M., Richardson, D.M., Le Maitre, D.C. & van Jaarsveld, A.S. (2008). Mapping ecosystem services for planning and management. *Agric. Ecosyst. Environ.*, **127**, 135-140.

Hausmann, A., Slotow, R., Burns, J.K. & Di Minin, E. (2015). The ecosystem service of sense of place: benefits for human well-being and biodiversity conservation. *Environ. Conserv.*, **43**, 117-127.

Hernández-Morcillo, M., Plieninger, T. & Bieling, C. (2013). An empirical review of cultural ecosystem service indicators. *Ecol. Indic.*, **29**, 434-444.

Klain, S.C., Satterfield, T.A. & Chan, K.M.A. (2014). What matters and why? Ecosystem services and their bundled qualities. *Ecol. Econ.*, **107**, 310-320.

Kruger, M. & Saayman, M. (2010). Travel motivation of tourists to Kruger and Tsitsikamma National Parks: a comparative study. *S. Afr. J. Wildl. Res.*, **40**, 93-102.

Lewicka, M. (2011). Place attachment: how far have we come in the last 40 years? *J. Environ. Psychol.*, **31**, 207-230.

MA. (2005). *Ecosystems and human well-being: synthesis*. Island Press, Washington, D.C.

Mace, G.M. (2014). Whose conservation? *Science*, **345**, 1558-1560.

Maes, J., Paracchini, M.L., Zulian, G., Dunbar, M.B. & Alkemade, R. (2012). Synergies and trade-offs between ecosystem service supply, biodiversity, and habitat conservation status in Europe. *Biol. Conserv.*, **155**, 1-12.

Martín-López, B., Iniesta-Arandia, I., García-Llorente, M. *et al.* (2012). Uncovering ecosystem service bundles through social preferences. *PLoS One*, **7**, e38970.

Martín-López, B., Montes, C. & Benayas, J. (2007). The non-economic motives behind the willingness to pay for biodiversity conservation. *Biol. Conserv.*, **139**, 67-82.

Mayer, M., Müller, M., Woltering, M., Arnegger, J. & Job, H. (2010). The economic impact of tourism in six German national parks. *Landsc. Urban Plan.*, **97**, 73-82.

Nahuelhual, L., Carmona, A., Laterra, P. & Barrena, J. (2014). A mapping approach to assess intangible cultural ecosystem services: the case of agriculture heritage in Southern Chile. *Ecol. Indic.*, **40**, 90-101.

Nelson, E., Mendoza, G., Regetz, J. *et al.* (2009). Modeling multiple ecosystem services, biodiversity conservation, commodity production, and tradeoffs at landscape scales. *Front. Ecol. Environ.*, **7**, 4-11.

Qiu, J. & Turner, M.G. (2013). Spatial interactions among ecosystem services in an urbanizing agricultural watershed. *Science*, **110**, 12149-12154.

Queiroz, C., Meacham, M., Richter, K. *et al.* (2015). Mapping bundles of ecosystem services reveals distinct types of multifunctionality within a Swedish landscape. *AMBIO*, **44**, 89-101.

R Core Team. (2014). R: A language and environment for statistical computing. R Foundation for Statistical Computing, Vienna, Austria. https://www.R-project.org/.

Raudsepp-Hearne, C., Peterson, G.D. & Bennett, E.M. (2010). Ecosystem service bundles for analyzing tradeoffs in diverse landscapes. *Science*, **107**, 5242-5247.

Reyers, B., Biggs, R., Cumming, G.S., Elmqvist, T., Hejnowicz, A.P. & Polasky, S. (2013). Getting the measure of ecosystem services: a social–ecological approach. *Front. Ecol. Environ.*, **11**, 268-273.

Rodriguez, J.P., Beard, Jr., T.D., Bennett, E.M. *et al.* (2006). Trade-offs across space, time, and ecosystem services. *Ecol. Soc.*, **11**, 28.

Russell, R., Guerry, A.D., Balvanera, P. *et al.* (2013). Humans and nature: how knowing and experiencing nature affect well-being. *Ann. Rev. Environ. Res.*, **38**, 473-502.

Satz, D., Gould, R.K., Chan, K.M.A. *et al.* (2013). The challenges of incorporating cultural ecosystem services into environmental assessment. *AMBIO*, **42**, 675-684.

Seppelt, R., Dormann, C.F., Eppink, F.V., Lautenbach, S. & Schmidt, S. (2011). A quantitative review of ecosystem service studies: approaches, shortcomings and the road ahead. *J. Appl. Ecol.*, **48**, 630-636.

Tao, C.-H.T., Eagles, P.F.J. & Smith, S.L.J. (2004). Profiling Taiwanese ecotourists using a self-definition approach. *J. Sustain. Tour.*, **12**, 149-168.

Tengberg, A., Fredholm, S., Eliasson, I., Knez, I., Saltzman, K. & Wetterberg, O. (2012). Cultural ecosystem services provided by landscapes assessment of heritage values and identity. *Ecosyst. Serv.*, **2**, 14-26.

Trentelman, C.K. (2009). Place attachment and community attachment: a primer grounded in the lived experience of a community sociologist. *Soc. Nat. Res.*, **22**, 191-210

Turner, K.G., Odgaard, M.V., Bøcher, P.K., Dalgaard, T. & Svenning, J.-C. (2014). Bundling ecosystem services in Denmark: trade-offs and synergies in a cultural landscape. *Landsc. Urban Plan.*, **125**, 89-104.

UNESCO. (2009). Report of decisions of the 33rd session of the World Heritage Committee. UNESCO, Seville, Spain. Available from: http://whc.unesco.org/archive/2009/whc09-33com-20e.pdf. Accessed 20 April 2016.

Uysal, M., McDonald, C.D. & Martin, B.S. (1994). Australian visitors to US National Parks and natural areas. *Int. J. Contemp. Hosp. Manage.*, **6**, 18-24.

van der Merwe, P. & Saayman, M. (2008). Travel motivations of tourists visiting Kruger National Park. *Koedoe*, **50**, 154-159.

Wolff, S., Schulp, C. & Verburg, P.H. (2015). Mapping ecosystem services demand: a review of current research and future perspectives. *Ecol. Indic.*, **55**, 159-171.

PERMISSIONS

LIST OF CONTRIBUTORS

Robin Abell
Global Water Program, The Nature Conservancy, Arlington, VA 22203, USA

Bernhard Lehner
Department of Geography, McGill University, Montreal, QC H3A 0B9, Canada

Michele Thieme
Freshwater Program, World Wildlife Fund, Washington, DC 20037, USA

Simon Linke
Australian Rivers Institute, Griffith University, Brisbane, QLD 4111, Australia

Brett G. Dickson and Luke J. Zachmann
Conservation Science Partners, Inc., 11050 Pioneer Trail, Suite 202, Truckee, CA 96161, USA
Landscape Conservation Initiative, Northern Arizona University, Box 5694, Flagstaff, AZ 86011, USA

Christine M. Albano
Conservation Science Partners, Inc., 11050 Pioneer Trail, Suite 202, Truckee, CA 96161, USA
John Muir Institute of the Environment, University of California - Davis, One Shields Ave., Davis, CA 95616, USA

Brad H. McRae
The Nature Conservancy, North America Region, 117 Mountain Ave, Suite 201, Fort Collins, CO 80524, USA

Jesse J. Anderson and David M. Theobald
Conservation Science Partners, Inc., 11050 Pioneer Trail, Suite 202, Truckee, CA 96161, USA

Thomas D. Sisk
Landscape Conservation Initiative, Northern Arizona University, Box 5694, Flagstaff, AZ 86011, USA

Michael P. Dombeck
College of Natural Resources, University of Wisconsin-Stevens Point, 800 Reserve St., Stevens Point, WI 54481, USA

Leejiah J. Dorward
Department of Zoology, University of Oxford Tinbergen Building, South Parks Road, Oxford OX1 3PS, UK

John C. Mittermeier
School of Geography and the Environment, University of Oxford, South Parks Road, Oxford, OX1 3QY, UK

Chris Sandbrook
United Nations Environment Programme World Conservation Monitoring Centre, 219 Huntingdon Road, Cambridge CB3 0DL, UK
Department of Geography, University of Cambridge, Downing Place, Cambridge CB2 3EN, UK

Fiona Spooner
Department of Genetics, Evolution and Environment, Centre for Biodiversity and Environment Research, University College London, Gower Street, London, WC1E 6BT, UK

Johan Ekroos
Centre for Environmental and Climate Research, Lund University, Ecology Building, Lund, Sweden

Julia Leventon and Jens Newig
Research Group Governance and Sustainability, Faculty of Sustainability, Leuphana Universität Lüneburg, Lüneburg, Germany

Henrik G. Smith
Centre for Environmental and Climate Research, Lund University, Ecology Building, Lund, Sweden
Department of Biology, Lund University, Ecology Building, Lund, Sweden

Joern Fischer
Institute of Ecology, Faculty of Sustainability, Leuphana Universität Lüneburg, Lüneburg, Germany

Mathew J. Hardy Sarah A. Bekessy & Ascelin Gordon
School of Global, Urban and Social Studies, RMIT University, Melbourne, VIC 3001, Australia

James A. Fitzsimons
The Nature Conservancy, Carlton, VIC 3053, Australia
School of Life and Environmental Sciences, Deakin University, Burwood, VIC 3125, Australia

Virgilio Hermoso
Centre Tecnològic Forestal de Catalunya (CEMFOR - CTFC), Crta. Sant Llorenc͵ de Morunys, Km 2, 25280, Solsona, Lleida, Spain
Australian Rivers Institute, Griffith University, Nathan, Qld 4111, Australia

Miguel Clavero
Estación Biológica de Doñana-CSIC, Américo Vespucio s.n., 41092, Sevilla, Spain

Dani Villero
Centre Tecnològic Forestal de Catalunya (CEMFOR - CTFC), Crta. Sant Llorenc͵ de Morunys, Km 2, 25280, Solsona, Lleida, Spain

Lluís Brotons
Centre Tecnològic Forestal de Catalunya (CEMFOR - CTFC), Crta. Sant Llorenc͵ de Morunys, Km 2, 25280, Solsona, Lleida, Spain
CREAF, Cerdanyola del Vallés, 08193, Spain
CSIC, Cerdanyola del Vallés, 08193, Spain

Carolyn J. Hogg
Zoo and Aquarium Association Australasia, Mosman, NSW, Australia

Catherine E. Grueber
Faculty of Veterinary Science, University of Sydney, Sydney, NSW, Australia
San Diego Zoo Global, San Diego, CA, USA

David Pemberton, Samantha Fox and Andrew V. Lee
DPIPWE, Save the Tasmanian Devil Program, Hobart, Tasmania, Australia

Jamie A. Ivy
San Diego Zoo Global, San Diego, CA, USA

Katherine Belov
Faculty of Veterinary Science, University of Sydney, Sydney, NSW, Australia

Martin Jeanmougin
Centre d'Ecologie et des Sciences de la Conservation (CESCO - UMR7204), Sorbonne Universités-MNHN-CNRS-UPMC, Muséum national d'Histoire naturelle, CP135, 43 rue Buffon, 75005, Paris, France

Camille Dehais
GERECO, 30 avenue Leclerc, F-38217, Vienne, France

Yves Meinard
Université Paris-Dauphine, PSL Research University, CNRS, UMR [7243], LAMSADE, 75016, FRANCE

Felix K. S. Lim & David P. Edwards
Department of Animal and Plant Sciences, University of Sheffield, Sheffield S10 2TN, UK

L. Roman Carrasco
Department of Biological Sciences, National University of Singapore, Singapore

Jolian McHardy
Department of Economics, University of Sheffield, Sheffield S1 4DT, UK

Alejandra Morán-Ordóñez
Quantitative & Applied Ecology Group, School of Biosciences, The University of Melbourne, Parkville, VIC 3010, Australia
Centre Tecnològic Forestal de Catalunya, Ctra. Antiga St. Llorenc͵ km 2, 25280 Solsona, Spain

Amy L Whitehead & Brendan A Wintle
Quantitative & Applied Ecology Group, School of Biosciences, The University of Melbourne, Parkville, VIC 3010, Australia

Gary W Luck
Institute for Land, Water and Society, Charles Sturt University, Albury, NSW 2640, Australia

Garry D Cook
CSIRO Ecosystem Sciences, Private Mail Bag 44, Winnellie, NT 0822, Australia

Ramona Maggini
ARC Centre of Excellence for Environmental Decisions, NERP Environmental Decisions Hub, Centre for Biodiversity & Conservation Science, University of Queensland, Brisbane, Qld 4072, Australia

James A Fitzsimons
The Nature Conservancy, Suite 2-01, 60 Leicester Street, Carlton, VIC 3053, Australia
School of Life and Environmental Sciences, Deakin University, 221 Burwood Highway, Burwood, VIC 3125, Australia

Guy Pe'er
Department of Conservation Biology, UFZ – Helmholtz Centre for Environmental Research, Permoserstr. 15, 04318 Leipzig, Germany
German Centre for Integrative Biodiversity Research (iDiv) Halle-Jena-Leipzig, Deutscher Platz 5e, 04103 Leipzig, Germany

Yves Zinngrebe
Department of Conservation Biology, UFZ – Helmholtz Centre for Environmental Research, Permoserstr. 15, 04318 Leipzig, Germany
Georg-August-University Göttingen,Department for Agricultural Economics and Rural Development, Platz der Göttinger Sieben 5, 37073 Göttingen, Germany

Jennifer Hauck
CoKnow Consulting – Coproducing Knowledge for Sustainability, Mühlweg 3, 04838 Jesewitz, Germany
Department of Environmental Politics, UFZ – Helmholtz Centre for Environmental Research, Permoserstr. 15, 04318 Leipzig, Germany

Stefan Schindler
Environment Agency Austria, Spittelauer Lände 5 (A-1090) Vienna, Austria
Department of Conservation Biology, Vegetation & Landscape Ecology, University of Vienna, Rennweg 14 (A-1030) Vienna, Austria

Andreas Dittrich and Christian Hoyer
Department of Computational Landscape Ecology, UFZ – Helmholtz Centre for Environmental Research, Permoserstr. 15, 04318 Leipzig, Germany

Silvia Zingg
Division of Conservation Biology, Institute of Ecology and Evolution, University of Bern, 3013 Bern, Switzerland
Bern University of Applied Sciences,School of Agricultural, Forest and Food Sciences, 3052 Zollikofen, Switzerland

Teja Tscharntke
Agroecology, Department of Crop Sciences, University of Göttingen, Grisebachstraße 6, 37077 Göttingen, Germany

Rainer Oppermann
Institute for Agro-ecology and Biodiversity (IFAB), Böcklinstr. 27, 68163 Mannheim, Germany

Laura M.E. Sutcliffe
Institute for Agro-ecology and Biodiversity (IFAB), Böcklinstr. 27, 68163 Mannheim, Germany
Plant Ecology and Ecosystems Research, University of Göttingen, Untere Karspüle 2, 37073 Göttingen, Germany

Clélia Sirami
Dynafor, Universite´ de Toulouse, INRA, INPT, INP-EI Purpan, Castanet Tolosan, France

Jenny Schmidt
Department of Environmental Politics, UFZ – Helmholtz Centre for Environmental Research, Permoserstr. 15, 04318 Leipzig, Germany

Christian Schleyer
Institute of Social Ecology, Alpen-Adria University Klagenfurt, Schottenfeldgasse 29, 1070 Vienna, Austria

Sebastian Lakner
Georg-August-University Göttingen,Department for Agricultural Economics and Rural Development, Platz der Göttinger Sieben 5, 37073 Göttingen, Germany

Adena R. Rissman
Forest and Wildlife Ecology, University of Wisconsin-Madison, 1630 Linden Drive, Madison, WI 53706, USA

Sean Gillon
Forest and Wildlife Ecology, University of Wisconsin-Madison, 1630 Linden Drive, Madison, WI 53706, USA
Food Systems and Society, Marylhurst University, 17600 Pacific Highway, Portland, OR, 97036, USA

Rebecca K. Runting & Jonathan R. Rhodes
School of Geography, Planning and Environmental Management, The University of Queensland, Brisbane 4072, Australia

ARC Centre of Excellence for Environmental Decisions, The University of Queensland, Brisbane 4072, Australia

Catherine E. Lovelock
School of Biological Sciences, The University of Queensland, Brisbane 4072, Australia

Hawthorne L. Beyer
ARC Centre of Excellence for Environmental Decisions, The University of Queensland, Brisbane 4072, Australia
School of Biological Sciences, The University of Queensland, Brisbane 4072, Australia

Anne H. Toomey
Department of Environmental Studies and Science, Pace University, New York, New York, USA
Center for Biodiversity and Conservation, American Museum of Natural History, New York, New York, USA

Andrew T. Knight
Department of Life Sciences, Imperial College London, Ascot, Berkshire, United Kingdom
ARC Centre of Excellence in Environmental Decisions, The University of Queensland, Brisbane, Queensland, Australia
Department of Botany, Nelson Mandela Metropolitan University, Port Elizabeth, South Africa
The Silwood Group, London, United Kingdom

Jos Barlow
Lancaster Environment Centre, Lancaster University, Lancaster, UK
Museu Paraense Emilio Goeldi, Belém, Brazil

Tungalag Ulambayar & Batbuyan Batjav
Mongolian Institute of Geography and Geoecology, Ulaanbaatar, Mongolia

María E. Fernández-Giménez
Department of Forest and Rangeland Stewardship, Colorado State University, Campus Mail 1472, Fort Collins, CO, 80523–1472, USA

Batkhishig Baival
Nutag Partners, Usnii St, Ulaanbaatar, Mongolia

Piero Visconti
Microsoft Research Computational Science Laboratory, 21 Station Road, Cambridge, CB1 FB, UK

Global Mammal Assessment Program, Department of Biology and Biotechnologies, Sapienza University of Rome, Viale dell'Universita` 32, Rome, 00185, Italy

Michel Bakkenes
PBL, Netherlands Environmental Assessment Agency, PO Box 303, 3720, AH, Bilthoven, The Netherlands

Daniele Baisero
Global Mammal Assessment Program, Department of Biology and Biotechnologies, Sapienza University of Rome, Viale dell'Universita` 32, Rome, 00185, Italy

Thomas Brooks
IUCN Species Survival Commission, International Union for Conservation of Nature, 28 rue Mauverney, CH-1196, Gland, Switzerland
World Agroforestry Center (ICRAF), University of the Philippines Los Ban˜ os, Laguna, 4031, Philippines
School of Geography and Environmental Studies, University of Tasmania, Hobart, TAS 7001, Australia

Stuart H. M. Butchart
BirdLife International, Wellbrook Court, Cambridge, CB3 0NA, UK

Lucas Joppa
Microsoft Research Computational Science Laboratory, 21 Station Road, Cambridge, CB1 FB, UK

Rob Alkemade
PBL, Netherlands Environmental Assessment Agency, PO Box 303, 3720, AH, Bilthoven, The Netherlands
Environmental Systems Analysis Group, Wageningen University, P. O. Box 47, 6700, AA, Wageningen, The Netherlands

Moreno Di Marco, Luca Santini, Luigi Maiorano, Luigi Boitani & Carlo Rondinini
Global Mammal Assessment Program, Department of Biology and Biotechnologies, Sapienza University of Rome, Viale dell'Universita` 32, Rome, 00185, Italy

Michael Hoffmann
IUCN Species Survival Commission, International Union for Conservation of Nature, 28 rue Mauverney, CH-1196, Gland, Switzerland
United Nations Environment Programme World Conservation Monitoring Centre, 219c Huntingdon Road, Cambridge, CB3 0DL, UK

Robert L. Pressey
Australian Research Council Centre of Excellence for Coral Reef Studies, James Cook University, Townsville, QLD 4811, Australia

Anni Arponen
Metapopulation Research Group, Department of Biosciences, University of Helsinki, P.O. Box 65, Helsinki, 00014, Finland

April E. Reside
Centre for Tropical Environmental & Sustainability Sciences, James Cook University, QLD, 4811, Australia

Detlef P. van Vuuren
PBL, Netherlands Environmental Assessment Agency, PO Box 303, 3720, AH, Bilthoven, The Netherlands Copernicus Institute of Sustainable Development, Department of Geosciences, Utrecht University, Heidelberglaan 2, 3584 CS, Utrecht, The Netherlands

Amy L. Whitehead, Heini Kujala, & Brendan A. Wintle
School of BioSciences, The University of Melbourne, Parkville, VIC 3010, Australia

Megan N. Dethier
Friday Harbor Laboratories and Biology Department, University of Washington, Friday Harbor, WA, 98250 USA

Jason D. Toft
School of Aquatic and Fishery Sciences, University of Washington, Seattle, WA 98195, USA

Hugh Shipman
Washington State Department of Ecology, Olympia, WA 98504, USA

Brendan Costelloe
Department of Life Sciences, Imperial College London, Silwood Park, Buckhurst Road, Ascot, Berkshire SL5 7PY, UK
Institute of Zoology, Zoological Society of London, Regent's Park, London NW1 4RY, UK
The Royal Society for the Protection of Birds, Potton Road, Sandy, Bedfordshire SG19 2DL, UK

Ben Collen
Centre for Biodiversity and Environment Research, Department of Genetics, Evolution and Environment, University College London, Gower Street, London WC1E 6BT, UK

E.J. Milner-Gulland
Department of Life Sciences, Imperial College London, Silwood Park, Buckhurst Road, Ascot, Berkshire SL5 7PY, UK

Ian D. Craigie
ARC Centre of Excellence for Coral Reef Studies, James Cook University, Townsville, QLD 4811, Australia

Louise McRae
Institute of Zoology, Zoological Society of London, Regent's Park, London NW1 4RY, UK

Carlo Rondinini
Global Mammal Assessment Program, Department of Biology and Biotechnologies, Sapienza University of Rome, Viale dell'Università 32, Rome 00185, Italy

Emily Nicholson
Department of Life Sciences, Imperial College London, Silwood Park, Buckhurst Road, Ascot, Berkshire SL5 7PY, UK
School of Botany, University of Melbourne, VIC 3052, Australia
Centre for Integrative Ecology, School of Life and Environmental Sciences, Deakin University, Burwood, Victoria 3125, Australia

Jeremy S. Brooks
School of Environment and Natural Resources, The Ohio State University, 2021 Coffey Rd., Columbus, OH, 43212, U.S.A

Reniel B. Cabral, Christopher Costello & Steven D. Gaines
Bren School of Environmental Science and Management, University of California, Santa Barbara, CA 93106, USA

Benjamin S. Halpern
Bren School of Environmental Science and Management, University of California, Santa Barbara, CA 93106, USA
Imperial College London, Silwood Park Campus, Buckhurst Road, Ascot SL57PY, UK
National Center for Ecological Analysis and Synthesis, 735 State St. Suite 300, Santa Barbara, CA 93101, USA

Judith M. Ament
Percy FitzPatrick Institute, DST/NRF Centre of Excellence, University of Cape Town, Rondebosch, Cape Town 7701, South Africa
Institute of Zoology, Zoological Society of London, Regent's Park, London NW1 4RY, UK
Centre for Biodiversity and Environment Research, Department of Genetics, Evolution and Environment, University College London, London WC1E 6BT, UK

Christine A. Moore
Percy FitzPatrick Institute, DST/NRF Centre of Excellence, University of Cape Town, Rondebosch, Cape Town 7701, South Africa

School of Geography and the Environment, University of Oxford, South Parks Road, Oxford OX1 3PY, UK

Marna Herbst
South African National Parks, Scientific Services, Private Bag X1021, Phalaborwa 1390, South Africa

Graeme S. Cumming
Percy FitzPatrick Institute, DST/NRF Centre of Excellence, University of Cape Town, Rondebosch, Cape Town 7701, South Africa
ARC Centre of Excellence for Coral Reef Studies, James Cook University, Townsville, QLD 4811, Australia

Index